How Green Is the City?

How Green Is the City?

Sustainability Assessment and the Management of Urban Environments

edited by Dimitri Devuyst
with Luc Hens and Walter De Lannoy

COLUMBIA UNIVERSITY PRESS
New York

Columbia University Press
Publishers Since 1893
New York   Chichester, West Sussex

Library of Congress Cataloging-in-Publication Data

How green is the city?: sustainability assessment and the management of urban environments /
edited by Dimitri Devuyst, Luc Hens, and Walter De Lannoy
p.   cm.
Includes bibliographical references and index.
ISBN 978-0-231-11802-6 (cloth : alk. paper) — ISBN 978-0-231-11803-3 (pbk : alk. paper)
1. Sustainable development 2. Urban economics—Social aspects. 3. Urban ecology.
I. Devuyst, Dimitri

HC79.E5 H676 2001
338.9'27'o91732—dc21   2001017379

Printed in the United States of America

Never doubt that a small group of thoughtful,
committed citizens can change the world:
indeed, it's the only thing that ever does.
—Margaret Mead

Contents

Boxes

Tables

Contributors

Carlo Aall received a university degree in agriculture (Cand Agric) from the Norwegian University of Agriculture. He has worked as a municipal environmental officer for two years. Since 1991 he has worked as a researcher at the Western Norway Research Institute, and is currently finishing a Ph.D. on municipal environmental policy.

Susanne E. Bruppacher has a master's degree (lic. phil. hist.) in developmental psychology from the University of Bern (Switzerland). She is currently a research assistant and Ph.D. student in psychology at the Interdisciplinary Center for General Ecology, University of Bern (Switzerland).

Walter De Lannoy is professor in human geography at the Vrije Universiteit Brussel (Belgium). He is specialized in urban geography and urban planning, transportation geography, and spatial planning, on which he has published extensively.

Dimitri Devuyst is a postdoctoral fellow of the Fund for Scientific Research-Flanders (Belgium) and coordinator of the EIA Center at the Vrije Universiteit Brussel. He is a biologist and holds a postgraduate degree and Ph.D. in human ecology from the Vrije Universiteit Brussel (Belgium). At the same university, Dimitri Devuyst has been coordinator of the Master's Program in Human Ecology and the Master's Program in Environmental Impact Assessment. He is recognized as an EIA expert by the Ministry of the Flemish Community in Belgium and has published on EIA in international journals such as *Environmental Management, Environmental Impact Assessment Review, Impact Assessment,* and *The Environmental Professional.* He is coeditor of *Environmental Management in Practice* (Routledge 1998). His current field of research is related to strategic environmental assessment and sustainable development in urban areas.

Paul Harland is a Ph.D. student at the Center for Energy and Environmental Research, Leiden University (the Netherlands). His research interests focus on the application of social psychological theories on environmentally relevant behavior. His dissertation will deal mainly with the influence of personal (moral) norms on behavioral decisions in the environmental domain.

Luc Hens graduated as a biologist and later received his Ph.D. in biology from the Vrije Universiteit Brussel (Belgium), where he is at present professor in human ecology and head of the Human Ecology Department. He also lectures at Antwerp University (Belgium), the Technical University of Sofia (Bulgaria), and Ankara University (Turkey). His specific area of research concerns environment and health-related issues and interdisciplinary instruments for sustainable development. Professor Hens acts as an expert in environmental policy in several councils in Belgium. He is the European editor for the *International Journal on Environmental Pollution.*

Elizabeth Kline is director of the Sustainable Communities Program at Tufts University in Massachusetts (USA) and is a professional planner, policy-maker, and applied researcher. She has a master's degree in urban planning and public administration and a second master's in geography, with a specialty in water resources. She was a high-level government environmental policy-maker, a town and city planner, and a program manager at an environmental nonprofit organization. For the past seven years, she has been a senior researcher at the Global Development and Environment Institute.

Lone Koernoev has an M.Sc. in planning from Aalborg University (Denmark) and is a doctoral candidate at the Department of Development and Planning at the same university. She has worked with life cycle assessment and environmental impact assessment before starting a research project about strategic environmental assessment and decision-making in 1996.

William M. Lafferty is professor of political science at the University of Oslo (Norway) and director of the Program for Research and Documentation for a Sustainable Society within the Research Council of Norway. He was also program chair for the 18th triennial World Congress of Political Science held in Quebec City, Canada, in August 2000. His most recent books in this area of research are *Towards Sustainable Development* (Macmillan: 1999, with Oluf Langhelle); *From the Earth Summit to Local Agenda 21* (Earthscan: 1998, with Katarina Eckerberg); and *Democracy and the Environment* (Edward Elgar: 1996, with James Meadowcroft).

William E. Rees is professor and director of the School of Community and Regional Planning at the University of British Columbia (Canada). He is internationally recognized for his groundbreaking work in ecological footprint analysis, on which he has published extensively.

Mark Roseland is director of the Community Economic Development Centre at Simon Fraser University (Canada) and is a professor in SFU's Department of Geography. He is the author of *Toward Sustainable Communities: Resources for Citizens and Their Governments* (1998) and the editor of *Eco-City Dimensions: Healthy Communities, Healthy Planet* (1997), both from New Society Publishers, as well as the North American editor of the international journal *Local Environment.* Dr. Roseland lectures internationally and advises communities and governments on sustainable development policy and planning.

Henk Staats is staff member of the Department of Social and Organizational Psychology at Leiden University (the Netherlands) and holds a doctoral degree. His teaching and research focus on the social psychology of human-environment relationships. He is coordinator of the Center for Energy and Environmental Research and supervisor of Paul Harland's Ph.D. research.

Sofie Van Volsem holds a degree in chemistry and a master's in environmental sanitation. During her two-year stay at the Human Ecology Department, Vrije Universiteit Brussel (Belgium), she was involved in the development of an environmental management system for secondary schools in Flanders. About 300 schools are currently using this system. She also has experience with life cycle analysis and sustainable food production. Currently she is working for the Flemish Government in Belgium to develop an action plan for the follow-up of the 4th North Sea Conference.

Thomas van Wijngaarden graduated in 1993 from the Agricultural University of Wageningen (the Netherlands) as an engineer in biology. He holds a postgraduate degree in environmental impact assessment from the Vrije Universiteit Brussel (Belgium). He was a scientific staff member at the Human Ecology Department, Vrije Universiteit Brussel (Belgium) from January 1997 until April 1999. Between October 1997 and April 1999 he was the coordinator of the Master's Program in Environmental Impact Assessment, Vrije Universiteit Brussel (Belgium). He is currently working on projects in Zambia.

Rodney R. White is professor of geography and director of the Institute for Environmental Studies at the University of Toronto, where he teaches courses on cities as ecosystems, risk analysis and management, and environment and health. Recent publications include *Urban Environmental Management* (John Wiley & Sons 1994) and *North, South and the Environmental Crisis* (University of Toronto Press 1993). He has combined an academic career with consulting assignments, mainly on infrastructure planning in West Africa.

Abbreviations

ASSIPAC	Assessing the Sustainability of Societal Initiatives and Proposing Agendas for Change
AT	Appropriate Technology
AVR	Average Vehicle Ridership
CEC	Commission of the European Communities
CED	Community Economic Development
CEQA	California Environmental Quality Act
CFC	Chlorofluorocarbon
CI	Consumers International
COICA	Coordinadora de las Organizaciones Indigenas de la Cuenca Amazonica (Spanish for Coordination of Indigenous Organizations of the Amazon Basin)
DMA	Durban Metropolitan Area
DOTIS	Duurzame Ontwikkeling in Tilburg, moderne Industrie-Stad (Dutch for Sustainable Development in Tilburg, modern Industrial City)
EA	Environmental Assessment
EC	European Commission
ECMT	European Council of Ministers of Transport
EFIEA	European Forum on Integrated Environmental Assessment
EIA	Environmental Impact Assessment
EIM	Environment in the Municipalities
EIS	Environmental Impact Statement
EMAS	Environmental Management and Audit Scheme
ESD	Ecologically Sustainable Development
EU	European Union
FAO	Food and Agriculture Organization
FEALWELL	Friendliest Environmental Alternative Without Excessive Losses in Lifestyle
FoE	Friends of the Earth

FRM	Family Resources Management
GAIA	General Agreement on the Integrity of (Ecological) Assets
GAP	Global Action Plan
GATT	General Agreement on Tariffs and Trades
GDP	Gross Domestic Product
GEO	Global Environment Outlook
GIS	Geographic Information Systems
GNP	Gross National Product
ha	hectares
HA21	Household Agenda 21
HIA	Health Impact Assessment
HOMES	Household Metabolism Effectively Sustainable
IA	Impact Assessment
ICLEI	International Council for Local Environmental Initiatives
IDM	Integrated Decision Making
IDRC	International Development Research Center
IISD	International Institute for Sustainable Development
IMF	International Monetary Fund
IMPACT	Integrated Method of Performance and Contribution Tracking
IUCN	The World Conservation Union - International Union for the Conservation of Nature
IULA	International Union of Local Authorities
LA21	Local Agenda 21
LCA	Life-Cycle Assessment
LETS	Local Employment Trading Systems
LGMB	Local Government Management Board
MEIR	Master Environmental Impact Report

MIT	Massachusetts Institute of Technology
MOE	Ministry of the Environment
NAFTA	North American Free Trade Agreement
NALRA	Norwegian Association of Local and Regional Authorities
NEPA	National Environmental Policy Act
NGO	Non-Governmental Organization
NIMBY	Not-In-My-BackYard syndrome
NOK	Norwegian Krone
NSESD	National Strategy for Ecologically Sustainable Development
OECD	Organization for Economic Cooperation and Development
ORTEE	Ontario Round Table on the Environment and the Economy
PEIR	Program Environmental Impact Report
RA21	Regional Agenda 21
REF	Reduced-Emission Fuel
SA	Sustainability Assessment
SAM	Sustainability Assessment Map
SDA	Sustainable Development Assessment
SDI	Sustainable Development Indicator
SEA	Strategic Environmental Assessment
SED	Strategic Environmental Discussion
SIA	Social Impact Assessment
SLA	Sustainable Lifestyle Assessment
SMMBL	Santa Monica Municipal Bus Line
SoER	State of the Environment Reporting
SPARTACUS	System for Planning and Research in Towns and Cities for Urban Sustainability
SSA	Strategic Sustainability Assessment
SUSA	Systemic User-driven Sustainability Assessment
SUSCOM	Sustainable Communities in Europe

TBC	Tuberculosis
TMP	Transportation Management Plan
TQM	Total Quality Management
UK	United Kingdom
UN	United Nations
UNCED	United Nations Conference on Environment and Development
UNCSD	United Nations Commission on Sustainable Development
UNEP	United Nations Environment Program
US	United States
VODO	Vlaams Overleg Duurzame Ontwikkeling (Dutch for Flemish Forum for Sustainable Development)
VUB	Vrije Universiteit Brussel
WBCSD	World Business Council for Sustainable Development
WCED	World Commission on Environment and Development
WHO	World Health Organization
WISC	Willapa Indicators for a Sustainable Community
WRI	World Resources Institute
WTO	World Trade Organization
WWF	World Wildlife Fund

Acknowledgments

We wish to thank our colleagues at the Human Ecology Department at Vrije Universiteit Brussel for their support, especially Michele Levoy for the thorough language reviews, and Razack Homaida "Robbie" Rabiah for assisting Michele in this job. Thank you to Glenn Ronsse for his computer assistance as well as to our other colleagues in the Human Ecology Department for their support.

We would also like to thank Holly Hodder and Jonathan Slutsky of Columbia University Press for their support of the book proposal, their encouragement, and their suggestions during the development of the book.

We want to express our sincerest appreciation to the other contributors to this volume for meeting deadlines and for their efficient collaboration.

Finally, we would also like to thank our families and friends for their support, suggestions, and encouragement. The idea for this book emerged during a walk on the beach with Youri Devuyst. Thank you for the inspiration. Also many thanks to Luc Devuyst, Johanna Verlinden, and Johan Leijsen for everything.

Dimitri Devuyst
Luc Hens
Walter De Lannoy

How Green Is the City?

Introduction to Sustainability Assessment at the Local Level

Dimitri Devuyst

A Human Ecological Perspective

This book is the result of a research project funded by the Fund for Scientific Research—Flanders (Belgium) that looks into strategic environmental assessment as an instrument to develop urban areas more sustainably. This research is carried out at the Environmental Impact Assessment Center, which is part of the Human Ecology Department at the Vrije Universiteit Brussel (VUB). In 1975 the VUB instituted a Master of Human Ecology educational program. As the interdisciplinary study of the relationships between the human species and its environment, human ecology is distinct from traditional animal, plant, or microbial ecology in that it recognizes the important role played by culture in shaping human society and behavior. In the course of evolution, humans have invented new methods for storing and transmitting accumulated knowledge, giving rise to an exclusive system of non-genetic transmission of acquired and stored information. While early humans became a dominant species in local ecosystems, modern technological society is now part of a global system of production and consumption, and an agent of worldwide dispersion of animals, plants, and microorganisms. The human species has been responsible for profound changes on the earth, comparable to that of no other species. But while humans are not the only animal capable of changing the physical and biotic environments, it is the only one capable of understanding and preserving the biosphere. Over the past decades there has been considerable evolution in thinking about human ecology. The increasing interest in the concept of sustainable development has shown a need for an interdisciplinary approach to environmental management. Clearly, human ecology has an essential role to play in defining what is meant by "sustainable development" and developing instruments for environmental management that can be used in the sustainable development of human societies. This is the goal of the Human Ecology Department and the Environmental Impact Assessment Center at VUB, and its international partners. As a result, this book aims to examine and develop so-called "sustainability assessment instruments," tools that can predict or estimate the impact of human activities on the sustainability of societies and measure progress made toward sustainable development. This book brings together contributions from an international group of European and North American scientists active in sustainable urban development, impact assessment, and measuring sustainability.

What Are the Long-Term Problems Facing the World?

The first issue of the United Nations Environment Program (UNEP) biennial Global Environment Outlook (GEO), "GEO-1 Global State of the Environment Report 1997," is a snapshot of an ongoing worldwide environmental assessment process. It was initiated in response to the environmental reporting requirements of Agenda 21 and to a UNEP Governing Council decision of May 1995, which requested production of a state of the environment report series in time for the next UNEP Governing Council in January 1997 (UNEP 1997). This report indicates that, despite both declining global birth rates since 1965 and recent policy initiatives toward more efficient and cleaner resource use in some regions, the large increase in world population, expanding economies in industrializing countries, and wasteful consumption patterns (particularly in developed countries of the world) will continue to increase global resource and energy consumption, generate burgeoning wastes, and spawn environmental contamination and degradation.

GEO-1 predicts additional major problems in the following fields:

✤ Increase in global carbon dioxide emissions and reversal of declining trends in acidifying sulfur and nitrogen concentrations leading to climate change

✤ Substantial expansion of agricultural activities into marginal lands at the expense of remaining wilderness and associated biodiversity

✤ The combination of increased pressure on land by expanding urbanization and losses of productive land through degradation and unsustainable management practices will lead to shortages in arable land and water, impeding development in several regions

In its 1998–1999 report the World Resources Institute (WRI 1999) makes the link between environmental, social, and health problems. In developing countries, environmental health threats stem mostly from problems such as a lack of clean water, sanitation, adequate housing, and protection from mosquitoes and other insect and animal disease vectors. According to WRI (1999), the following major global problems need to be considered:

✤ Water contaminated by feces—inadequate water, sanitation, and hygiene are responsible for an estimated seven percent of all deaths and disease globally. Diarrhea alone claims the lives of some 2.5 million children a year

❧ Overcrowding and smoky indoor air—this contributes to acute respiratory infections that kill 4 million people a year, mostly children under the age of five years

❧ Malaria—this disease kills 1 to 3 million people a year. Other mosquito-borne diseases, such as dengue and yellow fever, affect millions more each year and are on the rise

In addition, the AIDS epidemic is a serious global problem. AIDS is now the number one killer in Africa. In the hard-hit African countries, up to a quarter of all adults are infected and AIDS is wiping out development gains achieved over many decades (UNAIDS 1999). WRI (1999) links health problems to poverty. At the most fundamental level, many of the world's poorest poor, the 1.3 billion people who live on less than a (U.S.) dollar a day, are unable to secure the most basic necessities, such as adequate food, water, clothing, shelter, and health care. Another major cause of ill health globally is malnutrition, which is an issue of poverty and rarely an indicator of actual food shortages. The World Health Organization (WHO) confirmed in 1977 that poverty and inequality are two of the most important contributory factors to poor environmental conditions and poor health. Poverty can be considered one of the most pressing global problems. Reports on booming stock markets and economic growth often fail to indicate that poverty, as reflected in global income inequity, is increasing. In 1970, the richest 20 percent of humanity had roughly 30 times more income than the poorest 20 percent.

Today, this figure has doubled. The net worth of the world's 358 richest people is larger than the combined annual income of the poorest 45 percent of the world's population. In wealthier countries, a link can also be made between poverty in certain population groups, their health, and the environmental conditions they live in. Studies show that it is not a country's actual wealth, but rather how that wealth is distributed among its groups that is the best predictor of health status. For example, in 1970, Japan and the United States had similar income distribution and life expectancy. However, Japan currently has the longest life expectancy in the world, along with the most egalitarian income distribution. By contrast, the United States now ranks 21st out of 157 countries in life expectancy (WRI 1999), and income inequality has increased. GEO-1 expects that sharp regional differences will remain and poverty will be aggravated in several regions. If global economic gains are not accompanied more explicitly by investment in education, social development, and environmental protection, a move toward a more equitable, healthy, and sustainable future for all sectors of society will not be realized, and a new spate of urban and pollution-related health impacts may surface (UNEP 1997). A clear picture of the urgency of the problems facing the world is presented in Barney et al. (1993):

> Given the magnitude of the issues we face, we must expect that within the lifetime of a child born today, the world will change radically in one of two directions. If we continue with present beliefs, institutions, and policies, the world will become highly polarized, with a billion people in the wealthy industrialized countries of the North attempting to enjoy life and leisure a few decades longer while 10 billion plus people in the South spiral downward into increasingly desperate poverty exacerbated by global environmental deterioration. Ultimately the North spirals downward too, and the whole planet drifts off into a new dark age or worse.

> But there is another option open to us, one in which everyone comes to recognize that a healthy Earth is an essential prerequisite for a healthy human population. Under this option, the world could become less polluted, less crowded, more stable ecologically, economically, and politically if we humans would be willing to work together to:

> (a) create the religious, social, and economic conditions necessary to stop the growth of human population (b) reduce the use of resources (sources) and disposal capacity (sinks) by the wealthiest (c) ensure civil order, education, and health services for people everywhere (d) preserve soils and species everywhere (e) double agricultural yields while reducing both agricultural dependence on energy and agricultural damage to the environment (f) convert from carbon dioxide-emitting energy sources to renewable, non-polluting energy sources that are affordable even to the poor (g) cut sharply the emissions of other greenhouse gases (h) stop immediately the emissions of the chemicals destroying the ozone layer and (i) bring equity between nations and peoples of the North and South.

> We do not have generations or even decades to choose between these two directions because of the momentum inherent in population growth, capital investments, technological choices, and environmental changes. In fact, the choice of direction for Earth is being made today.

Not all people agree that environmental problems are urgent or that global environmental issues exist at all (e.g., Simon 1996). Such a wide range of opinions is fed by uncertainty in scientific evidence. Understanding a global problem such as climate change would, for example, require improved climate modeling. There are, however, issues that do not require more scientific research to be able to predict their catastrophic consequences.

The AIDS epidemic in Africa, for instance, has been documented and is a crisis of daunting proportions. Social and health-related problems affect people on a global scale, and violent conflicts and wars between and within countries erupt worldwide on a continuous basis. In other words, even if we do not agree on the technicalities of how to measure and explain all global environmental problems, we have evidence of major emergencies on Earth which are in need of remedial action. And even in those instances where scientific explanations are still inadequate, the precautionary principle should be applied. This principle holds that where there are threats of serious or irreversible damage, lack of full scientific certainty shall not be used as a reason for postponing cost-effective measures to prevent environmental degradation.

The current day-to-day decisions made by corporations and all levels of government, as well as those of private citizens, are determining which direction the planet will take. Decision-makers could be greatly helped by instruments that tell them clearly if their initiatives lead into the direction of more or less sustainability. This is what this book is all about. GEO-1 is very concerned that measures to solve existing global problems are introduced too slowly (UNEP 1997):

> GEO-1 substantiates the need for the world to embark on major structural changes and to pursue environmental and associated socioeconomic policies vigorously. Key areas for action must embrace the use of alternative and renewable energy resources, cleaner and leaner production systems worldwide, and concerted global action for the protection and conservation of the world's finite and irreplaceable freshwater resources.

> Nevertheless, despite [this] progress on several fronts, from a global perspective the environment has continued to degrade during the past decade, and significant environmental problems remain deeply embedded in the socioeconomic fabric of nations in all regions. Progress toward a global sustainable future is just too slow. A sense of urgency is lacking. Internationally and nationally, the funds and political will are insufficient to halt further global environmental degradation and to address the most pressing environmental issues—even though technology and knowledge are available to do so. The recognition of environmental issues as necessarily long-term and cumulative, with serious global and security implications, remains limited.

Most problematic on this planet is human population growth. *The State of World Population 1998* (UNFPA 1998) states that world population is growing by more than 80 million a year. In 1987, the total world population was five billion; it will pass six billion in 1999, and will continue to grow until at least the middle of the next century (UNFPA 1998). Clark (1989) makes an analogy between the exploding human population and the exuberant growth of bacteria introduced to a nutrient-rich petri dish. Since the petri dish is a closed system and contains only limited resources, bacterial populations will deplete available resources and submerge in their own wastes, resulting in stagnation or collapse of the population. While the bacteria do not have the ability to understand what is happening to them, human beings have an unprecedented understanding of how the planet works and how we are now on a path of unsustainable development. Can humanity do nothing more than race toward a catastrophe with their eyes wide open, fully aware of what is happening? Or can we collectively unite our state of mind and utilize strategies and technology to prevent doomsday theories from becoming a reality? Some experts, such as Lester Brown, president of the Worldwatch Institute, believe humanity will rise to the challenge and make the necessary changes. He believes the world may even be on the edge of an environmental revolution. The Worldwatch Institute sees indications of this by noting the environmental restructuring of our economies. For example, heads of major corporations and political leaders have started introducing measures—such as the development by petroleum corporations of solar and wind energy technology and their introduction in U.S. and European regions—long championed by environmentalists and non-governmental organizations (NGOs) are mobilizing millions of people for change. The big difference between the environmental revolution and previous revolutions, such as the agricultural and industrial revolutions, is the issue of time. The agricultural revolution was spread over thousands of years; the industrial revolution has been underway for two centuries. In contrast, the environmental revolution must be accomplished in a few decades (Worldwatch Institute 1999). Roodman (1999) envisions important roles for government, the corporate and nonprofit organization, as well as the private citizen in building a sustainable society. Each group can lead the others toward the goal of sustainability. For instance, government agencies must be committed to aggressively demarcate and defend environmental limits through legislation and enforcement. They will have to stimulate creativity within companies by encouraging corporate decisions that take advantage of the huge investment opportunities created by the eco-industrial revolution, e.g. in the energy sector. NGOs will need to push both government and corporate structures forward, with the help of voters, consumers, charitable organizations, and private owners of land and resources.

What Is Sustainable Development?

Since the publication in 1987 of "Our Common Future," the summary report of the World Commission on Environment and Development (WCED, 1987), *sustainable development* has become a popular term because it comes up in almost every discussion of environmental issues. The commission defined sustainable development as "development that meets the needs of the present generation without compromising the needs of future generations" (WCED 1987) or "the rearrangement of technological, scientific, environmental, economic and social resources in such a way that the resulting heterogeneous system can be maintained in a state of temporal and spatial equilibrium" (Hens 1996). There appears, however, to be much confusion and vagueness when we inquire in depth about exactly what "sustainable development" means. Agenda 21 is a report that helps clarify "sustainable development" and functions as an action plan for the 21st century. It is the result of the United Nations Conference on Environment and Development (UNCED), organized in Rio de Janeiro, Brazil, in 1992. *The Rio Declaration* is a document consisting of a preamble and twenty-seven articles reflecting the general principles of Agenda 21. This declaration explores ways of sustainable development that focus on human beings and their right to a healthy and productive life in harmony with nature (Hens, 1996). Important principles selected from *The Rio Declaration* are presented in the following box. Since the Rio conference, these principles have received wide attention from governments around the globe.

Selected Principles from the Rio Declaration on Environment and Development

The **public trust doctrine** means that governments must act to prevent environmental damage whenever a threat exists, whether it is covered by a specific law or not. This doctrine normally implies the right of the public to require that governments act in these circumstances on its behalf.

The **precautionary principle** holds that where there are threats of serious or irreversible damage, lack of full scientific certainty shall not be used as a reason for postponing cost-effective measures to prevent environmental degradation (or expressed more liberally, when in doubt about the impact of development, it will be managed according to the worst-case scenario of its impact on the environment). Politically, this principle is difficult to apply. In fact it is ignored in most countries. Choosing the side of caution is not an attractive option when considered against immediate projected economic benefits which can be spelt out in conventional development terms.

The principle of **inter-generational equity** is at the heart of the definition of sustainable development. It requires that the needs of the present are met without compromising the ability of future generations to meet their own needs. It depends on the combined and effective application of the other principles for sustainable development.

The principle of **intra-generational equity** requires that people within the present generation have the right to benefit equally from the exploitation of resources and that they have an equal right to a clean and healthy environment. This principle applies to the relationship between groups of people within a country and between countries. More and more, this principle is being applied in international negotiations. But within nations, it is particularly susceptible to cultural and socio-economic forces.

The **subsidiarity principle** is resurfacing world-wide after many decades of centralized planning and decision-making. In essence, this principle requires that decisions should be made by the communities affected or on their behalf, by the authorities closest to them. As appropriate, decisions should rest either at the national rather than international level or at the local rather than the national level. This has been the basic principle governing the devolution of planning systems world-wide, and it is intended to encourage local ownership of resources and responsibility for environmental problems and their solutions. The growing pressure for devolution on government needs to be balanced by a recognition that local areas are part of larger systems and cannot function in isolation. Often environmental problems may come from forces outside of local control such as "up-stream pollution from a neighboring country or community. In such cases the other principles for sustainable development would override the subsidiarity principle.

The **polluter pays principle (PPP)** suggests that the polluter should bear the cost of preventing and controlling pollution. The intent is to force polluters to internalize all the environmental costs of their activities so that these are fully reflected in the costs of the goods and services they provide. Problems would inevitably occur if an industry or a plant goes out of business through the rigorous enforcement of this principle. A community might decide that, for example, the employment benefits of keeping a factory open outweigh the health and other environmental costs of pollution. Environmental agencies in developed countries have usually taken a flexible approach with the continuation of government subsidies in special cases, and negotiations of individual programs have been undertaken to allow certain polluters to meet new environmental standards gradually.

The **user pays principle (UPP)** applies the PPP more broadly so that the cost of a resource to a user includes all the environmental costs associated with its extraction, transformation and use (including the costs of alternative or future uses foregone). The PPP and UPP can be expressed in similar ways through market systems and government regulation.

Source: Hens (1996) after Carew-Reid et al. (1994)

The concept of sustainable development can be defined in many different ways. One possible definition states that "sustainable development" is a societal project that aims to develop economic activities within the carrying capacity of the local ecosystem in such a way that the local population benefits as a whole, while preserving the well-being of future generations and people elsewhere." This definition shows the interlinking of economic, environmental, and social components. Indeed, as Redclift (1993) states:

> It soon becomes clear that we cannot achieve more ecologically sustainable development without ensuring that it is also socially sustainable. We need to recognize, in fact, that our definition of what is ecologically sustainable answers to human purposes and needs as well as ecological parameters. By the same token, we cannot achieve more socially sustainable development in a way that effectively excludes ecological factors from consideration.

What Is Sustainability Assessment?

The sustainability assessment instrument developed in this book is considered to be the premier tool in the VUB impact assessment family that includes: Environmental Impact Assessment (EIA), Strategic Environmental Assessment (SEA), Social Impact Assessment (SIA), and Health Impact Assessment (HIA). In other words, sustainability assessment is presented initially as a tool that can help decision-makers and policy-makers decide which actions they should and should not take in an attempt to make society more sustainable. While studying the link between instruments for impact assessment and sustainable development, it soon becomes clear that impact assessment on its own will not lead us to a sustainable society. There is clearly a need to study a variety of other instruments for environmental management (such as auditing and life cycle analysis), instruments to measure progress towards sustainability goals, and the link between these instruments and the planning, policy-making, and decision-making processes. Ultimately, after reading this book, there should be a much higher level of awareness of all these processes, instruments, and their interrelationships, as well the efforts being made to develop more sustainably. Therefore, this book deals with sustainability assessment not only as an instrument of the impact assessment family, but also how it pertains to other non-impact assessment-related tools such as indicators, targets, and ecological footprint analysis. As a result, this book has four major parts:

1. Part I discusses the problems of sustainable development at the local level, focusing specifically on urban areas. It appears that, as it currently exists, the structure of the typical modern city is not sustainable. However, many initiatives which

could lead to more sustainable urban areas are introduced. Sustainability assessment at the local level can function only when it is linked to policies and visions for urban sustainability. This part examines how the Local Agenda 21 and "eco-city" approaches can be used as a framework for assessing urban sustainability.

2. Part II examines tools for sustainability assessment to be used in the decision-making process where any policy, plan, program, or project approval by local officials could be subjected to a sustainability assessment before a decision is made. This approach makes it possible to check if a proposal is in line with the policies and visions for urban sustainability. Basically, the assessment methodology followed is comparable to that of other impact assessment instruments.

3. Part III deals with tools for setting a baseline and measuring progress in sustainability assessment. Indicators for sustainable development can be used to set the baseline situation and measure progress toward goals and targets set by local leaders. In this part the indicators-approach to sustainability assessment is examined. Ecological footprint analysis is also an instrument that can be used to examine the actual sustainability situation and monitor progress. Sustainability reporting is an aid that supports and helps other sustainability tools and processes. It brings together relevant information to provide a basis for effective, integrated policy-making, action planning and awareness raising.

4. Part IV, the final part of the book, looks into sustainability assessment at the most local level, which is the individual or household level. The sustainability assessment of lifestyles is discussed.

This book is not the first attempt to examine sustainability assessment; it is, however, the first work to give an overview of many different initiatives and activities in this field at the local level and build on these to develop state-of-the-art techniques in sustainability assessment. In 1987, in the wake of the publication of the results of the Brundtland Commission (WCED 1987), the Canadian Environmental Assessment Research Council decided to examine the link between EIA and sustainable development at two meetings. The results of the discussions were reported by Jacobs and Sadler (1988) and showed the need for a sustainability assessment:

> Research on improvements to EA processes in support of sustainable development should be linked to [these] broader issues of evaluation and decision-making. In this regard, there is an urgent need for "second generation" assessment processes, employing new and expanded concepts, methods, and procedures. Sustainable development assessments (SDAs) should explicitly address the economic, social, and ecological interdependencies of policies, programs, and projects. These must be coordinated with other

policy and management instruments as part of an overall approach to environment-economy integration. Some formidable scientific and institutional difficulties (and, no doubt, disciplinary foot-dragging) will be encountered in designing integrated tools for assessment, planning, and decision-making. Not to begin this process, however, is to miss the point of the paradigmatic change that is taking place in environment and development thinking. (Jacobs and Sadler 1988)

In response to the need for improved ways of assessing sustainable development, the International Institute for Sustainable Development (IISD) received support from the Rockefeller Foundation to bring together an international group of measurement practitioners and researchers to review progress to-date and to synthesize insights from ongoing efforts. The meeting took place in 1996 in the Italian municipality of Bellagio and resulted in the so-called Bellagio Principles for Assessment (Hardi and Zdan 1997). These principles are presented in the following box and are an important basis for any attempt at sustainability assessment. The Bellagio Principles should be kept in mind while reading this book and in the development of any tool for sustainability assessment.

The Bellagio Principles for Assessment Toward Sustainable Development

Principle 1: Guiding Vision and Goals
Assessment of progress towards sustainable development should:

> ✦ be guided by a clear vision of sustainable development and goals that define that vision

Principle 2: Holistic Perspective
Assessment of progress towards sustainable development should:

> ✦ include review of the whole system as well as its parts

> ✦ consider the well-being of social, ecological, and economic subsystems, their state as well as the direction and the rate of change of that state, of their component parts, and the interaction between parts

> ✦ consider both positive and negative consequences of human activity, in a way that reflects the costs and benefits for human and ecological systems, in monetary and non-monetary terms

Principle 3: Essential Elements
Assessment of progress towards sustainable development should:

> ✦ consider equity and disparity within the current population and between present and future generations, dealing with such concerns as resource use, over-consumption and poverty, human rights, and access to services, as appropriate

✤ consider the ecological conditions on which life depends

✤ consider economic development and other, non-market activities that contribute to human/social well-being

Principle 4: Adequate Scope
Assessment of progress towards sustainable development should:

✤ adopt a time horizon long enough to capture both human and ecosystem time scales thus responding to the needs of future generations as well as those current short-term decision-making requirements

✤ define the space of study large enough to include not only local but also long distance impacts on people and ecosystems

✤ build on historic and current conditions to anticipate future conditions—where we want to go, where we could go

Principle 5: Practical Focus
Assessment of progress towards sustainable development should be based on:

✤ an explicit set of categories or an organizing framework that links vision and goals to indicators and assessment criteria

✤ a limited number of key issues for analysis

✤ a limited number of indicators or indicator combinations to provide a clear signal of progress

✤ standardizing measurement wherever possible to permit comparisons

✤ comparing indicator values to targets, reference values, ranges, thresholds, or directions of trends, as appropriate

Principle 6: Openness
Assessment of progress towards sustainable development should:

✤ make the methods and data that are used accessible to all

✤ make explicit all judgments, assumptions, and uncertainties in data and interpretations

Principle 7: Effective Communication
Assessment of progress towards sustainable development should:

✤ be designed to address the needs of the audience and the set of users

✤ draw from indicators and other tools that are stimulating and serve to engage decision-makers

✤ aim, from the outset, for simplicity of structure and use of clear and plain language

Principle 8: Broad Participation
Assessment of progress towards sustainable development should:

✤ obtain broad representation of key grass-roots, professional, technical, and social groups, including youth, women, and indigenous people—to ensure recognition of diverse and changing values

✤ ensure participation of decision-makers to ensure a firm link to adopted policies and resulting action

Principle 9: Ongoing Assessment
Assessment of progress towards sustainable development should:

✤ develop a capacity for repeated measurement to determine trends

✤ be iterative, adaptive, and responsive to change and uncertainty because systems are complex and change frequently

✤ adjust goals, frameworks, and indicators as new insights are gained

✤ promote development of collective leaning and feedback to decision-making

Principle 10: Institutional Capacity
Continuity of assessing progress toward sustainable development should be assured by:

✤ clearly assigning responsibility and providing ongoing support in the decision-making process

✤ providing institutional capacity for data collection, maintenance, and documentation

✤ supporting development of local assessment capacity

Source: Hardi and Zdan (1997)

Assessment of Sustainability

Sustainability assessment makes sense only when linked to an assessment framework. This framework should clearly state what the assessors consider to be positive and negative evolutions for sustainable development of our societies. In this way sustainable development could become a frame of reference for public policy. Sustainability assessment initiatives should start from the knowledge that sustainable development is not a fixed state of harmony and that the concept is value-based. Hardi and Zdan (1997) rightly state that the design of a sustainable world will depend upon an operating set of values and that achieving progress toward sustainable development is clearly a matter of choices made by individuals, families, communities, and organizations, as well as the government. These choices should take into account long-term consequences and recognize our place within the ecosystem. Also in 1988, Gardner attempted to develop an assessment framework to be used in sustainability assessment (Gardner 1988). The idea was to use the basic principles of sustainable development, presented in the following box, as premises that have to be supported, or at least not contravened, in decision-making processes, in order to steer a course toward sustainable development. A distinction is made between substantive and process-oriented principles. *Substantive* principles are value-oriented; they describe the ends of decision-making. *Process* principles describe the structure, context, and processes of decision-making that are necessary in the pursuit of sustainable development. This assessment framework developed by Gardner (1988) sums up the important principles of sustainable development, and these can be used as criteria when assessing the sustainability of policies, plans, programs, and projects. Such an assessment framework should, however, be complemented by other assessment criteria which are a more detailed translation of sustainable development to the local situation and the specific problem studied. Gardner's criteria are an interesting basis from which to develop more advanced assessment frameworks.

Principles of Sustainable Development to be Used as Assessment Criteria in a Sustainability Assessment

1. **Ideology:** *Goal-seeking*
   Process-oriented
   + 1a proactive, innovative, generates alternatives
   + 1b considers range of alternatives and impacts
   + 1c based on convergence of interests
   + 1d normative, policy-oriented, priority-setting
   Substantive
   + A1 quality of life and security of livelihood
   + B1 ecological processes and genetic diversity
   + C1 equitable access to resources, costs and benefits
   + D1 individual development and fulfillment, self-reliance

2. **Analysis:** *Relational*
   Process-oriented
   + 2a focused on key points of entry into the system
   + 2b recognizing linkages between systems and dynamics
   + 2c recognizing linkages within systems and dynamics
   + 2d importance of spatial and temporal scales
   Substantive
   + A2 development as qualitative change
   + B2 awareness of ecosystem requirements
   + C2 equity and justice within and between generations
   + D2 endogenous technology and ideas

3. **Strategy:** *Adaptive*
   Process-oriented
   + 3a experimental, learning, evolutionary, responsive
   + 3b anticipatory, preventive, dealing with uncertainty
   + 3c moderating, self-regulating, monitoring

➤ 3d maintaining diversity and options for resilience

Substantive

➤ A3 (growth for) meeting a range of human needs

➤ B3 maintenance, enhancement of ecosystems

➤ C3 avoid ecological limits and associated inequity

➤ D3 culturally-appropriate development

4.   **Organization:** *Interactive*

Process-oriented

➤ 4a collaborative for the synthesis of solutions

➤ 4b integration of management processes

➤ 4c integration of societal, technical, and institutional interests

➤ 4d participatory and consultative

Substantive

➤ A4 organizations must respond to societal change

➤ B4 ecological principles guide decision-making

➤ C4 democratic, political decision-making

➤ D4 decision-making locally initiated, participatory

Source: Gardner (1988)

As said before, the development of a strategic or long-term vision for the future is very much linked to the moral values of our societies, and these are to a large degree influenced by religious and/or non-religious philosophies of life. The U.S. Millennium Institute made a courageous and rare attempt to call on the world's spiritual leaders to become involved in a sustainable future. The Millennium Institute grew out of the well-known "Global 2000 Report to the President," which was commissioned by former U.S. President Jimmy Carter and developed by Gerald O. Barney in 1980 (Barney 1980). "The Global 2000 Revisited: What Shall We Do?" report (Barney et al. 1993) was prepared in 1993 for the Parliament of the World's Religions. It gives an overview of the critical issues with which we are faced, examines choices and tasks ahead, and looks into the role of faith as a partner in sustainable development. The following box is a summary of urgent questions posed to spiritual leaders regarding sustainable development. Clearly, religious leaders, as well as atheists, moralists, and philosophers, are a major force in the way human society will continue to develop. Through their inspirational leadership, they can influence heads of state, leaders of corporations and other institutions, and individuals to a changed way of thinking and living. By reminding us of our moral obligation to be stewards of the earth, they can play a powerful role in the development of visions for sustainable development and consequently, in the development of assessment frameworks to be used in sustainability assessment.

Questions for the World's Spiritual Leaders

What are the traditional teachings—and the range of other opinions—within your faith on how to meet the legitimate needs of the growing human community without destroying the ability of Earth to support the community of all life?

　✤ What does your faith tradition teach about how the needs of the poor are to be met as human numbers continue to grow? What does your faith teach about the causes of poverty? What trends and prospects do you see for the poor?

　✤ How are the needs and wants of humans to be weighed relative to the survival of other forms of life? What trends and prospects do you see for other forms of life?

What are the traditional teachings—and the range of other opinions—within your faith on the meaning of "progress" and how it is to be achieved?

　✤ What does your faith tradition teach about the human destiny? Is the human destiny separable from that of Earth?

　✤ What is your destiny, the destiny of the followers of your faith tradition? What does your tradition teach concerning the destiny of followers of other traditions?

　✤ How are we to measure "progress?" Can there be progress for the human community without progress for the whole community of life?

　✤ How is personal "success" related to "progress" for the whole?

What are the traditional teachings—and the range of other opinions—within your faith tradition concerning a proper relationship with those who differ in race or gender (conditions one cannot change), or culture, politics, or faith?

　✤ Much hatred and violence is carried out in the name of religion. What teachings of your faith tradition have been used—correctly or not - in an attempt to justify such practices?

　✤ Discrimination and even violence by men toward women is often justified in the name of religion. Which, if any, of the teachings of your faith have been used - correctly or incorrectly—in this way?

　✤ How does your faith tradition characterize the teachings and followers of other faiths? Do some adherents of your tradition hold that the teachings and followers of other faiths are evil, dangerous, misguided? Is there any possibility that your faith tradition can derive wisdom, truth, or insight from the teachings of another faith?

What are the traditional teachings—and the range of other opinions—within your faith on the possibility of criticism, correction, reinterpretation, and even rejection of ancient traditional assumptions and "truth" in light of new understandings or revelations?

　✤ Does your faith tradition envision new revelation, new understanding, new interpretation, new wisdom, and new truth concerning human activity affecting the future of Earth?

Source: Barney et al. (1993)

Sustainability Assessment and Decision-Making

One of the most important aims of EIA, when it was first introduced in the United States, was to change the decision-making process. The signing into law of the National Environmental Policy Act by President Richard Nixon on January 1, 1970, institutionalized a new direction in public policy that was to use science in an integrated interdisciplinary way to redress excessive weighting of agency decisions on the side of narrowly conceived economic and engineering considerations (Caldwell 1982). Moreover, EIA had to open up the decision-making process, make it transparent and available to external control. This was accomplished by introducing public participation and freedom of information legislation. Similarly, sustainability assessment should have the important function of encouraging decision-makers to give the appropriate attention to the sustainability characteristics of their policy, plan, program, and project intentions. Sustainability assessment should also clarify how planners, policy-makers, and decision-makers take into account the goals of sustainable development in the realization of their initiatives. This information should then become open and available to the public. Integration of environmental, social, and economic development issues is one of the top priorities of sustainable development. Therefore, Chapter 8 in Agenda 21 examines how environment and development considerations can be integrated in the decision-making process. The overall objective of Agenda 21 on this issue is to improve or restructure the decision-making process so that consideration of socioeconomic and environmental issues become fully integrated and a broader range of public participation is assured. To support a more integrated approach to decision-making, Agenda 21 proposes, among other things, the adoption of comprehensive analytical procedures for prior and simultaneous assessment of the impacts of decisions, including the impacts within and among the economic, social, and environmental spheres (UNCED 1992). Agenda 21 also calls for the introduction of Strategic Environmental Assessment (SEA) or impact assessment, which extends beyond the project level to policies and programs (UNCED 1992).

Ideally, sustainability assessment should be fully integrated into the decision-making process. One approach to decision-making in which sustainability assessment could be integrated is the "ACTION" approach, which tries to incorporate environmental planning and decision-making into one process. ACTION (Ryding 1994) stands for:

A   =   Anticipate all of the different environmental, technical, and economic consequences (insofar as humanly possible).

C   =   Conclude the most plausible economic and environmental development.

T   =   Tabulate possible advantages and disadvantages with all potential decisions.

I   =   Initiate a thorough evaluation of the advantages and disadvantages pertaining to the different alternative decisions.

O   =   Omit all less-useful alternatives, and highlight the preferable ones.

N   =   Negotiate with relevant parties to gain acceptance for the preferred action.

Moreover, Chapter 40 of Agenda 21 encourages us to see sustainability assessment as an instrument that leads to the production of information usable for decision-making. Therefore, mechanisms should be strengthened or established for transforming scientific and socioeconomic assessments into information suitable for both planning and public information (UNCED 1992).

Local Level Aspects of Sustainability Assessment

This book will examine sustainability assessment at the local level. Focus on the local level is linked to the sustainable development motto, "Think Globally, Act Locally." The importance of introducing sustainable development at the local level is reflected in Chapter 28 of Agenda 21, which deals with the introduction of Local Agenda 21 initiatives by local communities:

> Because so many of the problems and solutions being addressed by Agenda 21 have their roots in local activities, the participation and cooperation of local authorities will be a determining factor in fulfilling its objectives. Local authorities construct, operate, and maintain economic, social and environmental infrastructure, oversee planning processes, establish local environmental policies and regulations, and assist in implementing national and subnational environmental policies. As the level of governance closest to the people, they play a vital role in educating, mobilizing and responding to the public to promote sustainable development.

(Source: UNCED 1992)

In practice, Agenda 21 has proved to be very difficult to implement at the national and international levels. At the local level, however, many success stories can be told on the creative introduction of Local Agenda 21 processes. Moreover, indications can be found in which those municipalities that are most advanced with the development of a Local Agenda 21 (for example, the Belgian region of Flanders) are also the municipalities that mostly feel the need for the introduction of one or another form of Strategic Environmental Assessment (SEA) and sustainability assessment (Devuyst et al. 1998).

Sustainability assessment initiatives currently originate mostly in cities or municipalities. At the local level, sustainable development can be translated into tangible, action-oriented and smaller scale initiatives. Because the methodology of sustainability assessment is still in its infancy and needs further development, the local level is the best and most effective point to start working with sustainability assessment.

Although this book deals with sustainability assessment at the local level, initiatives at the national level also exist. For example, Threshold 21 is a computer model developed by the U.S.-based Millennium Institute that helps decision-makers to both plan for development and build consensus among key groups. The model projects the impact of policy alternatives graphically and provides an analysis of different scenarios into the future. Threshold 21 uses more than 1,500 equations to assess the impacts of changes in education, goods production, demographics, health care, nutrition, trade, agriculture, energy, forests, water, pollution, and technology. The model can be tailored to analyze a region, a sector, or a specific policy, and it enables the user to trace each impact back to its root causes. The resulting integrated assessment is a visually compelling presentation (Millennium Institute 1997). Sustainability assessment at higher levels of government should also be studied in the future, hopefully, learning from experience at the local level. The earth is becoming increasingly urbanized. In 1900, only 160 million people, one tenth of the world's population, were city dwellers. In contrast, shortly after the year 2000, half of the world (3.2 billion people) will live in urban areas, a 20-fold increase (O'Meara 1999). Moreover, the typical lifestyle of the modern European or North American urban or suburban inhabitant is a threat to sustainable development. For example, in the period of 1969–1988, the energy requirements of all households in the Netherlands increased by about 30 percent (Wilting and Biesiot 1998); and in addition to the growth in the use of the car in 1995, energy demand for passenger transport was almost six times higher than in 1960 (Noorman and Schoot Uiterkamp 1998). Another example is that the ecological footprint of the 2.4 million residents of Toronto at seven hectares per capita imposes roughly the equivalent load on Earth as that of the nearly 18 million residents of Bombay, India (India has an average per capita footprint of approximately one hectare) (see Rees in this volume). Therefore, focusing on sustainability assessment at the local level often means focusing on urban issues. This does not, however, mean that we should neglect smaller municipalities and rural communities. On the contrary, the survival of the city is closely linked to its hinterland and should be taken into account. The Habitat Agenda, the result of the United Nations Conference on Human Settlements which took place in Istanbul in 1996, also stresses the importance of urban issues (UNCHS 1996). The economic, social, political, and environmental futures of the planet will most certainly depend on how urban issues are addressed. These issues include the increasingly important role of government decentralization and of non-public actors, how local and national institutional and financial capacities need to be strengthened to address urban issues, and finally, that most financial and technical resources have to originate from countries, cities, and communities themselves (Mega and Pedersen 1998).

In the end a more sustainable society will only become a reality when each and everyone of us participates. Therefore, this book also looks at the most local level of households and individuals. Here we must take into account the trends in the changing composition of households. Most European countries have, for example, experienced similar trends toward smaller households due to decreasing levels of fertility, changes in the age structure of the population, and the diversification in living arrangements. Consequently, the numbers of households have increased in many countries. This has far-reaching consequences for the future demand for services such as housing, energy, and water (van Diepen 1998).

Sustainable Urban Development and Sustainable Community Development

Mike Davis's work on the city of Los Angeles (Davis 1991) shows that contemporary urban development in southern California not only results in typical environmental problems such as the excessive use of energy and water, overproduction of waste, and air pollution, but also that it has severe societal consequences. People in wealthier neighborhoods isolate themselves behind walls and shop in malls guarded by private security officers. The demand by the wealthy for increased spatial and social insulation leads to the destruction of accessible public space. Public amenities are radically shrinking, parks are becoming derelict and beaches more segregated, libraries and playgrounds are closing, and streets are becoming more desolate and dangerous. Davis's (1995) description of the development of Las Vegas is another example of how environmental, social, and economic aspects of city development are closely interlinked and related to urban planning, urban policy and decision-making, and sustainable development issues. When

Davis wrote his article on Las Vegas in 1995, the economy of the city was booming, resulting in nearly a thousand new residents arriving each week. Las Vegas shows the same problems as other cities in the southwest of the U.S., including: overconsumption of water, weak local government, limited availability of public space, refusal to use "hazard zoning" to mitigate natural disasters, dispersed land use, dictatorship of the automobile, and extreme social and racial inequality (Davis 1995).

Although Las Vegas is located in the fragile southern Nevada desert ecosystem, overconsumption of water is part of the Las Vegas lifestyle. Habitat for desert species, such as the desert tortoise, is threatened by recreational vehicles such as dune buggies and dirt bikes and by urban sprawl. The city limits encompass barely one-third of the metropolitan population. Poverty, unemployment, and homelessness are disproportionately concentrated within the boundaries of Las Vegas. Tax resources are separated from regional needs. The lack of public parks mostly affects low-wage service workers who do not have the means to go, for example, jet-skiing across Lake Mead or lounge by the hotel pools. Summer flash-flood problems are a result of not taking into account floodplains in developing the city, lack of open space, and the paving of large surfaces. Urban functions are not centralized, but widely dispersed throughout the city as a result of the use of cars. Consequently, walking in the city to go working, shopping, and so forth is no longer a viable option. Money goes to the construction of new highways, not to public transportation. Clark County (the municipal division in which Las Vegas is located) has the "lowest vehicle occupancy rate in the country" and the "longest per person, per trip, per day ratio." As a result, there is an increasing amount of days with unhealthy air quality. Hypergrowth has also increased social inequality, with jobless immigrants far outpacing the supply of new jobs. Clark County has witnessed soaring welfare caseloads, crime, mental illness, child abuse, and homelessness (Davis 1995). Davis's analysis of Las Vegas shows how urban decision-making can simultaneously lead to environmental problems and social exclusion.

Sustainable urban development must take a broad view of urban issues and attempt to solve urban problems by integrating environmental, social, and economic components. In other words, merely solving problems of air or water pollution will not bring us much further. What is needed is an inclusive approach that determines how a variety of pressing urban problems are linked, while taking into account social problems such as poverty, homelessness, the spreading of tuberculosis in major cities, HIV, or lack of access to potable water in cities in developing countries. For example, research in the U.S. indicates that there is a pattern of disproportionate exposure to environmental hazards and degradation among marginalized groups, including racial and ethnic minorities and poor, less educated, and politically powerless people. This phenomenon is referred to as "environmental racism" and has led to a movement for environmental justice (WRI 1999). Understanding the underlying mechanisms of environmental racism is important in the framework of sustainable urban development. It becomes clear that to reach its goals, sustainable urban development should be closely linked to sustainable community development. Maser (1997) defines sustainable community development as a community-directed process of development based on: a) transcendent human values of love, trust, respect, wonder, humility, and compassion; b) active learning, which is a balance between the intellect and intuition, between the abstract and the concrete, between action and reflection; c) sharing that is generated through communication, cooperation, and coordination; d) a capacity to understand and work with and within the flow of life as a fluid system, recognizing, understanding, and accepting the significance of relationships; e) patience in seeking an understanding of a fundamental issue rather than applying band-aid-like quick fixes to problematic symptoms; f) consciously integrating the learning space into the working space into a continual cycle of theory, experimentation, action, and reflection; and g) a shared societal vision that is grounded in long-term sustainability, both culturally and environmentally. Rudlin and Falk (1999) state that creating sustainable communities means creating neighborhoods in which change will take place naturally and gradually over time. Sustainable neighborhoods should be like sustainable forests: they should develop naturally and should be constantly renewed by new growth and contain a rich variety of species. Sustainable neighborhoods should enhance the quality of social and economic life of their citizens and be a joy to live in, to work in, or to visit. Places that are popular will attract people and investment and will be constantly renewed. Rudlin and Falk (1999) feel that the most important challenge of the sustainable urban neighborhood is to engender in its residents a feeling that they belong, a pride in the area, and a sense of responsibility for it. An example of sustainable community development is the introduction of furniture recycling centers in Belgian cities. These projects consist of using abandoned buildings in run-down neighborhoods as workshops where people can bring furniture that they no longer need. The furniture is renovated by long-term unemployed people who are often part of marginalized groups. The employees receive professional training, gain work experience, and may eventually find better paying jobs in the private sector. The renovated furniture may be purchased by the general public and the proceeds go to the running of the workshops. Additional funds are received from local, national, or international sources. From a sustainable development point of view, this type of project is vitally important because it creates employment opportunities, brings life into formerly abandoned buildings, saves resources, and in this case, reduces waste by reusing furniture.

Sustainable Development and Different Levels of Authority

Obviously, sustainable development cannot be limited strictly to the local level. Initiatives for sustainable development at this level should work synergistically with initiatives at higher levels of government. Local leaders cannot solve sustainability issues solely on their own and must coordinate and cooperate with other policy makers at all levels to reach a more sustainable state. The following hypothetical example, which deals with mobility issues in Brussels, Belgium, demonstrates this process. The Belgian capital city of Brussels is flooded by 322,000 commuters every working day (BIM 1997) who must commute from the Brussels suburbs, as well as communities in the Flemish and Wallon regions. Since 63.7 percent of commuters use their own automobiles, the result is heavy traffic on the roads leading to Brussels daily. Several measures could be introduced to make the home-to-work travel more in line with sustainable development principles. At the local level, the Brussels regional government and the 19 local authorities, which together make up the Brussels Region, can only take limited measures and would not be able to solve the problems on their own. In this example, collaboration would be necessary between the different regional government agencies, which ideally should make agreements to develop an interregional public transportation system of buses and light rail systems. Regional authorities could also collaborate to make it more difficult for people to build new houses in rural areas, stop the urban sprawl, and encourage people to live in already urbanized areas, preferably nearby their job. Moreover, in this example, federal officials could also play an important role because the Belgian railway system remains a federal institution. Better railway services between Brussels and its outlying areas would encourage drivers to leave their cars at home and travel to work by train. The Brussels region or the individual municipalities could take some important measures on their own, such as reducing the availability of parking spaces in the city center, making streets more friendly to bicycle and pedestrian traffic, and giving priority lanes and priority traffic lights to public transportation. A sustainability assessment would help determine whether this type of commuting would be possible through the introduction of measures (following unilateral agreement at all levels of municipal, provincial, regional, and federal government). The "GEO-1 Global State of the Environment Report 1997" of UNEP (1997) points out that the continued preoccupation with immediate local and national issues and a general lack of sustained interest in global and long-term environmental issues remain major impediments to environmental progress internationally.

As O'Connor (1998) puts it, we not only need to "think globally, act locally," but also to "think locally, act globally," and ultimately, to "think and act both globally and locally":

> "Acting globally" implies the awareness of strategic thinking and actions not only against ecologically and socially disastrous practices of a particular corporation or industry, but also the global institutions whose decisions affect the lives of hundred of million of people. The key targets are the International Monetary Fund (IMF), the World Bank, and the General Agreement on Tariffs and Trade (GATT) and the new regional linkages such as the European Commission, the North American Free Trade Agreement,   or NAFTA), and Japan's informal financial and industrial empires in Asia. Their policies regarding Third World debt and "economic adjustments," infrastructure investment, and the rules governing world and regional trade have created immeasurable ecological harm and human misery. "Act globally" means to make the IMF and other undemocratic world economic bodies accountable for their policies and programs, and to demand that future policies be geared to the needs of the people of the world and the globe's fragile ecologies, rather than to the interests of central banks, treasury ministries, and privately owned financial monopolies. (O'Connor 1998)

The focus of sustainable development on the local level should be counterbalanced with a clear commitment of local authorities towards cooperation with different levels of government and a global outlook on problems facing the earth. The 1990s are characterized to a large extent by signs of "Balkanization," tribalism, and the dissolution of states. The disintegration of Yugoslavia and its ethnic conflicts, the genocide in Rwanda, and the constitutional revisionist movement in the U.S. (which wants to throw federal power back to the states) are only a few examples of this. Frey (1995) also indicates trends towards Balkanization in population redistribution in the U.S., such as uneven urban revival, regional race divisions, regional divisions by skill and poverty, the baby boom and elderly realignments, and suburban dominance and city isolation. While on the one hand, sustainable development strives for self-reliance and empowerment of local communities, attention should be given equally to cultural, educational, and informational exchanges between communities, nations, and continents. The focus of sustainable development within any given locale should not be misused for the breeding of tension and animosity between neighboring communities. The emphasis on local characteristics in sustainable development has both ecological and sociological reasons. From an ecological perspective, development should not exceed the carrying capacity of the local ecosystem. On the other hand, sociological issues deal with fighting inequality between population groups in a local community and on a global scale. It also means making local communities more independent from negative outside influences. We should see this as fighting against exploitation of local populations by outsiders. It means, for example, standing up against outsiders who want to take advantage of a relatively weak, untrained local labor force, simply to maximize profits and to contribute little or no investment in local community development. Sustainable development also means fighting against uneven development of our urban areas. Sustainable development attempts to deal with overcoming these problems, and sustainability assessment could be used to check whether any given proposal takes into account the need for building bridges between disparate communities or groups. In other words, we need to find a delicate equilibrium between a local and global outlook on sustainability issues.

This is, for example, exactly what The Council of Europe tries to do with the introduction of the European Charter of Local Self-Government. The Council of Europe has been active for many years in the development of a charter with the goal of encouraging local autonomy. In the light of the discussion on sustainable development, this charter becomes increasingly important and could be a catalyst in the attainment of a more sustainable state at the local level. The Council of Europe is the custodian of human rights and the upholder of the principles of democratic government in Europe. As far back as 1957, the Council of Europe showed appreciation of the importance of local authorities by establishing a representative body for them at the European level, which has since become the Standing Conference of Local and Regional Authorities of Europe. The European Charter of Local Self-Government as a convention was opened for signature by the member states of the Council of Europe on October 15, 1985. It came into force on September 1, 1988. This charter is the first and only international convention to define and protect the principles of local self-government in Europe. The aim of the charter is to offset the lack of joint European standards to protect and further develop the rights and fundamental freedoms of local authorities, which is the level closest to the citizens, enabling them to effectively participate in taking the decisions which affect their everyday environment. The charter requires states that have ratified it to implement fundamental rules guaranteeing the political, administrative, and financial independence of local authorities (Council of Europe 1998). Inevitably, it comes down to a correct application of the subsidiarity principle. This means that problems are best solved in the subsystem in which they arise. For instance, within the European Union, the Maastricht Treaty lays down limits for actions at the level of the European Community by stating: "In areas which do not fall within its exclusive competence, the Community shall take action, in accordance with the principle of subsidiarity, only if and in so far as the objectives of the proposed action cannot be sufficiently achieved by the member states."

In practice this means that the European-level authorities:

✦ Must not rule its citizens' lives

✦ Must act only in areas of common concern and where states cannot act effectively on their own

✦ Must highlight the diversity while respecting the specific characteristics of the states, regions, and professions

✦ Must be close to the citizens, entrusting the implementation of programs to local authorities and national civil services

✦ Must be transparent, clear and accessible to all citizens (Ministère des Affaires Etrangères 1996)

In other words, higher officials should only intervene when there is a need to; they should not be too forceful, but rather let local authorities play their appropriate role.

Sustainability at the Local Level: Dream or Reality?

In the eyes of many people, sustainable development is a goal for dreamers. They do not consider it a viable option. Examples show us, however, that important gains toward more sustainable societies are possible, and that these are often the result of an alternative way of urban planning:

✦ In U.S. cities, less than one percent of all urban trips are made by bicycle; increasing this to 30 percent may seem impossible. This is, however, an everyday reality in Dutch cities, where approximately 30 percent of trips are made by bike (O'Meara 1999)

✦ In Stockholm, the city council allows homes to be built only a short walk from offices and stores, in so-called "transit villages" around suburban rail stations. This urban planning initiative prompted car trips to fall by 229 kilometers per person between 1980 and 1990 (O'Meara 1999)

✦ Planning for an efficient transit system in Curitiba, Brazil, has resulted in a car traffic decline by 30 percent since 1974, even as the population doubled (O'Meara 1999)

Urban planning should be carried out with cultural considerations of those who may be affected. Urban design can make or break community relations, and can, for example, incorporate the specific requirements of working mothers, children, and people with specific disabilities. For example, work by Dolores Hayden on non-sexist urban design looks into the typical suburban block and examines how it can be redesigned to accommodate the lifestyles of working women (Hayden 1996). Well-known experts such as Jan Gehl (1971), Jane Jacobs (1961), and Kevin Lynch (1981) have examined the planning and designing of livable urban neighborhoods. More recently, planners and architects such as David Rudlin and Nicholas Falk (1999) have been looking into the link between the sustainable urban neighborhood, the local community, and the lifestyle of individual residents of the area. These examples that lead cities to a more sustainable future are the result of a specific way of development, based on creative planning and decision-making of local officials. Sustainability assessment can be a catalyst in steering planning and decision-making away from unsustainability. The role of urban planning in sustainable development should therefore be taken seriously.

Town and country planning is one of the major responsibilities of local authorities. The way people live their daily lives is determined to a large extent by the planning and design of settlements. For example, someone living in a city in southern California is most likely to use the car for going to work, shoppig, and recreation because local planning there promotes mono-functional areas, low-density housing, and very limited public transport (De Weerdt et al. 1996).

Someone living and working in downtown Paris or New York, on the other hand, is likely to travel by public transport because of the compactness of the city, the mix of residential and other functions, an extensive metro system, and the difficulty of finding a parking space. In Amsterdam and Beijing many people travel by bicycle. But this is not possible in Los Angeles where the distances to be covered are long; it is quite dangerous in Paris and New York because of the aggressive stream of motorized traffic. However, the inhabitants of Paris or Amsterdam, who travel by bus or bike, might not be any more environmentally conscious or interested in living more sustainably than the inhabitants of Los Angeles, but their respective ways of life are, to a large extent, a result of many decades of urban planning (De Weerdt et al. 1996). Sustainability thus implies different solutions for different places. It implies that the use of energy and materials in urban areas should be in balance with what the region can continuously supply through natural processes such as photosynthesis, biological decomposition, and the biochemical processes which support life. The immediate implication of this principle is a vastly reduced energy budget for cities and a smaller, more compact urban pattern interspersed with productive areas to collect energy, grow crops for food, fiber and energy, and to recycle wastes. New urban technologies should be less dependent on fossil fuels and rely more on information and a careful integration with biological processes. This should create cities of far greater diversity of design than we have today, with each region developing in a unique urban form based on its own regional characteristics. This diversity has been overridden and ignored in the past by the availability of cheap energy, the great leveler of regional diversity and character. A sustainable community exacts less from its inhabitants in time, wealth, and maintenance, and it demands less from its environment for land, water, soil, and fuel (Van der Ryn and Calthorpe 1986). In 1986, WHO launched its Healthy Cities project. This project makes use of an inclusive approach to solving the health problems of urban populations. The basic assumption is that future improvements in health will not be due principally to medical interventions, but will arise from improving the environment and promoting lifestyles conducive to health (Davies and Kelly 1993). For WHO, a healthy city is one that is engaged in a process of creating, expanding, and improving the physical and social environments and community resources that enable people to mutually support each other in performing all the functions of life and developing their maximum potential. A healthy city in the future would have a clean, safe, high-quality physical environment, and would operate within its ecosystem. The basic human needs of the city's inhabitants (food, water, shelter, income, safety, work) would be met. The community would be strong, mutually supportive, and non-exploitative, participating actively in community governance. Individuals would have access to a wide variety of experiences and resources with the possibility of multiple contacts and interactions with other people. The city would have a vital, diverse economy and its people would have a strong sense of connectedness with their biological and cultural heritage, as well as with other groups and individuals within the city. The city's physical and governmental design would be compatible with and support all of these circumstances. There would also be an optimum level of public health and appropriate health care services accessible to all (WHO 1988).The WHO Healthy Cities project uses an approach that is in line with principles of sustainable urban development in which health problems are considered in a wider context, and more healthy people will be the result of actions in a wide range of different fields, including measures for environmental, social, and economic health.

Smart Growth: Another Term for Sustainable Urban Development?

Is sustainable development of our urban areas too expensive? Is sustainable development incompatible with economic growth? Making our cities more compact is an example in which both the environmental and the financial factors improve. O'Meara (1999) states that changes in the layout of urban neighborhoods can lower energy demands from transportation by a factor of 10. Low-density neighborhoods require more water and sewer pipes, power lines, roads, and building materials. Economic benefits from compact growth have been calculated. For example, the State of New Jersey would save 1.3 billion (U.S.) dollars if a compact growth development option was followed instead of the sprawl-as-usual. Similarly, if rapid suburban development in the State of Maryland continues at its current pace between 1995 and 2020, it is predicted that new sewers, water pipes, schools, and roads will cost about $10 billion more than if population growth were accommodated by more condensed development (O'Meara 1999). In other words, building compact cities not only saves on energy and building materials, but also on transportation costs and taxpayers money. Moreover, "Smart Growth" initiatives show that the economy can grow without jeopardizing the state of the green areas outside the cities. The State of Maryland is a major advocate of Smart Growth with the 1997 passage of a wide-reaching package of legislation to strengthen the state's ability to direct growth and to enhance older developed areas. Smart Growth in Maryland is introduced in five different and mutually reinforcing ways:

- ✦ Smart Growth Areas Act: a law limiting most state infrastructure funding related to development to existing communities or to those places designated by state or local governments for growth
- ✦ Rural Legacy: a grant program to create greenbelts to protect geographically large rural areas from sprawl through purchase of easements and development rights
- ✦ Brownfields: three programs to facilitate clean-up of contaminated areas and commercial/industrial development of those sites
- ✦ Job Creation Tax Credit: income tax credits to businesses that create new jobs within designated areas to promote redevelopment
- ✦ Live Near Your Work Demonstration Program: state, employer, and local government matching cash grants to home buyers who purchase homes near their workplace

Maryland is not alone in this. Voters in the 1998 U.S. elections sent a strong message about land use. After decades of growth that destroyed forests and fields for the development of new suburbs, some 200 state and local ballot measures aimed at conservation and more compact development got citizen approval (Johnson 1999).

Performance Measurement by Government Authorities

An important trend in government functioning of the 1990s is to measure and report on performance. Performance measurement by government authorities seeks to promote improved government performance through better planning and reporting of the results of government programs. This is also what sustainability assessment and sustainability reporting aim to do. Many examples can be given from governments from every corner of the world that introduce initiatives for performance measurement. For example, in 1993, the U.S. Congress passed and the President signed into law a bipartisan initiative known as the Government Performance and Results Act (GPRA), also known as the Results Act. The purpose of the Results Act is to improve the efficiency and effectiveness of Federal programs by establishing a system to set goals for program performance and to measure results. In 1998 the requirements of the Results Act for five-year strategic planning, annual program performance plans, and annual program performance reports should have come into force government-wide (Congressional Institute 1999). The Act aims to shift the focus of federal management and accountability from a preoccupation with staffing, activity levels, and completed tasks to a focus on results. It wants public officials to look into the real difference that federal programs make in people's lives (USGAO 1997). Another example is the Flemish Region in Belgium where a decision was made in December 1995 to introduce a strategic planning process. This process includes a cycle of annually returning activities. Strategic plans are developed for many different fields, such as the environment, the Flemish scientific research policy, the urban problems, employment, mobility, and land-use. Highlights of the Flemish strategic planning process for each step are (Administratie Planning en Statistiek 1995):

➻ External factor analysis: this analysis gives insight in the relationship between the government and the external actors and factors which have a role to play in the particular subject for which a strategic plan is developed

➻ Strategic goal-oriented analysis: in this phase it is determined which results the government wants to reach within a certain period of time. Strategic goals and Critical Success Factors (CSFs) have to be developed. Continual measurement is necessary to evaluate if expected results are met. Therefore, we need to develop indicators

✤ Efficiency-oriented analysis: all activities, performances, and expected results in relation to the strategic plan should be evaluated against their pre-set goals. This type of analysis should always be part of the planning process and should lead to an action plan indicating priorities

✤ Resource analysis: this analysis should look into the resources needed to reach the goals and should be presented in a long-term budget

✤ Consolidation: the consolidation phase results in a coherent and balanced strategic plan to which a correct long-term budget is linked. It should take into account all rational elements which are the result of the efficiency-oriented analysis and the resource analysis

Performance measurement has also been widely used in the U.K. since the 1980s to improve the quality and responsiveness of the public service, and in New Zealand, the 1974 Local Government Act requires local authorities to report annually against performance targets which must address quality, quantity, timeliness, location, and cost (Productivity Commission 1997). In Australia, the Commonwealth Minister for Local Government asked the Productivity Commission to review the value and feasibility of developing national performance indicators for local authorities. The key question for the review was how best to use the tool of performance measurement to facilitate better performance by local government. Interest in developing performance indicators for local government has been evident in Australia at commonwealth, state, and local government levels since the 1980s. In the early 1990s, the national drive for microeconomic reform and associated attempts to improve efficiency and effectiveness in the public sector provided further impetus to the Local Government Ministers' Conference to develop national performance indicators for local government. In 1994, a manual was developed to help councils to benchmark their performance (Productivity Commission 1997). At local level in the U.S., the city of Sunnyvale, California, is at the forefront of performance measurement. One underlying reason for the success achieved in Sunnyvale is the fact that every program manager uses the system to plan, manage, and assess progress on a day-to-day basis. Each year the city submits a detailed annual performance report to the mayor and city council, indicating how well the performance objectives have been achieved. The objectives are tied into twenty-year strategic plans covering 28 areas of city service, showing long-term goals for the city. The development of sustainability assessment systems, sustainability reporting, indicators, and targets for sustainable development arise logically from this worldwide trend of focusing on government performance measurement. Therefore, this book also looks more closely into the use of reporting, indicators, and targets. The real use of sustainability indicators in measuring urban sustainability performance is currently still in an initial stage. Indicators are considered more useful if they can be judged against specified benchmarks and thresholds. Indicators are meaningless without specified objectives, and they cannot contribute to the improvement of the urban quality of life if there is no policy framework giving directions for change linked to objectives and targets to be attained (Mega and Pedersen 1998).

Concluding Remarks: Impact Assessment in Societies Around the World

This book has been written from the perspective of several European and North American societies, and the majority of examples come from this region as well. However, the basic principles of sustainability assessment are applicable anywhere. The main difference between a sustainability assessment in the Northern and Southern hemispheres would be found in the topics examined and in the vision for sustainable development which would serve as assessment framework. Whatever development phase a local community is in, the use of sustainability assessment will be helpful in reaching a more sustainable state. Sustainability assessment can be useful any time decisions have to be made on the introduction of new policies, plans, or other initiatives.

It is impossible to single out regions of the world where sustainability assessment would be more important in comparison with others. All societies should reach out to a more sustainable state, each in their own way. Nations that are already highly developed should focus on the undoing of (and alternatives to) harmful activities, whereas still=developing regions would have the option to try and reach a more sustainable society without first copying the western, unsustainable style of development. The more changes or development initiatives that take place in a society, the more important the instrument of sustainability assessment becomes.

Sustainability assessment can be used by all governments and decision-makers. However, in general, when we look at the concept of "impact assessment," it implies a number of activities that require governments and decision-makers to be open to public interference, input from experts and scientists, and control by independent bodies. Therefore, one can question the feasibility of introducing impact assessment instruments in non-democratic nations. The basic goals of impact assessment should lead to the opening up of the decision-making process, the participation of the public, the introduction of scientific information in the decision-making process, and the reduction of the number of irrational decisions or decisions that are not for the good of society in general. These effects of introducing an impact assessment system do not seem to be compatible with the operations of non-democratic regimes. But, then again, introducing even an incomplete impact assessment system in non-democratic nations may start a slow and gradual process of an evolution towards a more democratic way of government.

Impact assessment, and therefore also sustainability assessment, will only work when authorities are committed to it and want to make it work. There are many examples of authorities that introduce EIA systems because they are legally bound to do so, but do not really believe in its usefulness. They develop minimal EIA systems without guarantees for quality of the reports produced. Since these authorities do not provide sufficient support, money, or motivation, their EIA systems do not work well, which gives these authorities added proof of its uselessness (Devuyst 1994).

It should be clear that sustainability assessment is more than an instrument to be applied by governments and public agencies. It could, for example, also be performed by "watchdogs" (commissions of independent experts) or by Non-Governmental Organizations (NGOs). These organizations could scrutinize the policy-making and planning practices of local and other authorities, implementing the sustainability assessment methodology. Businesses could also make use of sustainability assessment, especially multinational companies that have widely diversified activities around the globe. When a society transforms to a more sustainable state, the most successful industries will be those that anticipate the transition and position themselves to exploit the huge investment opportunities created (Roodman 1999). Sustainability assessment could guide companies in an internal process of policy-making and planning in line with sustainable development principles. As is shown in the last part of this book, sustainability assessment is also applicable at the household and individual level, opening up the instrument of sustainability assessment to a wider public. Additional research should be carried out to further develop sustainability assessment methodologies adapted to the many different roles it could play in society. It is clear, for example, that a sustainability assessment performed by one of the children of a household, as part of a school assignment, needs to satisfy different requirements than a sustainability assessment for the planning department of a major city. Obviously, one of the main challenges for the future of sustainability assessment is its applicability not as a technocratic and rigid instrument, but as a flexible set of tools that can be naturally integrated into all corners of society.

References

Administratie Planning en Statistiek. 1995. Overzicht van het strategisch Planningsproces. Strategisch plan voor Vlaanderen. Brussels: Ministerie van de Vlaamse Gemeenschap.

Barney, G. O. 1980. *The Global 2000 Report to the President, Volumes 1, 2 and 3.* Washington: US Government Printing Office.

Barney, G.O., J. Blewett, and K. R. Barney. 1993. *Global 2000 Revisited: What Shall We Do? A Report on the Critical Issues of the 21st Century Prepared for the 1993 Parliament of the World's Religions.* Arlington, VA: Millennium Institute.

BIM, Brussels Instituut voor Milieubeheer. 1997. *Bescherm het leefmilieu: mobiliseer uw bedrijf! Richtsnoer voor de invoering van bedrijfsvervoerplannen in het Brussels Hoofdstedelijk Gewest.* Brussels: Brussels Instituut voor Milieubeheer.

Caldwell, L. K. 1982. *Science and the National Environmental Policy Act. Redirecting Policy through Procedural Reform.* University, AL: University of Alabama Press.

Carew-Reed, J., R. Presott-Allen, S. Bass, and B. Dalal-Clayton. 1994. *Strategies for National Sustainable Development. A Handbook for their Planning and Implementation.* London, UK: International Institute for Environment and Development and Gland, Switzerland: World Conservation Union.

Clark, W. C. 1989. Managing Planet Earth. *Scientific American* 261, 3: 47-54.

Congressional Institute. 1999. Government Performance and Results Act of 1993 Report of the Committee on Governmental Affairs United States Senate to Provide for the Establishment, Testing, and Evaluation of Strategic Planning and Performance Measurement in the Federal Government, and for Other Purposes. The Results Act. Website http://server.conginst.org/congist/results/index%21. html (18 Jan 1999).

Council of Europe. 1998. *European Charter of Local Self-Government and Explanatory Report.* Strasbourg: Council of Europe Publishing.

Davies, J. K. and M. P. Kelly. 1993. *Healthy Cities. Research and Practice.* London: Routledge.

Davis, M. 1991. *City of Quartz: Excavating the Future in Los Angeles.* London: Verso.

Davis, M. 1995. House of cards. Las Vegas: Too many people in the wrong place, celebrating waste as a way of life. *Sierra Magazine* 80, 6: 36-43.

Devuyst, D. 1994. *Instruments for the Evaluation of Environmental Impact Assessment.* Ph.D thesis. Brussels: Human Ecology Department, Vrije Universiteit Brussel.

Devuyst, D., T. van Wijngaarden, G. De Beckker, and L. Hens. 1998. *Duurzame ontwikkeling in steden en gemeenten van het Vlaamse Gewest: een stand van zaken.* Research Report 1 of the Environmental Impact Assessment Center, Human Ecology Department, Vrije Universiteit Brussel. Brussels: Environmental Impact Assessment Center.

De Weerdt, H., J. Van Assche and D. Devuyst. 1996. The role of local authorities in achieving sustainable development. In B. Nath, L. Hens, and D. Devuyst, eds., *Sustainable Development.* Brussels: VUBPress.

Frey, W. H. 1995. The new geography of population shifts: Trends toward Balkanization. In R. Farley, ed., *State of the Union: America in the 1990s. Volume 2: Social Trends:* 271-336. New York: Russell Sage Foundation.

Gardner, J. E. 1988. The elephant and the nine blind men: An initial review of environmental assessment and related processes in support of sustainable development. In P. Jacobs and B. Sadler, eds., *Sustainable Development and Environmental Assessment: Perspectives on Planning for a Common Future.* Hull, Canada: Canadian Environmental Assessment Research Council.

Gehl, J. 1971. *Livet mellem husene.* Copenhagen: Arkitektens Forlag.

Hayden, D. 1996. What would a non-sexist city be like? Speculations on housing, urban design, and human work. In R. T. LeGates and F. Stout, eds., *The City Reader*. London: Routledge (first published in 1980).

Hardi, P. and T. Zdan. 1997. *Assessing Sustainable Development. Principles in Practice*. Winnipeg: International Institute for Sustainable Development.

Hens, L. 1996. The Rio conference and thereafter. In B. Nath, L. Hens and D. Devuyst, eds., *Sustainable Development*. Brussels: VUBPress.

Jacobs, J. 1961. *The Death and Life of Great American Cities*. New York: Vintage Books.

Jacobs, P. and B. Sadler. 1988. *Sustainable Development and Environmental Assessment: Perspectives on Planning for a Common Future*. Hull, Canada: Canadian Environmental Assessment Research Council.

Johnson, C. 1999. The new metropolitan agenda. Sprawl is out. Sustainable neighborhoods are in. *Urban Age* Winter 1999.

Lynch, K. 1981. *A Theory of Good City Form*. Cambridge, Mass.: MIT Press.

Maser, C. 1997. *Sustainable Community Development. Principles and Concepts*. Delray Bay: St. Lucie Press.

Mega, V. and J. Pedersen. 1998. *Urban Sustainability Indicators*. Dublin: European Foundation for the Improvement of Living and Working Conditions.

Millennium Institute. 1997. Threshold 21. *Millennium Institute*. Website http://www.igc.org/millennium/t21/index.html (13 Nov 1998).

Ministère des Affaires Etrangères. 1996. The Principle of Subsidiarity. *French Ministry of Foreign Affairs*. Website http://www.france.diplomatie.fr/frmonde/euro/eu05.gb.html (26 Febr 1999).

Noorman, K. J. and A. J. M. Schoot Uiterkamp. 1998. Diagnosing and evaluating household metabolism. In K. J. Noorman and T. Schoot Uiterkamp, *Green Households? Domestic Consumers, Environment and Sustainability*, pp. 236-53. London: Earthscan.

O'Connor, J. 1998. Think globally, act locally? Toward an International Red Green Movement. In J. O'Connor, *Natural Causes. Essays in Ecological Marxism*, pp. 302-03. New York: The Guilford Press.

O'Meara, M. 1999. Exploring a new vision for cities. In L. Brown, C. Flavin and H. F. French, eds., *State of the World 1999. A Worldwatch Institute Report on Progress Toward a Sustainable Society*. New York: W.W. Norton & Company.

Productivity Commission. 1997. Performance measures for councils—Improving local government performance indicators. *Productivity Commission*. Website http://www.indcom.gov.au/inquiry/localgov/final/index.html (18 Jan 1999).

Redclift, M. 1993. Sustainable development: Needs, values, rights. *Environmental Values* 2, 1: 3-20.

Roodman, D.M. 1999. Building a sustainable society. In L. Brown and C. Flavin, eds., *State of the World 1999. A Worldwatch Institute Report on Progress Toward a Sustainable Society*. New York, London: W.W. Norton & Co.

Rudlin, D. and N. Falk. 1999. *Building the 21st Century Home. The Sustainable Urban Neighbourhood*. Oxford: Architectural Press.

Ryding, S. O. 1994. An environmental management initiative. In Ryding, ed., *Environmental Management Handbook*. Amsterdam, Oxford: IOS Press and Boca Raton: Lewis Publishers.

Simon, J. L. 1996. *The Ultimate Resource 2*. Cambridge: Princeton University Press.

UNAIDS. 1999. What is the International Partnership Against AIDS in Africa. *International Partnership Against AIDS in Africa*. Website http://www.unaids.org/africapartnership/whatis.html#aids (30 Nov 1999).

UNCED, United Nations Conference on Environment and Development. 1992. *Agenda 21: Program for Action for Sustainable Development.* New York: United Nations Conference on Environment and Development.

UNCHS, United Nations Conference on Human Settlements. 1996. *Report of the United Nations Conference on Human Settlements (Habitat II).* Istanbul: United Nations.

UNEP, United Nations Environment Programme. 1997. "United Nations Environment Programme Global State of the Environment Report 1997." *Global Environmental Outlook-1.* Website http://www.grid.unep.ch/geo1/exsum/ex1. htm (3 Mar 1999).

UNFPA, United Nations Population Fund. 1998. *The State of World Population 1998.* Website http://www.unfpa.org/swp/1998/pdffiles.htm (5 Mar 1999).

USGAO, United States General Accounting Office. 1997. Measuring Performance. Strengths and Limitations of Research Indicators. Washington, DC: General Accounting Office.

Van der Ryn, S. and P. Calthorpe. 1986. *Sustainable Communities: A New Design Synthesis for Cities, Suburbs, and Towns.* San Francisco: Sierra Club Books.

van Diepen, A. M. L. 1998. Developments in household composition in Europe. In K.J. Noorman and T. Schoot Uiterkamp, eds., *Green Households? Domestic Consumers, Environment and Sustainability,* pp. 82-100. London: Earthscan.

WCED, World Commission on Environment and Development. 1987. *Our Common Future.* Oxford: Oxford University Press.

WHO, World Health Organization. 1988. *Promoting Health in the Urban Context.* Copenhagen: WHO Healthy Cities Papers no. 1.

WHO, World Health Organization. 1997. *Health and Environment in Sustainable Development: Five Years After the Earth Summit.* Geneva: World Health Organization.

Wilting, H. C. and W. Biesiot. 1998. Household energy requirements. In K. J. Noorman and T. Schoot Uiterkamp, eds., *Green Households? Domestic Consumers, Environment and Sustainability,* pp. 64-82. London: Earthscan.

Worldwatch Institute. 1999. World may be on edge of environmental revolution. *Worldwatch Institute.* Website http://www.worldwatch.org/alerts/990225.html (4 Mar 1999).

WRI, World Resources Institute. 1999. 1998-1999 World Resources. Environmental Change and Human Health. Washington, DC: World Resources Institute.

The Conundrum of Urban Sustainability

William E. Rees

By early in the coming decade, the majority of humankind will be living in cities. For the first time in the two-million-year history of our species, the immediate human environment will be the "built environment." By some conventional accounts, accelerating urbanization is just another piece of evidence that the human economy is "decoupling" from the environment, that humanity is finally leaving nature and the rural countryside behind. This is a perceptual error. Even as we urbanize, human beings remain dependent on the environment, and cities become the major drivers of global ecological change. Moreover, the sustainability of our cities is inextricably linked to the integrity and sustainability of the rural hinterland. It follows that cities are simultaneously part of the problem and a "key to [global] sustainability" (Rees and Wackernagel 1996).

Why are some observers misled by urbanization? Part of the answer lies in the fact that urbanization is usually regarded solely as a demographic or economic phenomenon, as just one in a series of progressive steps in human socio-cultural evolution that takes us away from our origins in hunting and gathering. Indeed, many people see urbanization as a transition to a higher plane of civilization. Western industrial society tends to associate "rural" and "agricultural" with general underdevelopment and a presumptively inferior peasant culture. After all, it is the city that serves as the seat of government, the engine of economic growth, the center of culture, the wellspring of new knowledge, and the repository of cumulative learning (Jacobs 1985).

It is true, of course, that cities can be all of these things—our greatest cities are among the most magnificent of human achievements. But there is another side to cities and urbanization to which the modern technological eye is blind. Urbanization represents an unprecedented—but oddly ambiguous—human ecological transformation. The massive migration of people to cities produces a dramatic shift in the spatial relationships of human populations to "the land," but this shift occurs without in any way reducing the material dependence of people on the biophysical products and services of that land.

This quasi-paradox has serious implications for sustainability. First, urbanization distances people both physically and psychologically from the ecosystems that support them. This immediately decreases the typical city dweller's sense of dependence on the land, a trend that is exacerbated by their increased reliance on trade. Because it eliminates any negative feedback on their own population or lifestyle, trade reduces the incentive for urban populations to conserve local ecological resources. For example, agricultural land near sprawling cities is frequently lost to food production so that it can rise to its economically "highest and best use." Also, people who live on imports are blind to any far-off negative ecological and social consequences of production processes in the exporting areas. Meanwhile, social pressures and the generally higher income associated with the urban lifestyle increases both the propensity and capacity of city dwellers to consume. Thus, urbanization, combined with trade, has the potential to accelerate the loss of both local and global carrying capacity (Rees 1994; Rees and Wackernagel 1994).

Herein lies the urban sustainability conundrum: even as urbanites lose their sense of direct connection to nature, the city demands ever-greater quantities of food, material commodities, and energy—which must often be shipped great distances—to sustain the increasingly consumer lifestyle of itsr inhabitants. Meanwhile, the processing and consumption of these resources returns an equal volume of degraded and often toxic waste back into the environment at great cost to local ecosystems and ultimately, the global commons. In short, whether through commercial trade or natural flows, modern cities are able to draw on resources everywhere and dump their garbage all over the world.

Upping the Urban Ante

While "50 percent urban" is an important milestone on the road to an urban world, the process is still accelerating in many countries. From 1950 to 1990, the world's urban population increased by 200 million to about 2 billion, adding an average of just 1.1 percent of the 1950 level each year. In the 1990s, the pace of urbanization picked up. The world's urban population is expected to have grown by about 50 percent in this decade to almost three billion by the end of the year 2000. By 2025, the UN projects that about 5.1 billion people will reside in cities, an increase of 70 percent in the first quarter of the century (UN 1994; UNDP 1998). If this projection holds true, it means that in the next 27 years, *the urban population alone will grow by the equivalent of the total human population in the early 1930s.*

Not only are there more cities, but existing cities are getting bigger. In 2015, almost 17 percent of urban dwellers will live in large cities of over 5 million, and there will be 71 such megacities by 2015. Significantly, most urban growth will occur in the less-developed countries, which are by definition less-well equipped to cope with it. The Third World is projected to have 114 cities of over 4 million people in 2025, up from only 35 cities in 1980 (UNDP 1998); of the 44 cities globally with between 5 and 10 million inhabitants in 2015, some 36 to 39 will be in the developing world. Significantly, in 1950, the only city to exceed 10 million people (New York) was in a developed country. However, by 2015, 23 of the 27 cities expected to reach this size will be in less developed countries (UN 1994).

Many cities in the developing world are woefully ill-prepared to accommodate this population tidal wave. In the mid 1990s, as many as 25 percent of urban dwellers in the developing world did not have access to safe potable water supplies and 50 percent lacked adequate sewage facilities. Accordingly, the World Bank estimates that developing countries alone will need to invest 200 billion U.S. dollars a year in basic infrastructure in the period to 2005, most of it for urban regions. Given anticipated urban population growth and material demand to 2025, "...it would be reasonable to expect the total volume of investment [in infrastructure] to reach six trillion [U.S.] dollars by that time" (NRTEE 1998:11).

Keep in mind that the world must construct adequate new physical accommodations to support an urban population increase for just the next 27 years, which will be the equivalent of the total accumulation of people in all of history up to the 1930s. Thus, in addition to basic infrastructure, the world's cities will have to construct millions of new dwelling units, stores, and offices; thousands of new schools, hospitals, and water and waste-water treatment plants; countless square kilometers of new roads and parking facilities for tens of millions of additional motor vehicles; and all manner of supportive transportation, communications, and related urban infrastructure. (In effect, we will be doubling the 1970s urban presence on the planet.) In the face of such data, it is little wonder that most people expect cities to be "using much more energy,[materials,] water, and land than ever before and doing so in more concentrated, land-, capital-, knowledge-, and technology-intensive ways" (NRTEE 1998:2).

Cities and Sustainability

The prevailing pace and pattern of urbanization obviously has enormous potential to severely dim the prospects of global sustainability. At the same time, the economies of scale and agglomeration economies associated with the city might well be exploited to enhance the future of humanity (Girardet 1992; Rees 1992; Rees and Wackernagel 1996; Rees 1997). The critical question is: How can we reconcile the expected use of so much more of everything with the accumulating evidence that the human "load" on the ecosphere already exceeds global carrying capacity? (see Goodland 1991)

Despite the potentially pivotal role of cities, and despite two UN Habitat conferences on urban prospects (Vancouver in 1976 and Istanbul in 1996), the future role of the city has been all but ignored in the mainstream sustainability debate. For example, the World Conservation Strategy of 1980, which apparently first explicitly used the term "sustainable development," paid no special attention to accelerating urbanization. The Brundtland report did raise the issue, but kept the main emphasis on the "urban crisis in developing countries" (WCED 1987: 8), effectively letting more prosperous nations off the sustainability hook.

Overlooking the future impact of the city on sustainability—particularly rich cities—is difficult to reconcile with physical reality. Up to 80 percent of the populations of high-income countries already live in cities and, as noted, half of humanity will be urbanized by early in the twenty-first century. High-income cities in particular impose ever greater burdens on nature. Since the wealthiest 25 percent of the human population consumes 80 percent of the world's economic output (WCED 1987), approximately 64 percent of the world's economic production/consumption and pollution is associated with cities in *rich* countries. Ten percent or less is tied to cities in the developing world. In short, approximately "half the people and three-quarters of the world's environmental problems reside in cities; and rich cities, mainly in the developed Northern Hemisphere, impose by far the greater load on the ecosphere and global commons" (Rees 1997: 304). "Today, with [increasing numbers of] megacities feeding off a global hinterland, new popular ideas are required to ensure the sustainability of our cities" (Girardet 1992: 40). A sustainability assessment of urbanization is clearly long overdue.

References

Girardet, H. 1992. *The Gaia Atlas of Cities: New Directions for Sustainable Urban Living.* London: Gaia Books (also: Anchor/Doubleday).

Goodland, R. 1991. The case that the world has reached limits. In R. Goodland, H. Daly, S. El Serafy and B. Von Droste, eds., *Environmentally Sustainable Economic Development: Building on Brundtland.* Paris: UNESCO.

Jacobs, J. 1985. *Cities and the Wealth of Nations: Principles of Economic Life.* New York: Vintage Books.

NRTEE. 1998. *Canada Offers Sustainable Cities Solutions for the World.* Discussion Paper for a Workshop. Ottawa: National Round Table on the Environment and the Economy.

Rees, W. E. 1992. Ecological footprints and appropriated carrying capacity: What urban economics leaves out. *Environment and Urbanization* 4, 2: 121-130.

Rees, W. E. 1994. Pressing global limits: Trade as the appropriation of carrying capacity. In T. Schrecker and J. Dalgleish, eds., *Growth, Trade and Environmental Values.* London, ON: Westminster Institute for Ethics and Environmental Values.

Rees, W. E. 1997. Is 'sustainable city' an oxymoron? *Local Environment* 2: 303-310.

Rees, W. E. and M. Wackernagel. 1994. Ecological footprints and appropriated carrying capacity: Measuring the natural capital requirements of the human economy." In A-M.

Jansson, M. Hammer, C.Folke and R. Costanza, eds., *Investing in Natural Capital: The Ecological Economics Approach to Sustainability.* Washington, DC: Island Press.

Rees, W. E. and M. Wackernagel. 1996. Urban ecological footprints: Why cities cannot be sustainable and why they are a key to sustainability. *EIA Review* 16: 223-248.

UN, United Nation. 1994. *World Urbanization Prospects: The 1994 Revision.* New York: The United Nations.

UNDP, United Nations Development Programme. 1998. *Urban Transition in Developing Countries.* New York: United Nations Development Programme.

Wackernagel, M. and W. E. Rees. 1996. *Our Ecological Footprint: Reducing Human Impact on the Earth.* Gabriola Island, BC and New Haven, CT: New Society Publishers.

WCED, World Commission on Environment and Development. 1987. *Our Common Future.* Report of the [UN] World Commission on Environment and Development. New York and Oxford: Oxford University Press.

Part I

Sustainable Development in Urban Areas
Dimitri Devuyst

Since this book deals with assessing sustainability at the local level, Part I looks into the concept of "sustainable urban development," how to make this concept more operational, and how to put it into practice. Contributions in this part deal primarily with the following questions: *Is sustainable urban development realizable? What approaches can be used to attain a more sustainable state in urban areas? How can we deal with the uniqueness of each city (and take into consideration the differences between poorer and more wealthy urban areas) in sustainable development initiatives?*

In Chapter 1 Rodney R. White traces the history and gives an overview of current issues of sustainable development in urban areas. Sustainable development of our modern urban areas is a major challenge. Because we do not know whether it is in fact attainable, sustainability should be regarded as a social aspiration; certainly it is a feature we all would like our societies to have. White is optimistic about the future and sees ample opportunities for making our cities more sustainable. This chapter also deals with organizational and physical problems inherent in trying to develop a sustainable society and delves into the question of how sustainable urban development might work technically. White expects that a move toward a more sustainable society will not occur spontaneously under the present "rules of engagement" with the biosphere. Therefore, a set of new key rules are proposed and examples are given.

One of the ways for local authorities to attain a more sustainable state is through the introduction of a Local Agenda 21 process. In Chapter 2 William M. Lafferty reviews the textual and operational basis for Local Agenda 21. Local Agenda 21 is described in Chapter 28 of Agenda 21 and is an appeal to local authorities to engage in a dialogue for sustainable development with the members of their constituencies. In other words, the development of a Local Agenda 21 is a process that should lead cities and municipalities to a more sustainable state. Updated information is provided on the status of Local Agenda 21 activities in Europe, with specific focus on the Norwegian situation. Lafferty also develops criteria for best available practice in Local Agenda 21. The aim is that these criteria will gain a wide enough acceptance so that they can be used as benchmarks for identifying "best cases" for Local Agenda 21 and for annual reports on overall progress with respect to the Local Agenda 21 process.

The Local Agenda 21 approach to sustainable urban development is mostly process-oriented and pragmatic. This contrasts to the eco-city approach, which is much more visionary and therefore also very important in the creative thought process essential in the development of scenarios for future sustainable urban management. In Chapter 3 Mark Roseland discusses the rapidly growing interest in the practical application of the eco-city approach. This approach brings together a collection of apparently disconnected ideas about urban planning, transportation, health, housing, energy, economic development, natural habitats, public participation, and social justice. Roseland examines how various authors, categorized as "Designers," "Practitioners," "Visionaries," or "Activists," have different views on the "eco-city." Examples from the United States, Canada, and Brazil show how components of eco-cities are being put into place, although they still lack some form of synthesis. Roseland concludes that the emerging eco-city vision offers the possibility of such a synthesis, which is also a requirement of a sustainable urban development.

Thomas van Wijngaarden discusses "Urban Agriculture" as an example of an eco-city development in urban areas of industrial and developing countries. The author looks into the contribution of urban agriculture in food and income security, pollution prevention and waste reduction, and improving the living conditions in urban environments.

Different cities each have their unique set of sustainability problems and solutions. Therefore, it is impossible to prescribe one standard solution for the sustainable development of all cities. In Chapter 4 Dimitri Devuyst examines how the concept of sustainable development has different meanings for urban areas at different levels of development. The city of Durban is used as a case study: since extreme forms of poverty and affluence co-exist in this city, it is considered an ideal testing ground for research on sustainable urban development. The author stresses the need for creative solutions for sustainability problems in developing regions. The Filipino community tradition of Bayanihan is presented as an example of how non-Western societies can take advantage of their own culture in the sustainable development of the city. Taking ancient habits and traditions seriously should, however, not prevent the introduction of new and innovative ideas which may also help in attaining a more sustainable state.

Part I illustrates that urban sustainability issues should be of major concern to all of us and that there is ample room for introducing innovative ideas in city planning and management that can lead to a more sustainable state. Although certain local authorities take some action in sustainable urban development, most initiatives remain limited and a lack of a thorough sustainability approach becomes apparent.

Sustainability assessment, the main theme of this book, should help planners, decision-makers, and policy-makers decide whether or not their initiatives lead to a more sustainable society. Consequently, it should first be clear what sustainable urban development means and how it can be put into practice. Part I of this book, therefore, also aims to provide theoretical and practical background information on the meaning of sustainable urban development.

Chapter 1

Sustainable Development in Urban Areas: An Overview

Rodney R. White

Summary

As a concept, the notion of sustainable urban development is simple, appealing, and an essential component of global sustainability. On a technical level we already have many of the pieces that we need to construct communities that would be much more sustainable than those we inhabit today. We can even identify working examples of urban activities that are more sustainable than the typical ones. However, the incentive structure of regulations, prices, and taxes does not encourage the diffusion of these improved practices. We need to put into place the regulations—such as a tax on carbon emissions—that will create a marketplace that rewards those people who behave in a more sustainable way. Despite the fact that national governments sign agreements—such as the Convention on Climate Change—that form the framework for new national regulations, the best hope for implementing more sustainable practices is at the level of local government and its communities, as well as individual households. In order to encourage change at this level, we should provide the information that people need to measure their impact on the planet and encourage them to develop strategies to reduce that impact.

Sustainable Development

Sustainable development was initially associated with sustainable environmental development in response to the degradation and destruction of ecosystems and species that have occurred as a side-effect of the growth of the human economy and population over the last 200 years. No sooner was the term coined when critics pointed out that a strictly physical interpretation of the crisis seemed to imply assigning a lower priority to human needs, both economic and social. In fact, this assumption is counterproductive to the goals of sustainable development and is very rarely implied by its advocates because the growing weight of human poverty demonstrates the continuing need for the expansion of economic and social opportunity. Thus, most seekers of sustainable development assume that these opportunities should expand in order to become available to all, while remaining within the physical bounds of what can be maintained indefinitely. We must keep in mind that there is a very important distinction between the social, economic, and physical aspects of sustainability. Whereas poor decisions regarding social and economic aspects might be terminal for an individual, or smaller groups of individuals, they are rarely so for the entire human species. This is not true for the physical aspect; our global society is making irreversibly destructive choices every day, such as allowing the loss of biodiversity (and hence the extinction of species) and by changing the composition of the atmosphere in ways that cannot be reversed in the lifetime of people living today.

Sustainable development became an issue of widespread concern with the publication in 1987 of "Our Common Future," the summary report of the World Commission on Environment and Development (the Brundtland Commission), which had been holding meetings around the world for the previous two years (WCED 1987). The commission produced a carefully balanced assessment that attempted to determine the severity of the world's deepening environmental predicament without sounding alarmist or triggering despair. The phrase "sustainable development" itself is characteristic of that balancing act and therein creates an apparent paradox. The development path charted by today's wealthier countries is anything but sustainable on a global scale, so how can we conceive of "development" that operates in a radically different way? That is a question that this book explores at the local level.

For many years—indeed, since the earliest civilizations—observers have commented on the destructive side-effects of many human endeavors, including major economic activities such as mining, agriculture, transportation, and industry. Some side-effects were slow-moving processes such as soil erosion, deforestation, depletion of fisheries, acid deposition, and eutrophication of water bodies. Locally, these processes might occur in accelerated form such as landslides and forests fires, but on the global scale the changes were incremental and rarely triggered widespread alarm.

This complacency was shattered in the 1980s by the realization that human activity was changing the composition of the atmosphere quite rapidly with potentially irreversible impacts that might be highly detrimental to the health of all members of the human species, whether rich or poor (White 1994). It became clear that the depletion of the stratospheric ozone layer and the accumulation of greenhouse gases in the atmosphere have the potential to destabilize humanity's planetary niche. At the very least, this kind of documented atmospheric change has injected a large amount of uncertainty into human affairs and the negative side-effects can no longer be safely assigned to just the poorer members of society (as could many previous environmental stresses, such as polluted drinking water). It has become increasingly apparent that not even the advanced technology and ingenuity of modern science can find substitute products to compensate for the increased penetration of ultra-violet radiation or the loss of reasonably predictable climatic regimes. The prospect of these kinds of atmospheric changes that cannot be reversed in the lifetime of this generation have finally seized the popular imagination as a potentially very serious problem for us all.

It was this fear that was the impetus behind the conception of a development path that was sustainable, defined by the Commission as, "Development that meets the needs of the present generation without compromising the needs of future generations." This simple aspiration is understandable enough as a goal, but it remains unclear whether we—as a global community of human beings—can put it into operation. Despite the uncertainty regarding the practicality of the goal, the search for "sustainable development" has become a concept with a life of its own. Why is it only now just beginning to be part of our consciousness?

Origins of the Concept of Sustainable Development

An idea as fundamental as this must have its precursors and indeed it does. Throughout human history, there have always been writers and artists who have imagined a world more harmonious than the one they inhabited. Such harmony is not confined to the physical environment, but also embraces the cultural environment and the concept of justice. One of the most famous books on this theme, *Utopia*, was written by Thomas More (originally published in Latin in 1516) to describe another country, one most unlike the England in which he lived. King Henry VIII, the monarch at the time, so strongly objected to More's opinions that More was beheaded. Although most of *Utopia* is devoted to social, religious, and political matters, it also describes physical details of everyday life, including buildings, the provision of food and water, and the importance of farming, which is practiced by every member of this idyllic society, on a rotational basis. These ideas resonate through the ages, down to Ernest Callenbach's *Ecotopia* (1975).

It says a great deal for human aspirations and imagination that writers such as these can portray living conditions so unlike anything they could have experienced. Before the Industrial Revolution, except for deforestation, the land had not been so extensively ravaged. Living conditions, however, were appalling compared with the conditions enjoyed by most families in industrialized nations today. Ironically, it was the Industrial Revolution—and the agricultural revolution which accompanied it—that gradually improved the standard of living for the common man, not just the elite. Yet it was these modern production processes that have also inflicted the deepest and most long-lasting damage on the biosphere on which humans and all other species depend.

The Industrial Revolution

The Industrial Revolution is a simple phrase that encapsulates a complex range of interlinked processes, some of which were extremely beneficial for the human species and others which were detrimental to the environment in the longer term. Consequently, people reached only for the benefits and ignored the depletion of natural, or ecological, capital. In physical terms, the revolution included a dramatic switch from the reliance on organic materials and energy sources to inorganic sources—that is, from wood and thatch for construction to bricks and iron; from human, water, wind, and animal power to fossil fuels. This had profound implications for the accumulation of waste in the water, on the land, and in the air. The subsequent development of chemical industry also made the waste stream from human activities more complex and problematical.

Not only did the mix of wastes change, but also the volume of waste increased exponentially. Rivers that had once broken down human and animal wastes naturally became sewers. Village rubbish dumps grew into vast pits of nondegradable materials. Spoil heaps from small scale, artisanal mines became towering slag heaps and major features of the industrial landscape. Economist Herman Daly describes this transformation as increasing "throughput" of the materials, energy, and water that people now required for their daily needs (Daly and Cobb 1990).

After the more acute deficiencies of the Industrial Revolution (labor exploitation, lack of sanitation, and so on) were addressed, the human population in the industrial regions began to soar. People had larger families and they began to live much longer. As they became wealthier, they required more "throughput" and, consequently, they created more waste.

In retrospect, from an end-of-the century perspective, this process of human enrichment based on fossil fuels, chemically controlled agriculture, deforestation, depletion of marine resources, and so on, was not without its dangers. Yet until quite recently such concerns would have been dismissed by the population at large and nearly all of their elected leaders. As noted recently by the World Bank, "These losses were once viewed as the price of economic development" (World Bank 1998). What most people saw was the positive side of these economic changes—literacy, longer life, higher incomes, and more consumer goods. In the 1960s, the accumulation of these goods was celebrated as a goal in itself; indeed Walter Rostow viewed it as the ultimate stage of economic growth, when all consumers would enjoy the "age of high mass consumption" (Rostow 1969). Society's goal had shifted from a modest "chicken in every pot" to a television in every room and three automobiles in every garage.

Can the alarming side effects of this level of accumulation be controlled? The possibilities are diminished as the population in the richest countries move from urban living to lower density suburban living. One negative effect of living at lower densities is that under our present incentive structure, public transportation is not convenient or economically self-sufficient; hence, owning an automobile is currently a necessity for all suburban residents. With a way of life so ingrained, how can we begin to implement sustainable urban development?

Sustainable Urban Development: The Lingering Paradox

Modern cities in wealthy nations epitomize the benefits of the fossil fuel-based development path; they offer the best access to high-paying jobs, entertainment, health care, and educational opportunities. They are also profligate in their use of energy, water, and materials; the built environment absorbs heat during the day and thereby creates its own "heat island effect," which requires more energy for cooling homes, offices, and hotels. Even in winter, in a northern city like Toronto, large buildings use energy for cooling. In the words of a colleague, the modern "built environment" is nothing but a "heated rock pile" (Hansell 1997). None of this is sustainable, in terms of the situation we will bequeath to the next generation.

The term "sustainable urban development"—like its parent, "sustainable development"—represents an aspiration only (Stren et al. 1992; Atkinson and Atkinson 1994; Nijkamp and Capello 1998). We do not know if it is attainable or not, but clearly it is a desirable goal. If we observe the modern industrial and post-industrial city, we see an assemblage of immensely wasteful processes dedicated to a range of activities, many of which are seemingly futile. The use of the automobile as a means of mass transportation is the most obvious case of wastefulness and futility. In virtually every major modern city, "rush hour" has now become the "crush hour," consisting of dense packs of slowly moving, or stationary, vehicles. New freeways bring temporary relief and then become as clogged as the streets they replaced. The management of the hydrological cycle is scarcely more efficient, as we all use far more water than we need, and then we poison it with the pollutants produced in our cities, thereby making it much more expensive to clean and return to use.

Although today's richer cities are much cleaner now than they were when in the throes of the nineteenth-century Industrial Revolution, opportunities for reducing the production of troublesome residuals, or pollutants, abound as we shift from an industrial economy to an information economy. Information technology now makes the possibility of "telecommuting"—working from home—a reality for some businesses. Yet the proliferation of the use of automobiles for non-commuting trips has negated any reduction in transportation demand this shift in the employment profile has produced.

Is the modern city a hopeless case in our search for sustainability? On the contrary, the very wastefulness of our current activities implies that opportunities for making our cities more sustainable do exist in abundance. Thus, we can imagine making future decisions that move us toward sustainability by reducing our impact on the natural, or ecological, capital on which future generations depend.

There still remains an inherent ambiguity in the search for sustainability at the local level, such as an urban system. Does usage of the term, and declaring it as desirable societal goal, mean that each local area, each rural and urban system must now be self-sufficient? Does this mean that each city must now identify a surrounding "bioregion" from which it must draw all its supplies and treat all its wastes? It has been calculated that we will need the resources of the equivalent of three planet earths, if the world's current population continues to live at the OECD average of today (Wackernagel and Rees 1996). Given the ambiguous, even contradictory, nature of the term "sustainable urban development," it may be useful to put it aside for the time being and replace it with a more readily operational concept such as "the ecological city," which can be described as "a built environment that operates in harmony with the ecosystems on which it depends." Even if for the time being we replace the difficult goal of sustainability with the even more ambiguous term, "harmony," there still remain some very difficult proposals to consider. For example, is the ultimate goal to establish self-sufficient bioregions, each with an "ecological footprint," greatly reduced from those currently found in the more prosperous nations? Would there be no trade in goods—or export of residuals, such as air pollution—permitted across the boundaries of these regions? How big would they be? Could they change in size or composition over time? Would we have to assume that the world population was down to replacement level to even think about such a scheme as viable?

What Does "Sustainable Urban Development" Mean Organizationally?

The physical problems inherent in trying to devise a sustainable society are daunting, yet the organizational challenge poses even more difficult questions, at every level from individual behavior to global government. The organizational issues will be discussed here, and the physical issues in the next section.

As the human impact on the biosphere has increased, environmental issues have steadily climbed up the hierarchy of human organizations. Environmental issues such as air and water pollution were not dealt with at the local level until the 1950s when national governments first enacted clean air legislation. In the 1960s and 1970s the problems of regional scale acid deposition and the pollution of international waterways led to the first international agreements, in Western Europe and North America, for example. In 1988 at the conference "The Changing Atmosphere," it was recognized that the composition of the atmosphere was a problem of the global commons and that all sovereign nations would have to work toward an agreement (Environment Canada 1988). The Earth Summit, at Rio de Janeiro in 1992, was the first time that all countries of the world met together, whatever their level of wealth or ideological persuasion. The perceived urgency of the global environmental challenge had driven the leaders of nearly all the world's nations to consult one another and attempt to map out a future path for human development on Earth. The outcome from Rio included a set of conventions and an ambitious handbook for environmental reform, Agenda 21, a blueprint for the twenty-first century (Grubb, et al. 1993).

It is commendable and encouraging that the world's leaders got this far and were able to maintain a regular schedule of subsequent meetings on the Convention on Climate Change. International rhetoric is, not surprisingly, very rarely matched by positive action; this level of dialogue is only the most visible part of what is nothing less than a revolution in the way all the world's people live. At its simplest level, this means, first, as the Brundtland Commission (WCED 1987) recognized, that the commitment to eliminate world poverty must be accepted as a priority, while simultaneously the wealthier people of the world must radically change their lifestyle to reduce the volume of the "throughput" of materials, energy, and water that they currently use. All of this implies the involvement at every level of society, from national governments through regional and municipal governments, down to smaller neighborhoods and individual households.

The need for this kind of multi-level participation was anticipated at the Rio Conference where a commitment was made to develop "Local Agenda 21s" at the local government level. This recognition is a logical continuation of the injunction to "think globally and act locally," proposed at the United Nations Conference on Human Development, held at Stockholm in 1972. Admirable as that slogan is, the intervening years have seen little organized action to put it into practice. The emergence of a new type of global governance based on the recognition of the need for simultaneous social and environmental reform now provides hope that significant action at the local level can form a valuable—if not *the* most valuable—contribution to implementation of the conventions that have been signed at the international level. For example, at the second United Nations Conference on Human Settlements (Habitat II) in Istanbul in 1995 many representatives of urban governments and NGOs presented contributions to Local Agenda 21. Organizations such as the International Council for Local Environmental Initiatives have helped to build networks of reform-minded urban groups, including many municipalities and some regional governments. At the Kyoto Conference of the Parties to the Convention on Climate Change in 1997 municipal governments played an official role for the first time; they also demonstrated their support to the venture by producing a ranking of cities that had gone furthest to reduce greenhouse gas emissions within their jurisdictions.

This is no small achievement; indeed, several significant barriers had to be overcome simply to get this far. First, these global negotiations between sovereign states are the exclusive responsibility of central governments, not provinces or municipalities; up until this point central governments had retained exclusive control. Second, political pressure exists at the opposite end of the scale by voters in local elections who are generally reluctant to fund international activities by their representatives. It is often construed as a duplication of effort and a waste of scarce local revenues by voters. Thus, we can assume that considerable momentum lies behind the recent steps toward forming an environmentally active international network of local governments.

We do not yet know if this emerging infrastructure of global environmental governance will one day be capable of implementing the physical changes that will be necessary to build our sustainable cities, or ecologically benign, settlements. But at least an encouraging start has been made.

How Might Sustainable Urban Development Work Technically?

Although sustainability will require agreements between many organizations in order to become operational, the fundamental building block for its survival is household and individual behavior. Much of the change may happen voluntarily, at or least willingly, following effective leadership at the local level. A first step toward this change is knowledge of the environmental implications of our current behavioral modes. This may be accomplished through a number of channels of communication simultaneously.

First, environmental education must occur at every level, formal and informal. Many people still regard the environment as something "technical" or "scientific" that they cannot, or need not, understand. This assumption immediately closes the individual off from any personal engagement with the challenge of sustainability. As quickly as possible people need to understand that the "environment" is what keeps us alive, through breathing, eating, drinking, and so on. It is not separate or extraneous; it permeates every aspect of our lives and is something for which we not only have a responsibility, but a vital need. It should be made clear through this type of education that the environment is as important for our well-being as shelter from the elements and money in the bank. As a society, we need to take "the environment" out of the realm of science and re-install it as common knowledge, just as it was for our pre-industrial forbears.

Second, each individual and household should be made aware of their impact on the environment through the measurement of their "ecological footprint," which estimates the amount of land it takes to support the current lifestyle. Invariably, people in prosperous nations are astonished to learn just how much land and energy they do require. It is this kind of awareness that encourages them to assume personal responsibility for sustainability (Onisto, et al. 1997; Wackernagel and Rees 1996).

Third, in order to provide incentives to reduce that footprint, we need visible indicators around us to monitor our ongoing use of certain resources in the house, such as water, electricity, and natural gas. All of these uses are (or can be) metered, but the meters are usually placed in out-of-the-way corners of the house; instead, they should be located in a prominent position, such as on the kitchen wall, where everyone would have a visible incentive to use less and keep the system running at the minimum usage necessary. Rate structures should also be changed so that the unit cost increases above a certain threshold of reasonable use.

These knowledge gaps can be summarized as:

✦ An understanding of the system as a whole

✦ An understanding of each particular individual's impact on that system

✦ Visible indicators of that impact at the household level

The technical challenges for an ecological city (as defined above) fall into three broad categories, among which there is always some overlap—these are water, energy, and materials. Although water is rarely seen as a critical factor in many wealthy countries of the Organization for Economic Cooperation and Development (OECD), it is already at a critical point in the poorer nations of the world. It is just a matter of time before even in the most developed countries it will become more salient. Just as living plants have factors that limit their growth such as temperature, water, and nutrients, so does human society; of these factors, water is probably the most critical because there is no substitute for it, and water, in turn, is a determining factor in the world's food production.

The following ideas are suggestive only, as it is too early in our search for sustainability to be prescriptive. There may be better alternatives that may emerge that we cannot even imagine today. However, the prescriptive terms "should" and "must" are used as a stimulus for debate until those better alternatives are identified.

Household Water Usage

The improvement of household water quality was probably the most important factor in reducing mortalities during the Industrial Revolution, although water-related diseases like typhoid, cholera, and yellow fever continued to be an intermittent problem even for the more prosperous countries throughout the nineteenth century. Once the water quantity/quality problem was solved, people in these countries lost sight of its critical importance to their well-being. Water was cheap and abundant, with many households paying only a token amount for an unlimited supply. Water use was, and continues to be, abused under this system. While it is important for public health reasons to make good quality water readily available for every household, the current system must be revamped from the benign neglect into which it has fallen in the public consciousness. All households should have their water supply metered and higher prices should be charged for non-essential uses such as washing automobiles and maintaining water-demanding yards and gardens. As an industrial material, it should be charged at the full price for delivery and restoration for all commercial uses in industry and agriculture.

Household Energy Usage

Energy has also been made available very cheaply through the widespread use of fossil fuels, with no forethought of the environmental cost, such as acid deposition, climate change, ozone impacts on health, and so on. Cheap energy, like cheap water, was often seen as a key component of economic growth, and there has been an unwillingness to take a comprehensive view of the costs of this policy. The ramifications of this uncritical provision of cheap energy have been apparent since the 1950s, but even now there is a resistance to stand back and see what our policies have wrought. As a result, our modern settlements are comprised of thousands of separately heated and cooled buildings that are supplied by independent power plants requiring enormous amounts of water for cooling, while this warmed water is then dissipated into a nearby cold water body. Ideally, district or neighborhood-wide heating and cooling systems should be used for efficiency, but not within a system that removes choice of temperature from a particular building or room. We also need to replace the automobile as a mass transportation system with a public rail and bus system wherever it can be supported by the density of population. Because density is the key to efficient public transport, higher density must be maintained as an essential characteristic of our settlements.

As previously noted in an earlier section, the Industrial Revolution brought about a shift from the predominant use of organic materials to the use of the inorganic. This meant, at the simplest level, that human waste was no longer "carried away" by microorganisms that broke them down and used them in their own biological cycles. Instead, our solid wastes began to accumulate, requiring transportation (and energy) to take them away and land in which to bury them. Other wastes were discharged directly to the ground, to water bodies, and into the air. Waste management is still viewed in a piecemeal fashion, with various authorities carrying away solid waste, others in charge of restoring water quality, and still others trying to restore air quality. None of these various agencies have time or incentive to get together and review the whole implication of the growing impact of human waste products on the natural recycling processes of the biosphere. Ideally, the search for sustainability should include a review of the activities and the re-use of human wastes in a strictly biological context.

Implications of Sustainable Urban Development

None of the ideas presented so far are likely to occur spontaneously under the present "rules of engagement" with the biosphere. Clearly, these rules will have to change significantly if any improvement is to be expected. This section will introduce only a few key rules, while the detailed implications will be found later in this book.

Development along the material path stimulated by the Industrial Revolution is epitomized by the capture of stored energy from fossil fuels. Gradually, this discovery has freed humanity from the constraints of growing wood for fuel and timber, which was a land-use in competition with every other land-use, such as agriculture. Once energy could be excavated, this tight constraint of competing land-uses was broken. The Industrial Revolution made many individuals enormously wealthy; it also reduced the human mortality rate to the extent that the species multiplied at a rate biologists typically associate with a pest outbreak. Pest outbreaks occur when a species suddenly finds itself in a situation where it can reproduce much more effectively, either because the conditions that support its life have improved (like humans discovering coal), or a predator has suffered a reversal in numbers. The pest then rapidly exploits its expanded niche, until ultimately, it crashes. In other words, the population falls to its lowest historical level or possibly even becomes extinct. Like all pest outbreaks, the very rapid expansion of our species will come to an end. We may be unique among other species on this planet in that we have the capacity to examine our situation as we go along; we also have the ability to regulate our situation if we can achieve consensus as to how to do that. The attempts to develop a global consensus on our predicament are an indication that this examination and regulation process has begun. If we are going to move toward sustainability, the next step is to change the current incentive structure, which has produced the current crisis.

The guideline for changing this incentive structure is that we should pay more, through pricing and taxation, for the activities that harm the planet and pay less for activities that make a positive contribution. A tax on carbon is one sweeping measure that could have immediate and far-reaching consequences. In the short run, it would drive the economy toward more efficient use of fossil fuels and simultaneously discourage non-essential or substitutable uses. This kind of tax would raise the cost of transportation, thereby giving local products (or substitutes) a pricing advantage over international trade. The world would become a more bioregional sort of place, as entrepreneurs would begin to re-examine local opportunities. The increased cost of transporting solid wastes far from the point of production would be yet another long-term beneficial side-effect of such a tax. These kind of shifts would not happen easily because people would more than likely object to tax increases and overall price increases, along with diminution in their choice of goods. Another probable reaction by the public to overcome is the typical "not-in-my-backyard" (NIMBY) reactions to dealing with wastes locally.

A second change concerns the cost of labor relative to the cost of materials. Under the present incentive structure there is very little reason to conserve materials because they are inexpensive. On the other hand, labor is relatively expensive. This incentive structure has produced a "throwaway society" with a growing problem of solid waste disposal and a seemingly permanent body of unemployed; it runs counter to the recognition of the need to reduce, re-use, and recycle. If the full costs of waste disposal, including polluted water and polluted air, were paid for by the users of the goods and services that produced the wastes, then those costs would rise dramatically. It then makes economic sense to employ people to use, and re-use, materials more efficiently.

This shift would put people back to work to repair the damage they have done, and are continuing to do, to the planet. In this way, people could take care of the planet on which they depend, instead of continuing to abuse it with their thoughtless use of its resources. Positive initiatives to develop this opportunity should be encouraged at the local level.

At first glance, it may seem improbable and unrealistic to discuss such massive changes in incentive structures, yet several changes at similar societal scales have taken place very recently. Examples include the dismantling of apartheid in South Africa, the (fairly) peaceful disaggregation of the Soviet Union, the apparent end of the Cold War, and the successful suing of tobacco companies in North America to pay for their 100-year assault on public health. If anyone had predicted any of these events even twenty years ago, they would not have been taken seriously. Compared with these changes, the ones proposed above are relatively benign; there would be many winners and very few losers.

Examples of Sustainable Urban Development

Only a few, partial examples will be offered here—few, because the remainder of this book will provide many examples of themes that have been introduced in this chapter; partial, because there is no working example of an urban system that is "sustainable" in the sense of the Brundtland Commission's definition. Even the most ecologically benign settlements are depleting nature's capital.

One of the most comprehensive attempts to design a sustainable urban settlement was made by Ebenezer Howard with his plans for "garden cities" (Howard 1985). Paradoxically, Howard's motivation was social improvement and his principal concern for sustainability was financial, for which he proposed a form of cooperative ownership. Nevertheless, his design included details of land-use, transportation, and economy that would have been substantially greener than anything we have today. Since he worked in the pre-automobile age, everyone traveled by foot, bicycle, or train; his design called for plenty of green space within the urban area and surrounding agricultural land, which would provide most of the food and material needs of the city—a bioregion, in fact. Sadly, only the low-density land-use parts of the design were put into effect in his lifetime, and even those communities became absorbed into London's suburban sprawl. His concept was revived in the post-World War II development of new towns in Britain and continental Europe, but these communities involved a high percentage of automobile use and the local employment intentions were partly lost as the towns were drawn into the commutersheds of the major cities. Nevertheless, the proposal remains on the table and it could be implemented today, as designed.

The town of Curitiba, Brazil, is frequently held up as an example of what could be achieved even in adverse conditions. In the 1970s, when the city embarked on its reform, there was already an unpromising mix of widespread poverty and reliance (among those who had a well-paying job) on the automobile (Rabinovitch 1992). The city invested in a high-capacity public transport system based on buses running in dedicated lanes, hence providing a rapidity of travel that enabled it to compete with the automobile. The gradual extension of this service eventually won 75 percent of commuters to the bus system. Finding even one such example proves that the drift towards a deteriorating modal split, losing public transport commuters to the automobile, can be reversed if the will is there. In Dorchester, England, there is an large extension of the town being built on a greenfield site called Poundbury (Krier, 1989). A driving principle behind the design of Poundbury is the provision of well-serviced neighborhoods to which residents can walk to shops, and in which some of them would work, hence eliminating the need for most automobile-based trips. It is too early to know which activities will evolve in this experimental community, but at least the land-use design permits a very different travel behavior option from the norm in the wealthier cities of the world.

In terms of solid waste reduction through recycling, there are many examples around the world where schemes have been put in place which substantially reduce the volume of household waste that goes to landfill. In Toronto, for example, a combination of composting and curbside collection of garden waste, paper, metals, and plastics can reduce household solid waste generation by 75 percent. However, the financial viability of these efforts depends on the value of the recycled materials and the markets for them are not yet mature. After several years as a voluntary measure, Toronto is now introducing compulsory metering of water use. On average, it is found that metering reduces household use by as much as 20 percent.

These are only a few examples and nowhere do we see, yet, a determined move to make our urban settlements more sustainable. However, every successful change provides a demonstration effect that can be emulated wherever the political will can be found.

Conclusion

The notion of sustainable urban development is simple and appealing. It is also a necessity if human beings are to develop a sustainable lifestyle because more than half the world's population is already urban, and the trend toward further urbanization is strong, especially in the poorer countries of the world. However, we are a very long way from knowing how to turn the concept into practice.

The difficulty is based on the historical fact that the prosperity of wealthier nations has been achieved through the widespread use of fossil fuels and other non-renewable sources of energy. This energy path now can be seen as a dead-end in that the modern expansion of human activities has resulted in a massive loss of biodiversity and a changing composition of the atmosphere which will alter the climate in ways we cannot yet predict.

There is no question that we already have the technology to handle or to greatly reduce our negative impacts on the biosphere and to move toward a more sustainable way of life (von Weizsacker et al. 1998). We can also see the emergence of new forms of global governance through networks of urban communities willing to implement the necessary changes, and it is probably these local initiatives that will provide the momentum and determination to meet the commitments that are being signed between national governments.

However, the implementation of the more environmentally benign technology will require substantial changes in lifestyle, and this will, in turn, require changes in the incentive structures that govern contemporary urban life in the richer countries of the world. This alternative lifestyle needs to be demonstrated as a viable choice so that the poorer countries do not emulate the environmentally destructive practices that are the norm in richer cities today, such as widespread reliance on the automobile as a means of mass transportation. In the most general terms, we need to develop an incentive structure that will encourage society to reduce, re-use, and recycle. Through regulation, markets, and taxes we could substantially reduce our use of energy, water, and materials, and with technology that is available now, reduce the production of solid, liquid, and gaseous wastes. We can identify examples around the world where some of the correct steps have been taken and sustained. However, there is, as yet, no strong movement towards making these practices a worldwide standard.

References

Atkinson, A. and F. K. Atkinson, eds. 1994. "Sustainable Cities," special issue of *Third World Planning Review* 16, 2.

Callenbach, E. 1975. *Ecotopia. The Notebooks and Reports of William Weston.* New York: Bantam Books.

Daly, H. E and J. B. Cobb. 1990. *For the Common Good.* London: Green Print.

Environment Canada. 1988. *The Changing Atmosphere: Implications for Global Security.* Ottawa: Environment Canada.

Grubb, M., M. Koch, A. Munson, F. Sullivan and K. Thompson. 1993. *The Earth Summit Agreements: a Guide and Assessment.* London: Earthscan.

Hansell, R. I. C. 1997. Personal communication.

Howard, E. 1985 (originally published in 1898). *Garden Cities of Tomorrow.* Builth Wells: Attic Press.

Krier, L. 1989. Master plan for Poundbury development in Dorchester. *Architectural Design Profile*: 46-55.

More, T. 1974 (originally published in 1516). *Utopia*. New York: Washington Square Press.

Nijkamp, P and R. Capello, eds. 1998. "Sustainable Cities," special issue of *International Journal of Environment and Pollution* 10, 1.

Onisto, L.J., E. Krause and M. Wackernagel. 1997. *How Big Is Toronto's Ecological Footprint? Using the Concept of Appropriated Carrying Capacity for Measuring Sustainability*. Toronto: Centre for Sustainable Studies and the City of Toronto.

Rabinovitch, J. 1992. Curitiba: Towards sustainable urban development. *Environment and Urbanization* 4, 2: 62-73.

Rostow, W. 1969. *The Stages of Economic Growth*. Cambridge: Cambridge University Press.

Stren, R. E., R. R. White and J. B. Whitney, eds. 1992. *Sustainable Cities: Urbanization and the Environment in International Perspective*. Boulder: Westview.

von Weizsacker, E., A.B. Lovins and L. H. Lovins. 1998. *Factor Four: Doubling Wealth—Halving Resource Use*. London: Earthscan.

Wackernagel M. and W. E. Rees. 1996. *Our Ecological Footprint: Reducing Human Impact on the Earth*. Gabriola Island, BC: The New Society Publishers.

White, R. R. 1994. *Urban Environmental Management: Environmental Change and Urban Design*. Chichester: John Wiley & Sons.

World Bank. 1998. *World Development Report 1998/99: Knowledge for Development*. Oxford: Oxford University Press.

WCED, World Commission on Environment and Development. 1987. *Our Common Future*. Oxford: Oxford University Press.

Chapter 2

Local Agenda 21: The Pursuit of Sustainable Development in Subnational Domains

William M. Lafferty

Summary

The pursuit of sustainable development at the local and regional level of governance has received a powerful impetus from the follow-up to Chapter 28 of Agenda 21, the "action plan" adopted at the Rio Earth Summit in 1992. "Local Agenda 21" (LA21) has become an international symbol for new initiatives for sustainable local and regional communities. With strong support from the United Nations Commission on Sustainable Development (UNCSD), the International Council for Local Environmental Initiatives (ICLEI), and the European Commission, Local Agenda 21 is one of the major success stories from the Earth Summit. The purpose of this chapter is to document the growing consensus as to what the idea of a "Local Agenda 21" involves, and to present data and analytical perspectives from a comparative research project sponsored by the European Commission (SUSCOM). The article concludes that, while there is considerable progress toward realizing the intentions of Agenda 21 in this area, there is also an emerging conflict over the nature of LA21 with respect to its oppositional vs. integrationist function in local politics for sustainable development.

Introduction

One of the most distinctive features of the idea of sustainable development as expressed in the Brundtland Report (WCED 1987) is the need for new forms of social mobilization and decision-making. The achievement of environmental, economic, and social sustainability requires different modes of collective insight and competence than those prevalent in advanced liberal-capitalist societies. This chapter focuses on these requirements that come directly from the Rio action plan, with specific reference to subnational political domains. The chapter reviews the textual and operational basis for Local Agenda 21 (LA21), and provides updated information on the status of LA21

activities in Europe. With reference to ongoing comparative research sponsored by the European Commission and supplemented by data from applied
research projects in Norway, the chapter aims to provide a more consensual
point of departure for further implementation and evaluation.

What Is Local Agenda 21?

Chapter 28 of the Rio Action Plan[1]

Seven years after the Rio Earth Summit, relatively few traces of the Summit's
ambitious program for change are in place. Two major exceptions are progress
on the Framework Convention for Climate Change and the implementation
of Chapter 28 of Agenda 21. While the former has been widely publicized
through the recurring "Conferences of the Parties" where representatives of
the leading climate-degrading countries have been seriously pursuing a new
international regime for the control of greenhouse gases, the Chapter 28 of
Agenda 21 has evolved in a less visible but no less significant manner.

Chapter 28 of Agenda 21 is the shortest in the forty-chapter action plan, a
fact which may account for its relative success. While other chapters are
packed full of guidelines, recommendations, and ambitious goals regarding
substantive problem areas, Chapter 28 is a relatively simple appeal to "local
authorities" to engage in a dialogue for sustainable development with the
members of their constituencies. More specifically, Chapter 28 is addressed
to local authorities as one of several major groups which the Agenda singles
out as particularly relevant for achieving the aims of the overall Agenda
itself. Because "...so many of the problems and solutions being addressed by
Agenda 21 have their roots in local activities...," the participation and
involvement of local authorities is viewed as "a determining factor" in fulfilling the objectives of the action plan. As the level of governance closest to the
people, local authorities "...play a vital role in educating, mobilizing and
responding to the public to promote sustainable development..." (United
Nations 1993: para. 28.1).

It is within this focus that Chapter 28 then goes on to stipulate only four
major objectives, three of them with very specific deadlines:

1. By 1993, the international community should have initiated a consultative process aimed at increasing cooperation between local authorities.

2. By 1994, representatives of associations of cities and other local authorities should have increased levels of cooperation and coordination with
   the goal of enhancing the exchange of information and experience
   among local authorities.

3. By 1996, most local authorities in each country should have undertaken a consultative process with their populations and achieved a consensus on a Local Agenda 21 for the community.

4. All local authorities in each country should be encouraged to implement and monitor programs which aim at ensuring that women and youth are represented in decision-making, planning, and implementation processes.

What we learn from these objectives is: (1) that the aim of Chapter 28 is an identifiable result of a consultative process—a "Local Agenda 21"; (2) that the effort in question should be both cooperative and coordinated across national boundaries, with the specific assistance of the international community and transnational NGOs; and (3) that the subprogram should make a particular effort to bring women and youth into the change processes.

These concepts are then followed up more specifically in three subparagraphs on "activities." Generally, they serve to fill out the objectives, adding a number of specific ideas and identifying the types of international organizations and transnational bodies that could become involved in the coordinating activities.

Based on these relatively sparse documentary sources, it can be said that a "Local Agenda 21" is a strategic program, plan, or policy that has emerged from a consultative process initiative by local authorities with both local citizens and representatives of relevant local stakeholders, with a particular interest in involving women and youth. The purpose of the strategic program is to implement Agenda 21 at the local level, which, by implication, means that the purpose is to entrust local authorities with a particular responsibility for achieving sustainable development within their particular subnational domains.

ICLEI and the Aalborg Charter [2]

Several international and regional organizations have played a major role in following up, and filling out, the documentary signals provided by Chapter 28. Foremost among these has been the International Council on Local Environmental Initiatives (ICLEI). Established two years prior to the Earth Summit, ICLEI played a major role in preparing and coordinating Chapter 28 of the Agenda. Working closely with organizations such as the United Nations Environment Program (UNEP), the International Union of Local Authorities (IULA), and the European Commission, ICLEI has taken a clear and forceful lead after the summit in sponsoring and promoting Local Agenda 21 initiatives. Two areas of major importance for an understanding of what "Local Agenda 21" has come to mean are the European "Sustainable Cities and Towns Campaign" and the survey conducted by ICLEI as part of the reporting exercise for the 1997 special session of the UN General Assembly, "Earth Summit +5."

Beginning with a resolution by the European Council in 1991 to establish an "Expert Group on the European Environment," a multi-organizational effort gradually coalesced on "Sustainable Cities and Towns in Europe," the name of a conference held in Aalborg, Denmark, in May 1994. The conference resulted in the so-called "Aalborg Charter," a three-part document outlining basic values and strategic options for sustainable development in European urban areas, and launching a broad-based campaign for sustainable cities and towns in Europe. Part III of the Charter made a specific commitment to following up Chapter 28 of the Rio Agenda, and also made a direct connection between the Charter, Agenda 21, and the European Union's Fifth Environmental Action Program, "Towards Sustainability."

Most importantly in the present context, however, the Aalborg Charter also proposed what it considered to be an eight-stage model for preparing and implementing a local action plan (that is, a Local Agenda 21):

1. Recognition of the existing planning and financial frameworks as well as other plans and programs.

2. The systematic identification, by means of extensive public consultation, of problems and their causes.

3. The prioritization of tasks to address identified problems.

4. The creation of a vision for a sustainable community through a participatory process involving all sectors of the community.

5. The consideration and assessment of alternative strategic options.

6. The establishment of a long-term local action plan toward sustainability which includes measurable targets.

7. The programming of the implementation of the plan including the preparation of a timetable and statement of allocation of responsibilities among the partners.

8. The establishment of systems and procedures for monitoring and reporting on the implementation of the plan. (Aalborg Charter 1994)

These stages can be interpreted as a logical practical extension of the objectives and activities put forth in Chapter 28 of the Rio Agenda. Taken together with Part I of the Charter (which outlines the notion and principles of "sustainability"), we arrive at a relatively concise understanding of what a "Local Agenda 21" is all about:

➹ A Local Agenda 21 is a local action plan for the achievement of sustainable development

➹ It is to be worked out through a broad consultative process between local authorities, citizens, and relevant stakeholder groups, and eventually integrated with existing plans, priorities, and programs

➹ The "consultation" in question is clearly meant to be a separate process from existing protective and remedial environmental activities

➹ The process has a clear strategic intent. Though the actual content of a Local Agenda 21 is not spelled out, there is a clear presumption of both change and instrumental rationality with respect to a realization of the Earth Summit goals

➹ The action plan should be implemented with due provision for ongoing input, monitoring, and revision underway, and it should make special efforts to engage women and youth in all phases of the implementation process

➹ Chapter 28 of Agenda 21 is specifically addressed to "Local Authorities." The responsibility of national governments is primarily facilitative with respect to the LA21 process

➹ The substance of any particular Local Agenda 21 will be *relative to the specific nature of the local community* in question (its geography, demography, economics, society, and culture), and it should be expected to *evolve dynamically over time*

The Implementation of LA21 in Europe: Development of the SUSCOM Project

Trying to attach a specific connotation to the notion of a "Local Agenda 21" is important for a number of reasons. Empirical overviews of the implementation process in several European countries (Lafferty and Eckerberg 1998) indicate that there is considerable confusion as to just what the concept means. While a certain amount of conceptual openness is surely necessary for applying the Agenda across a broad spectrum of local and regional communities, there is also a need for a common understanding as to what is different and significant with the idea. If LA21 is to be understood as anything resembling "environmental policy," the idea will quickly deteriorate into a catchall category with little potential for either evaluation or cross-national comparison and analysis. In addition, therefore, to the differentiating material just presented, next we discuss evaluative criteria in a more systematic way. The initial attempt to come to grips with the evaluation problem resulted in a set of six substantive criteria (Lafferty and Eckerberg 1998: 5-6):

1. A more conscious attempt to relate environmental effects to underlying economic and political pressures (which, in turn, derive from political decisions, non-decisions, and markets).

2. A more active effort to relate local issues, decisions, and dispositions to global impacts, both environmentally and in regard to global solidarity and justice.

3. A more focused policy for achieving cross-sectoral integration of environment-and-development concerns, values, and goals in planning, decision-making, and policy implementation.

4. Greater efforts to increase community involvement; that is, to bring both average citizens and major stakeholder groups, particularly business and labor unions, into the planning and implementation process with respect to environment-and-development issues.

5. A commitment to define and work with local problems within: (a) a broader ecological and regional framework, as well as (b) a greatly expanded time frame (that is, over three or more generations).

6. A specific identification with (reference to) the Rio Summit and Agenda 21.

In the first phase of our project, these criteria were employed to monitor implementation in eight different countries: Norway, Sweden, Finland, the Netherlands, Great Britain, Ireland, Germany, and Austria. The results from this analysis were then channeled into a second phase, where the project achieved status as a "concerted action" within a research program for Climate and the Environment under Directorate General XII of the European Commission. The research network has since been expanded to include Italy, Spain, Denmark, and France, in addition to the original eight countries. The project title has been shortened to SUSCOM: "Sustainable Communities in Europe."[3]

In applying the preceding criteria to specific country analyses, the research team originally differentiated between three different types or levels of environment-and-development activity.

The first level refers to policies and initiatives that are primarily designed to either *conserve nature* or *improve and redress the environment*. These initiatives could have been taken prior to the publication of the Brundtland Report (WCED 1987) and address environmental concerns in a relatively narrow, more technical, and more "natural-science" type of perspective.

Such activities are simply referred to as "environmental initiatives," and are not presumed to reflect any of the preceding six characteristics.

The second level refers to policies and initiatives that *specifically refer to the concept of "sustainable development"* as expressed in the Brundtland Report or those that use broad concepts such as "global ecology," reflecting the general concerns of the Brundtland Commission without specifically using the terms and categories of the report itself. Such activities should reflect most or all of criteria 1 through 5, and can be referred to as "initiatives for sustainable development." Most of these local initiatives would have been instigated in the period following the publication and dissemination of the Brundtland Report; that is, between 1987 and 1992. [4]

Finally, at the third level, we identify *activities which make specific reference to the Rio Summit and/or Agenda 21.* Only these activities qualify, in the strict sense of the term, as a "Local Agenda 21." Such activities should reflect all six of the previous criteria, and they should do so as a conscious attempt to implement the intentions of Chapter 28 of the Agenda. It should be stressed that these brief guidelines were intended as a sensitizing device to facilitate a common understanding of, and approach to, Local Agenda 21.

Making the differentiations between the environmental initiatives in question proved very necessary for systematic reporting and comparative analyses. The original findings revealed that there was considerable confusion as to what a Local Agenda 21 signifies, particularly on the part of national governments in their reports to the United Nations Commission on Sustainable Development (Lafferty and Eckerberg 1998: Chapter 10). If everything that has to do with either improving the environment or achieving sustainability is to be categorized as a Local Agenda 21, there is clearly no way to meaningfully monitor and evaluate the impact of the Rio accords. The major differentiation in this regard is between, on the one hand, the first level of more traditional environmental activities, and, on the other— the second and third levels—both of which express the values and goals of sustainable development.

Preliminary Findings

The original assessments, particularly of the total number of local authorities that had embarked on LA21 processes at the time of the "Earth Summit +5" special session of the General Assembly (in June, 1997), indicated three different levels of implementation. These were identified (Lafferty and Eckerberg 1998) as "pioneers," "adapters," and "latecomers" (Table 2-1).

We found the results of this comparison surprising with respect to the variation of local government policy revealed in other areas (see Lafferty and Eckerberg 1998). There is no clear pattern that can be detected at first glance.

TABLE 2-1 *The Status of LA21 in Eight North European Countries*

Pioneers	Adapters	Late-comers
Sweden	Finland	Germany
United Kingdom	Norway	Ireland
The Netherlands		Austria

For example, the Nordic countries, which in most other respects are quite similar in their political and administrative traditions, here prove to be very dissimilar. The spectrum runs from Sweden, with (officially) a very large number of LA21 initiatives (peaking in 1994), to Norway and Finland, which showed very little LA21 activity prior to 1996. Similarly, countries with reputedly well-established environmental policies can be found in all of the three categories: with the Netherlands and Sweden among the pioneers, Norway in- between, and Germany considerably lagging behind. Finally, the UK and Ireland appear at the two extremes, despite their geographical and cultural affinity. In the UK, LA21 began as a major success story, while the Irish had hardly applied the concept. Austria was also shown to be a clear laggard. It should be pointed out here, however, that the positioning of a particular country among the "pioneers" related only to its relative placement; that is, it was labeled a "pioneer" only in comparison with the other countries canvassed. Being a "pioneer" in LA21 implementation does not imply that the majority of municipalities in the country are indeed involved in LA21. Nor is it correct to say that there are numerous municipalities in these countries that have implemented LA21 according to all the dimensions and tasks covered by the criteria outlined above. We were mainly relying at this stage, on the reports submitted by the countries themselves and as "cleared" by either the UNCSD or ICLEI. Closer analyses indicated that, even among the "pioneers," local authorities who claimed to be far advanced in the LA21 process, seldom reflected advanced practices for local sustainability.

Having said this, however, the division of the countries into three groups provided a first indication of the overall success of LA21 in relative terms. One might conclude that the pioneers had, at this stage, a larger share of municipalities embarked on LA21 and sustainable development processes compared to the adapters and latecomers, who were mostly involved in more traditional environmental activities.

Original attempts to explain the difference uncovered were made on a very ad hoc basis. We were primarily interested in uncovering relatively simple patterns of correlation at this stage of the analysis. The more detailed descriptive results of the original analysis are reported in Lafferty and Eckerberg (1998) and need not be repeated here. The study also generated, however, two other perspectives which warrant repetition: first, a series of factors that seemed to play a significant role in promoting LA21 implementation in one or more of the countries studied; and, second, a set of four tentative generalizations as to more systematic explanatory perspectives.

Significant Factors in the Implementation of Local Agenda 21 in Europe

Each of the following factors appeared to have a positive effect in one or more of the original eight countries studied, so that any combination of the factors should have a cumulative positive effect:

* A previous involvement on the part of representatives of local authorities in the UNCED process
* An active positive attitude on the part of responsible central-government officials to the LA21 idea
* Central-government initiatives and campaigns to disseminate information on Local Agenda 21
* The availability of central government financial resources to subsidize LA21 initiatives
* Enough local government autonomy to render the LA21 idea interesting and possible
* Membership in cross-national environment-and-development alliances, charters, and the like
* A previous history of international "solidarity" orientations and activities at the local level
* Previous municipal involvement in environmental and sustainable-development pilot studies
* Previous experience with "cooperative management regimes" (among social partners and stakeholders)
* Active individual "firebrands" for LA21 at the local level
* Perceived possibilities for coupling LA21 with the creation of new jobs
* Perceived conditions of "threat" to local environment-and-development conditions from external sources

Key Aspects in the Search for Underlying Explanatory Processes

From the point of view of moving from descriptive assessment to a more explanatory perspective, we concluded the eight-nation study by identifying four key aspects that seemed to have general implications across several of the country reports. For each of these, we identified what we believe to be key questions and dimensions for future model building and analysis.

Interpretation of the Implementation Task

This is a question of "defining the situation," with a need for documenting both the history of meaning related to the task and the current power structure for implementation. Which national "agent" has responsibility for bringing the particular aspect of the global program "home"? How is the particular task integrated into the overall national strategy for following up and implementing the global program for sustainable development, and what is the relationship between the responsible agent and the broader constellation of forces affecting policy priorities and the distribution of implementation resources?

Domains and Levels of Implementation Responsibility

How is the implementation task viewed with respect to central-local government? What is the current mix of constitutional authority, policy responsibility, and resource allocations among national, regional, and local authorities? Is the facilitator role of national authorities accepted and acted upon, and how are nongovernmental organizations—particularly umbrella organizations for local and regional authorities—integrated into the implementation process?

The Availability of Resources and the Inertia of Reforms

Are resources—in the form of information, guidelines, and earmarked funding—made available for specific acts of implementation directed toward change? What is the current economic and fiscal situation of local communities, and what is the status of existing local environmental policy responsibility? Is there compatibility between the general characteristics of the LA21 idea and existing official environmental policy, or does the history, priorities, and institutionalization of current policy create barriers against new initiatives?

Mobilization of Local Citizens and Stakeholders

To what degree have individual citizens and representatives of the major groups previously been involved in environment-and-development activities at the local level? Are there legal provisions in place for involving affected parties in planning and implementation, and have NGOs (including business and labor) been active or passive in the specific policy area? How do local politicians and personnel within the responsible local administration view the prospect of increased popular involvement? As a welcome and constructive supplement? Or as a threat to established positions and hard-won sectoral achievements?

These benchmark perspectives and questions provided a solid point of departure for the next stage of the analysis, which is currently in the process of assessing implementation reports from all twelve countries in the expanded EU network. The results of this analysis will then be used to structure the third stage of the project, whereby each country team will focus on those aspects of the LA21 process which are most relevant for the project's more general analytic themes. As currently defined, these consist of: (1) implications for the study of policy implementation in general; (2) issues related to the interaction between community planning and democratic participation; and (3) the larger issue of subsidiarity within the European Union. A report on these issues will be completed by the end of 1999, employing in the process a "dialogue conference" between key stakeholders, civil servants, and social scientists to clarify the broader implications of the findings.

Recent Developments in Norway

As indicated above, Norway was a relatively late "adapter" to the LA21 process. It was not until the fall of 1995, more than three years after the Rio Summit, that the idea was seriously placed on the political and administrative agenda. Though it is difficult to get public officials to speculate as to why Norway was so reticent in pursuing Chapter 28, two reasons seem to stand out. First, Norway likes to consider itself as very advanced in the pursuit of sustainable development within global forums, and, in the eyes of many officials, Agenda 21 did not go far enough in stipulating targets and allocating resources. There was thus considerable disappointment in the wake of the Rio Summit, which led to a general feeling that Norway had somehow already progressed beyond the prescriptions in the Agenda. It is interesting to note that Agenda 21 has still not been translated into Norwegian, nor does there exist any coherent national strategy for pursuing the Agenda (other than—as we shall see—Local Agenda 21).

Second, in 1992, Norway was in the process of completing a wide-ranging reform of environmental policy at the local level. The essence of the reform was to provide funds through the Ministry of the Environment for establishing a new administrative post, a "municipal environmental officer" in each of Norway's 435 municipalities. The new positions were phased in gradually. It begun with a pilot project in 1988, with the period 1992–94 being the most active period of the reform. The focus on carrying through with these activities apparently served to block the spread of the LA21 idea. The new environmental officers were simultaneously more bureaucratic and more narrowly focused on the technical aspects of environmental protection than was the intention of LA21. Many local politicians and administrative personnel felt that the new demands of LA21 were simply too much too soon with respect to the ongoing reform.[5] Norway's poor showing with respect to LA21 was clearly documented in a report published in the fall of 1995 (Armann et al. 1995). This was enough to move the issue more forcefully onto the political agenda of both the Ministry of the Environment and the Norwegian Association of Regional and Local Authorities (called *Kommunenes Sentralforbrund*—KS in Norwegian), so that today Local Agenda 21 is the single most publicized aspect of Agenda 21 in Norway. Two aspects of this heightened interest are of particular relevance for the question of sustainability assessment: (1) new criteria under development by the Ministry of the Environment for evaluating progress on LA21; and (2) the promotion of a "Regional Agenda 21" as an extension of the LA21 concept to the middle-level of Norwegian governance.

Criteria for LA21 "Best Cases"[6]

As part of an enhanced effort to promote and facilitate LA21 through central-government activities, the Ministry of the Environment has established a separate LA21 coordinating unit within the Division for Regional and Spatial Planning.[7] The unit has been active on a number of different fronts, most significantly: (1) working interactively with local authorities and relevant stakeholders to develop an overall LA21 strategy; (2) establishing a "resource and competence network" for improving documentation, communication and electronic resources available for LA21 activities, both nationally and internationally; (3) financing and supporting regional nodes for coordinating LA21 at the county level of Norwegian governance (see below); and (4) developing standardized benchmarks for assessing progress and promulgating "best cases."

With respect to the latter, the Program for Research and Documentation for a Sustainable Society (ProSus) has recently outlined for the Ministry a set of criteria that build on the dimensions mentioned above from the initial stage of the SUSCOM project. These criteria are designed to identify defining characteristics of LA21 activity at the municipal (urban and rural) level. The hope is that they will gradually gain enough wide acceptance so they can be used as national benchmarks for identifying "best cases" and for annual reports on overall progress with respect to the broader and more long-range intent of Chapter 28. Each criterion is identified with a short (and semantically similar) title and five subcategories for the purpose of identifying different types and degrees of activity within each of the five areas. Recognition of activity can be made separately for each of the five dimensions or cumulatively for that unit or units which "scores" highest across all five standards. The term "local authorities" means local politicians, civil servants, and other administrative personnel, and the term "stakeholder" is used as a substitute for the term "major group" as used in the Rio documents. As currently formulated, the five categories and subcategories are as follows:

A New Dialogue

Have local authorities:

✦ Passed a resolution in the Municipal Council in support of the "Fredrikstad Declaration" and of Local Agenda 21?

✦ Initiated efforts for a broadly based media campaign, providing information to citizens and other stakeholders on the commitment to Local Agenda 21?

✦ Arranged public hearings and meetings to disseminate information and to discuss and register suggestions for LA21 activities?

✦ Taken an initiative to develop methods for maintaining the dialogue through, for example, interactive information technology or other communication systems whereby citizen input can be channeled into LA21 processes?

✦ Taken concrete steps to develop a vision for the local community and moved to involve key stakeholders in goal-oriented efforts to achieve the vision?

A Sustainable Economy

Have local authorities:

✦ Made specific efforts to integrate the language and values of sustainable development into the municipality's key planning documents?

✦ Tried to promote a clearer understanding of the municipality's ecological setting in a regional context, with particular emphasis on biological diversity and the local resource base?

✦ Taken steps to chart the ecological impacts of the municipality's production and business structure (through "green accounting," environmental audits, etc.), and to integrate environmental impacts in long-term planning and budgeting?

✦ Initiated cooperative programs for addressing the environmental problems of households, and given direct support to stakeholder projects for sustainable production and consumption?

✦ Taken steps to introduce a method of "directional analysis" with respect to municipal policy, whereby all municipal activities are monitored with respect to values, goals, and targets for sustainable development?

Sectoral Integration

Have local authorities:

✦ Created a permanent administrative position with the responsibility of coordinating environmental policy and LA21 activities?

✦ Taken steps to establish a "stakeholder forum" for discussing, planning, and coordinating strategic initiatives for sustainable development?

✦ Tried to find routines for an effective application of "the precautionary principle" (within an ecosystem frame of reference) to trans-sector planning, budgeting, and implementation?

✦ Initiated a program for "putting one's own house in order" by the "greening" of municipal administrative routines, "green purchasing," and training of municipal employees in the values and principles of sustainable development?

✦ Taken the necessary procedural and administrative steps to establish a Local Agenda 21 plan as the primary strategic steering instrument for municipal activity?

Global Orientation

Have local authorities:

✦ Taken initiatives for direct contacts with local communities in less developed countries, through, for example, "twinning" programs or other types of cooperative arrangements between more or less developed nations?

✦ Charted existing points of contact and interaction between local stakeholders and governments, businesses, organizations, and communities in less-developed countries?

✦ Taken steps to document the effects of local production and consumption patterns (material flows, resource dependency, market impacts –

the "ecological footprint") with respect to less-developed countries, and to publicize the results?

✦ Made a conscious effort to integrate distributional issues among Northern and Southern hemispheres in municipal planning and initiatives?

✦ Initiated assessments of the global environmental impacts – particularly with respect to energy flows, climate change, and biodiversity – of local production and consumption?

Constructive Evaluation and Revision

Have local authorities:

✦ Invited local educational institutions of different levels to become involved in the evaluation and follow-up of LA21 efforts, both within their own institutional domains and through the application of specific types and levels of competence and local knowledge?

✦ Tried to involve citizens and stakeholder representatives in monitoring and evaluation, and establish procedures for channeling feedback and new initiatives back into municipal planning and revision?

✦ Taken steps to share information on LA21 experiences with other municipalities, and improve administrative competence in LA21 "best cases" and procedures?

✦ Introduced ongoing routines for monitoring and revising LA21 activities with respect to clearly formulated goals and indicators, and by promoting procedures such as environmental audits, EMAS-protocols, "directional analysis," and the like?

✦ Taken steps to combine active and regular reporting on LA21 progress with public meetings and hearings?

Regional Agenda 21 as County Planning for Sustainable Development[8]

Norway has 18 counties, varying in population from about 446,000 for Akershus to 75,000 for the most northern county, Finnmark. In addition, the cities of Oslo and Bergen administer their own county functions. The governing of the county is divided between, on the one hand, an electoral channel, with a county assembly (*fylkesting*) and an administrative unit (*fylkesrådmann*); and, on the other, a separate administrative unit (*fylkesmannskontor*) which is responsible for state functions at the county level. To confuse the matter even more, the electoral jurisdiction is known as the

"County Municipality," while the administrative arm of the state at the county levels is known as the "County Governor." All this is indicative of the shifting stages of centralization/decentralization in Norway's unitary state. The counties are caught in the middle, with a "bottom-up" tendency structuring the County Municipality and a "top-down" tendency maintaining its effect through the office of the County Governor.

With respect to the Regional Agenda 21, it is the administrative arm of the County Municipality that has taken the initiative. In Akershus County, this has resulted in a completely new version of the traditional sector plan for environmental protection for the jurisdiction. The plan has been completely altered in both form and substance, being now profiled as *"Regional Agenda 21: Global Environmental Challenges—Local and Regional Responsibility."* The plan has official status according to the legal procedures for planning at the county level of governance, and was adopted by a large majority of the County Assembly in January of 1998. Though the plan is a subplan of the over-arching "general plan" for the county, it is pointed out in the introduction to the document that the new sector plan "covers a broader spectrum of environmental challenges" than the general plan and that the sector plan is, in most areas, "more concrete" than the general plan.

The general tone and import of the Akershus RA21 plan is captured by an overview of the plan's six major sections:

1. Agenda 21—An Action Plan for the 21st Century

2. Challenges on the Way to the 21st Century

3. Sustainable Production and Consumption

4. Coordinated Land-Use and Transport Planning

5. Management of Biological Diversity and the Basis for Food Production

6. The Quality of the Environment along Coastal Areas and Watersheds

The plan also concludes with a separate "Action Plan" with specific tasks for each of the four priority areas (III through VI). All of the sub-points of the action plan are formulated as "shall-do" commitments, though there are no specific targets set.

In addition to the action plan, the RA21 also stipulates a number of more general initiatives. These include the establishment of a separate *Green Council*, which is to function as an "Agenda 21 forum." The purpose of the Council is to bring together representatives from the municipalities, regional (i.e., county) authorities, voluntary organizations, business and labor unions, "to focus on the local and regional responsibility for global environmental problems" (Akershus County 1998: 13). Of the other procedures adopted through the plan, two of the most significant are: (1) preparation of a yearly report on the progress of the plan, to be submitted to the County Assembly in connection with the yearly review of the General Plan; and (2) the adoption of "directional analysis" as a specific tool for target-setting and revision of the RA21 action plan.

Finally, it is important to point out that the procedure for adopting the RA21 for Akershus County was carried out in accordance with clear "dialogical" principles. The County produced and circulated (in cooperation with a leading informational NGO) a comprehensive discussion document which outlined the tasks and possibilities for creating a Regional Agenda 21, and conducted popular meetings and other procedures for focusing popular and group opinion on the issue. Each action-area had its own contact person in the Municipal County administration, and active efforts were made to solicit the opinions and suggestions of major target groups. Throughout the process, a major task was to identify the specific types of responsibilities adhering to a regional effort viewed as supplementary to, and supportive of, national and municipal efforts. [9]

RA21 in Nordland County: The Use of "Deliberative Democracy"

The second-most prominent RA21-process in Norway thus far is that of Nordland County in the north of Norway. Here, there was also an extensive preparatory phase in connection with an attempt to integrate the new consultative aspects of the effort with normal legal-political planning procedures. Though many aspects of the Nordland RA21 are similar to the Akershus model, they introduced one new feature in the form of a survey based on principles of "deliberative democracy" (Fishkin 1991 1995; Eriksen 1995). Given the fact that counties in the north of Norway are relatively large geographically, with widely dispersed population centers and great variety in occupational structure, the notion of regional authorities effectively conducting a "consultative process" with citizens and affected interests becomes very difficult to achieve in practice. It was decided, therefore, to carry out an experiment in deliberative procedures by, first, conducting a normal sample survey of the county's population on the specific issues identified for the plan; secondly, using the survey to identify a subgroup of the population who were willing to participate in a face-to-face gathering; and, thirdly, bringing together the subpopulation for a two-day "hearing" over the issues in question. The full results of this procedure have not yet been analyzed, but the most significant findings from the "hearing" were channeled directly into the decision-making procedure which led to the official adoption of the plan in June of 1998. The significance of the procedure is, of course, that the sample-survey methodology, combined with the actual face-to-face "deliberation," provides a technique for transferring the imagery and intent of the LA21 "consultation" to a higher level of political aggregation and responsibility. In comparison with the RA21 plan for Akershus, the RA21 for Nordland county is perhaps even more ambitious in its attempt to integrate principles of sustainable development directly into its specific recommendations.

Regional Assistance from the Center: County-based LA21 "Nodes"

Another recent initiative with direct implications for RA21 activity in Norway is the establishment of so-called "regional nodes" for the promotion of Local Agenda 21. The new effort is a cooperative project between the Ministry of the Environment and the Norwegian Association of Local and Regional Authorities (KS). The purpose of the effort is to provide county-based centers for providing information and coordinating LA21 activities at the county level. Funds have been provided by the Ministry (approximately three million NOK for 1998) to establish between four and six nodes in an initial phase of the project. The administrative responsibility for the operation of the new units lies with KS. Each node is to have representation from both the electoral (county municipality) and central-administrative (county governor) branches, as well as from the local KS-organization and one or more municipalities. The primary goal of the nodes is to coordinate the different activities related to both LA21 at the local level and the new RA21 activities at the county level.

In addition to, and parallel with, the nodes, the Ministry and KS have also taken a joint initiative to establish a national "Competence Network" for LA21. The principal aim of the network is to use modern communications technology (primarily Internet and the World Wide Web) to effectively spread information on LA21 activities both nationally and globally. Once the Competence Network is fully functioning, it will be tied directly into the nodal network. All three of these aspects serve to "operationalize" the LA21 idea at a regional (and, in Norway, intermediate) level of governance. In addition to providing new possibilities for promoting and realizing the UNCED program for sustainable development, the emergence of a separate Regional Agenda 21 program contributes strongly to two additional developments.

First, the initiatives come at a time when (in Norway, at least) there has been considerable discussion as to the viability of the county level of governance. The functions now performed at the county level are the result of a broad-based reform in governmental responsibilities that was carried out in the 1960 and 1970s. After nearly two decades of experience with the system, numerous commentators have found relatively little positive value in the reconstituted democratic role of the county municipality. The functions currently allocated to the level (coordination of secondary education, roads, and transport and health and social services) have not been politicized at the county level, and the involvement of citizens in county politics has been relatively low. The prospect of using RA21 procedures to focus issues of broader interest for sustainable development and to mobilize countywide, citizens, and interest in the process, offers a potential rejuvenation of regional politics and democracy.

Second, in a more substantive direction, the development of a Regional Agenda 21 perspective at the county level provides a potentially new function whereby the county can take the lead in reorienting county politics towards a more wholistic ecosystem perspective. Given the fact that the municipal boundaries have historical roots that have very little to do with either ecosystems or a fair allocation of natural-resource exploitation and sinks, the county can take on a vital role in coordinating and mediating municipality priorities and policies. In short, if ecosystem thinking is a prerequisite for physical sustainability and if a fair distribution of "ecological space" is a prerequisite for sustainable development in time and space, county-based Regional Agenda 21s can be a vital link in securing a middle-level of policy coordination and implementation.

Conclusion

In brief, it appears that the idea of Local Agenda 21 is the single most effective aspect of the Rio action plan. The activity generated by Chapter 28 of the Agenda, strongly supported by both the United Nations Commission on Sustainable Development (UNCSD) and the International Council for Local Environmental Initiatives (ICLEI), represents a relatively isolated case of Post-Rio success. It is primarily through LA21 that large numbers of local citizens at the local level of governance have become aware of the goals from Rio and the basic ideas and values of sustainable development. As the idea continues to spread, however, it is also clear that a strong potential for "derailing" the LA21 train exists. Underlying the diverse efforts and spontaneous interest is a potential conflict of considerable importance. Numerous groups and stakeholders, particularly from the more anarchistic wing of the environmental movement, have tended to view LA21 as a means of mobilizing against existing political institutions, whether at the central, regional, or local levels. The general feeling here is that LA21 is best expressed as an anti-bureaucratic and even anti-political-party movement: a general admonition to "let a thousand flowers bloom." Recently in Norway, leaders of the Conservative Party in Parliament joined lobbyists from the more decentralized elements of the environmental movement in a successful effort to shift funds that were originally earmarked for the regional nodes just mentioned, to a scheme of direct subsidies to the organizations themselves. Such an interpretation of Local Agenda 21 is, of course, directly at odds with both the original intent and actual follow-up of Chapter 28. The admonition (and responsibility) of Chapter 28 goes to *local authorities*. It is they who must initiate a new and different type of dialogue with citizens and stakeholders, *so that the results can then be channeled into long-range strategic planning and implementation*. The most succinct challenge to Local Agenda 21 is to join the top-down politics of democratic priorities and administrative responsibility, combined with the bottom-up politics of spontaneous initiative and grassroots knowledge and interests. If the "new politics" of LA21 are not joined with the "old politics" of democratic procedure and administrative responsibility, the momentum will surely die out; and the hope expressed in Rio that Agenda 21 should be viewed as a "new partnership for sustainable development" will surely be dashed.

Notes

[1]   The original document for the Rio Action Plan—"Agenda 21"—is here referenced as United Nations (1993). All of the Rio documents are available on-line at http://infoserver.ciesin.org/datasets/unced/unced.html

[2]   The complete text of the Aalborg Charter is available at http://www.iclei.org/europe/echarter.htm

[3]   The "work program" for the project is available at http://afux.prosus.nfr.no/la21/eu/program.html

[4]   As a concept for joining developmental with environmental concerns, the term "sustainable development" was apparently used by Barbara Ward in the mid-1970s (Holmberg and Sandbrook 1992: 19). Dennis Pirage contributed to spread the concept further in his anthology on *The Sustainable Society* in 1977, and the International Union for the Conservation of Nature (IUCN) made it a central idea of its *World Conservation Strategy* in 1980. (See Dahle, 1997: 197). It was not, however, until the idea was made the core concept of the Brundtland Report and the UNCED process that it gained its current prominence and legitimacy. (See Lafferty and Langhelle 1995 and Trzyna 1995)

[5]   See Carlo Aall, this volume, for further perspectives on the Norwegian reforms.

[6]   This section is based on work under progress for the Coordination Unit for Local Agenda 21 within the Ministry of the Environment in Norway. The list of criteria presented is a translated (and slightly revised version) from the list presented in the first draft to the Ministry (in December 1998). The author is grateful to the co-author of the report, Trygve Bjørnæs, and to the head of the Coordination Unit, Dagfinn Rivelsrud, for permission to use the material in the present context.

[7]   A profile of the Ministry's work and strategy in this area is available (in Norwegian) at http://odin.dep.no/md/publ/agend21/

[8]   This section is based on a paper presented to the Graz Symposium on "Regions—Cornerstones for Sustainable Development." (Graz, Austria, 28-30 October 1998) I am grateful to Michael Narodoslawsky and other participants at the symposium for valuable comments. The proceedings of the symposium are available in Gabriel and Narodoslawsky (1998).

[9]   It should be pointed out here, that there also has taken place an ongoing discussion and realignment of tasks within the County Governor's Office, both regionally with the different county offices, and as part of a centrally initiated procedure, to determine the correct balance between the responsibilities of the Ministry of the Environment and key central agencies on the one hand, and the County Governor's offices on the other. The clarification of this "regional" dimension has, in turn, given rise to further discussion and clarification as to the division of labor between the administrative offices of the Municipal County Council and the County Governor's Office.

References

Aalborg Charter. 1994. *The Charter of European Cities and Towns*. European Conference on Sustainable Cities and Towns. 27 May 1994. Aalborg.

Akershus County. 1998. [Akershus fylkeskommune]. *Regional Agenda 21: Globale miljøutfordringer – Lokal og regionalt ansvar*. ("Regional Agenda 21: Global Environmental Challenges – Local and Regional Responsibility"). Oslo: Akershus fylkeskommune.

Armann, K.A., J. Hille and O. Kasim. 1995. *Lokal Agenda 21 - Norske kommuners miljøarbeid etter Rio*. ("Local Agenda 21: Efforts of Norwegian Municipalities after Rio"). Oslo: Project for an Alternative Future, Report 95: 5.

Dahle, K. 1997. *Forsøk for Forandring? Alternative Veier til et Bærekraftig Samfunn*. ("Experiments for Change: Alternative Paths to a Sustainable Society"). Oslo: Spartacus Forlag.

Eriksen, E.O., ed. 1995. *Deliberativ Politikk: Demokrati i Teori og Praksis*. ("Deliberative Politics: Democracy in Theory and Practice"). Oslo: TANO.

Fishkin, J. 1991. *Democracy and Deliberation: New Directions for Democratic Reform*. New Haven: Yale University Press.

Fishkin, J. 1995. *The Voice of the People: Public Opinion and Democracy*. New Haven: Yale University Press.

Gabriel, I. and M. Narodoslawsky, eds. 1998. *Regions – Cornerstones for Sustainable Development*. Proceedings from the Graz Symposium, Graz, Austria, 28-30 October 1998. Graz: Austrian Network Environmental Research.

Holmberg, J. and R. Sandbrook. 1992. Sustainable development: What is to be done? In J. Holmberg, ed., *Policies for a Small Planet*. London: IIED/Earthscan.

Lafferty, W.M. and K. Eckerberg. 1998. *From the Earth Summit to Local Agenda 21: Working Towards Sustainable Development*. London: Earthscan.

Lafferty, W.M. and O. Langhelle. 1999. *Towards Sustainable Development: The Goals of Development – and the Conditions of Sustainability*. London: Macmillan.

Pirage, D. 1977. *The Sustainable Society*. New York: Praeger.

Trzyna, T.C., ed. 1995. *A Sustainable World: Defining and Measuring Sustainable Development*. Sacramento, CA: IUCN and California Institute of Public Affairs.

United Nations. 1993. Report of the United Nations Conference on Environment and Development, Rio de Janeiro, 3-14 June 1992. Vol. I: Resolutions Adopted by the Conference. (A/CONF.151./26/Rev.1 [Vol. I]: Containing the "Rio Declaration on Environment and

Development"; "Agenda 21"; and the "Non-Legally Binding Authoritative Statement of Principles for a Global Consensus on the Management, Conservation and Sustainable Development of All Types of Forests") New York: The United Nations.

WCED, World Commission on Environment and Development. 1987. *Our Common Future.* Oxford and New York: Oxford University Press.

Chapter 3

The Eco-City Approach to Sustainable Development in Urban Areas[1]

Mark Roseland

Summary

This chapter deals with the "eco-city," a relatively new concept that brings together ideas from several disciplines such as urban planning, transportation, health, housing, energy, economic development, natural habitats, public participation, and social justice. The author examines how "eco-city" principles can be introduced at the local level, how various dimensions of the eco-city vision are interlinked, and he helps the reader navigate the new literature which has recently appeared. Examples found in the United States, Canada, and Brazil are discussed. While many of these examples are impressive in scope or design, most have been adopted piecemeal rather than as part of a broader framework. In other words, the elements of eco-cities are being put in place but not, as yet, with the necessary synthesis.

Introduction

In this chapter I argue that a collection of apparently disconnected ideas about urban planning, transportation, public health, housing, energy, economic development, natural habitats, public participation, and social justice all hang upon a single framework that can be called "eco-city."

The term "eco-city" is relatively new, but it is based upon concepts that have been around for a long time. I begin by looking at eco-city origins, then examine other paradigms or movements which strongly influenced the development of the eco-city concept. From this broad base I explore the growing interest in how these ideas can be applied at the local level and consider the links between various dimensions of the eco-city vision.

Eco-City Origins

In 1975, Richard Register and a few friends founded Urban Ecology, in Berkeley, California, as a nonprofit organization to "rebuild cities in balance with nature." Since then, the organization has participated with others in Berkeley to build a "Slow Street," to bring back part of a creek culverted and covered eighty years earlier, to plant and harvest fruit trees on streets, to design and build solar greenhouses, to pass energy ordinances, to establish a bus line, to promote bicycle and pedestrian alternatives to automobiles, to delay and possibly stop construction of a local freeway, and to hold conferences on these and other subjects (Register 1994).

Urban Ecology started to gather real momentum with the publication of Register's *Eco-city Berkeley* (1987), a visionary book about how Berkeley could be ecologically rebuilt over the next several decades[2] and *The Urban Ecologist,* the organization's new journal. The momentum accelerated when Urban Ecology organized the First International Eco-City Conference, held in Berkeley in 1990. This conference convened with over 700 people from around the world to discuss urban problems and submit proposals toward the goal of shaping cities upon ecological principles. Among the outcomes were the Second International Eco-City Conference (held in Adelaide, Australia, in 1992) and the Third International Eco-City Conference (held in Yoff, Senegal, in 1996).

Shortly before the Second Eco-City Conference, David Engwicht, an Australian community activist, published *Towards an Eco-City* (1992), later reissued in North America as *Reclaiming Our Cities and Towns* (1993). In it, Engwicht illustrates how city planners and engineers have virtually eliminated effective human exchange by building more roads, taking commerce out of the cities into strip malls, gutting communities, and increasing traffic fatalities. A city, for Engwicht, is "an invention for maximizing exchange and minimizing travel." By that he means exchanges of all sorts: goods, money, ideas, emotions, genetic material, and the like. He advocates eco-cities where people can move via foot, bicycles, and mass transit and interact freely without fear of traffic and toxins.

Urban Ecology, now more than 20 years old, states that its mission is to create ecological cities by following these 10 principles (Urban Ecology 1996b):

1. Revise land-use priorities to create compact, diverse, green, safe, pleasant, and vital mixed-use communities near transit nodes and other transportation facilities.

2. Revise transportation priorities to favor foot, bicycle, cart, and transit over autos, and to emphasize "access by proximity."

3. Restore damaged urban environments, especially creeks, shore lines, ridgelines, and wetlands.

4. Create decent, affordable, safe, convenient, and racially and economically mixed housing.

5. Nurture social justice and create improved opportunities for women, people of color, and the disabled.

6. Support local agriculture, urban greening projects, and community gardening.

7. Promote recycling, innovative appropriate technology, and resource conservation while reducing pollution and hazardous wastes.

8. Work with businesses to support ecologically sound economic activity while discouraging pollution, waste, and the use and production of hazardous materials.

9. Promote voluntary simplicity and discourage excessive consumption of material goods.

10. Increase awareness of the local environment and bioregion through activist and educational projects that increase public awareness of ecological sustainability issues.

The Eco-City Context

Register, Engwicht, and Urban Ecology certainly deserve credit for popularizing the term "eco-city" in the last decade, but the eco-city concept is strongly influenced by other movements that were developing over the same period as Urban Ecology, and by a long line of thinkers and writers whose ideas were precursors to these concepts many decades ago.[3]

A survey of several paradigms or movements that have been floating around for the last 20 or so years may help readers understand the dimensions of eco-city concepts. These include, among others, healthy communities, appropriate technology, community economic development, social ecology, the Green movement, bioregionalism, native world views, and sustainable development.[4] Such a broad survey in a short space necessitates overgeneralization, but I believe the overview that emerges is worth that risk.[5]

Healthy Communities

Public health has been among the traditional responsibilities of local government. A century ago, municipalities were instrumental in improving public health by preventing the spread of disease through slum clearance, community planning, water treatment, and the provision of certain health services.

These early interventions were based on the view of health as the absence of disease and disease prevention as the main challenge for local government.

In the last two decades a new, broader conception of public health has been developed and adopted by municipal governments in Europe and North America. Although the name "healthy communities" implies a focus on medical care, the Ottawa Charter for Health Promotion (WHO 1986) recognizes that "the fundamental conditions and resources for health are peace, shelter, education, food, income, a stable eco-system, sustainable resources, social justice, and equity."

Local governments play a big role in all these areas through their impact on public hygiene (waste disposal and water systems), food handling and other public health regulations, recreational facilities, education, transportation, economic development, and land-use planning. In Europe, the World Health Organization has directed the successful creation of a 30-city network known as the Healthy Cities Project. In Canada, there have been approximately 100 active healthy community projects, and interest has been growing in Seattle and other U.S. cities.

Appropriate Technology [6]

E.F. Schumacher in 1973 coined the term "intermediate technology" to signify "technology of production by the masses, making use of the best of modern knowledge and experience, conducive to decentralization, compatible with the laws of ecology, gentle in its use of scarce resources, and designed to serve the human person instead of making him [*sic*] the servant of machines" (Schumacher 1973). The central tenet of "appropriate technology" (AT) is that a technology should be designed to fit into and be compatible with its local setting. Examples of current projects that are generally classified as AT include passive solar design; active solar collectors for heating and cooling; small windmills to provide electricity; roof-top gardens and hydroponic greenhouses; permaculture; and worker-managed craft industries. There is general agreement, however, that the main goal of the AT movement is to enhance the self-reliance of people on a local level. Characteristics of self-reliant communities that AT can help facilitate include: 1) low resource usage coupled with extensive recycling; 2) preference for renewable over nonrenewable resources; 3) emphasis on environmental harmony; 4) emphasis on small-scale industries; and 5) a high degree of social cohesion and sense of community (see, e.g., Darrow et al. 1981; Farallones Institute 1979; Mollison 1978, 1979; RAIN 1981). Communities that could be said to be practicing AT include the Amish of Lancaster County, Pennsylvania, and the Mennonites of southern Ontario (Foster 1987).

Community Economic Development

The concept of community economic development (CED), like some others here, suffers from an abundance of interpretation. At their finest, however, the distinguishing features of community economic development are characterized by the following working definition from the Community Economic Development Center at Simon Fraser University (1996):

> Community Economic Development is a process by which communities can initiate and generate their own solutions to their common economic problems and thereby build long-term community capacity and foster the integration of economic, social, and environmental objectives. [7]

Other observers describe CED in less flattering terms, arguing that in response to external funding priorities, community development organizations have lost their original focus on the creation of local employment opportunities and local control and generation of capital in low-income communities (Surpin and Bettridge 1986). Examples of CED range from small business counseling and import substitution ("buy local") programs to worker cooperatives, community development corporations, and community land trusts. Boothroyd (1991) argues that "[whether CED is practiced in hinterland resource towns, urban ghettos, obsolescent manufacturing cities, or native communities' reserves, the general objective is the same: to take some measure of control of the local economy back from the markets and the state]."

Social Ecology [8]

Social ecology focuses its critique on domination and hierarchy *per se:* the struggle for the liberation of women, of workers, of blacks, of native peoples, of gays and lesbians, of nature (the ecology movement) are ultimately all part of the struggle against domination and hierarchy. Social ecology is the study of both human and natural ecosystems, and in particular of the social relations that affect the relation of society as a whole with nature. Social ecology advances a holistic worldview, appropriate technology, reconstruction of damaged ecosystems, and creative human enterprise. It combines considerations of equity and social justice with energy efficiency and appropriate technology. Social ecology goes beyond environmentalism, insisting that the issue at hand for humanity is not simply protecting nature, but rather creating an ecological society in harmony with nature. The primary social unit of a proposed ecological society is the eco-community, a human-scale, sustainable settlement based on ecological balance, community self-reliance, and participatory democracy.

Social ecology envisions a confederation of community assemblies, working together to foster meaningful communication, cooperation, and public service in the everyday practices of civic life, and a "municipalist" concept of citizenship cutting across class and economic barriers to address dangers such as global ecological breakdown or the threat of nuclear war. Cooperation and coordination within and between communities is considered able to transcend the destructive trends of centralized politics and state power. The city can function, social ecology asserts, as "an ecological and ethical arena for vibrant political culture and a highly committed citizenry" (Bookchin 1987).

The Green Movement

The Greens believe in the "four pillars" of ecology, social responsibility, grassroots democracy, and nonviolence (Capra and Spretnak 1984).[9] These pillars translate into principles of community self-reliance, improving the quality of life, harmony with nature, decentralization, and diversity.[10] From these principles, the Greens question many ingrained assumptions about the rights of land ownership, the permanence of institutions, the meaning of progress, and the traditional patterns of authority within society. The Greens recognize that their movement will have to take different forms in different countries (Capra and Spretnak 1984). Starting in the mid-1970s in New Zealand (where it was called the Values Party), France (Les Vertes), and West Germany (Die Grünen), the Green movement soon spread to many other developed countries in Europe and North America. In countries with proportional representation, such as Germany, Green politicians have been elected to seats in the Bundestag. Indeed, in 1998 the German Greens became the junior partner in the coalition federal government. In North America, however, Greens admit their involvement in federal political campaigns is primarily a way to educate the populace and build the movement. Local campaigns may be considered more serious bids for power, as when the Arcata, California, Greens won a majority of seats on city council in 1996. Most North Americans still think Green simply means being pro-environment, but for Germans, being Green means being feminist, supporting civil liberties, working for solidarity with Third World peoples, and standing for an end to the arms race (Swift 1987).

Bioregionalism

The central idea of bioregionalism is *place*. Bioregionalism comes from *bio*, the Greek word for life, as in "biology" and "biography," and *regio*, Latin for "territory to be ruled." Together they mean "a life-territory, a place defined by its life forms, its topography, and its biota, rather than by human dictates;

a region governed by nature, not legislature" (Sale 1985). A *bioregion* is considered to be the right size for human-scale organization: often defined as a river basin or as a watershed, it is a natural framework for economic and political decentralization and self-determination. Bioregional practice is oriented toward resistance against the continuing destruction of natural systems, such as forests and rivers, and toward the renewal of natural systems based on a thorough knowledge of how natural systems work and the development of techniques appropriate to specific sites (Dodge 1981).

While bioregionalism as a movement is relatively new, its precursors date back at least a century.[11] Like social ecology, it is rooted in classical anarchism. The implications of bioregional social organization are clearly for local political control by communities on their own behalf combined with broader allegiance to an institutional structure that governs according to an ecological ethic. Bioregionalism considers people as part of a life-place, as dependent on natural systems as are native plants or animals. By virtue of the emphasis it places on natural systems, perhaps, bioregionalism may perhaps appear weak in terms of human systems; however, some "Green City" ideas (e.g., Berg et al. 1989) are rooted in bioregionalism.

Recent volumes edited by bioregionalist Doug Aberley explain how to do bioregional mapping for local empowerment (1993) and cover the history and theory of ecologically sound planning (1994). The "ecological footprint" analysis developed by Wackernagel and Rees (1996) is a bioregional tool which can consider the impact of cities on natural resources and ecosystems. Their work demonstrates that although some industrial cities may appear to be sustainable, they "appropriate" carrying capacity not only from their own rural and resource regions, but also from "distant elsewheres";that is, they import sustainability.

Native World View [12]

Although the subject of considerable debate, many observers (see, for example, McNeeley and Pitt 1985) argue that sustainable patterns of resource use and management have for centuries been reflected in the belief and behavior systems of indigenous cultures. These systems traditionally have been based in a world view that does not separate humans from their environment (Callicott 1982):

> The Western tradition pictures nature as material, mechanical, and devoid of spirit..., while the American Indian tradition pictures nature throughout as an extended family or society of living, ensouled beings. The former picture invites unrestrained exploitation of non-human nature, while the latter provides the foundations for ethical restraint in relation to non-human nature.

The World Commission on Environment and Development recognized how much industrialized cultures have to learn about sustainability from traditional peoples, and at the same time, how vulnerable the latter are to encroachment by the former (WCED 1987). As a native chief speaking at a symposium on sustainable development suggested, mainstream society would be wise to look at native "history, culture, and traditions and practices, and find out how they managed to survive for thousands of years before European contact" (Smith 1989).

Sustainable Development

In December 1983, amid growing concern over declining ecological trends and the seeming incompatibility of economic and environmental perspectives, the UN Secretary-General responded to a United Nations General Assembly resolution by appointing Gro Harlem Brundtland of Norway as Chairman of an independent World Commission on Environment and Development. For the next few years the Brundtland Commission, as it became known, studied the issues and listened to people at public hearings on five continents, gathering over 10,000 pages of transcripts and written submissions from hundreds of organizations and individuals. In April 1987 the commission released its report, *Our Common Future*. At the core of the report is the principle of "sustainable development." The commission's embrace of sustainable development as an underlying principle gave political credibility to a concept many others had worked on over the previous decade. The commission defined sustainable development as meeting "the needs of the present without compromising the ability of future generations to meet their own needs" (WCED 1987). This simple, but vague definition was also the foundation for *Agenda 21*, the document that emerged from the United Nations Conference on Environment and Development (also known as the "Earth Summit" held in 1992 in Brazil) as a sustainable development action plan for the 21st century (Voisey et al. 1996).

From Concept to Practice

Practitioners who turned to these paradigms or movements for direction in applying these concepts to the communities where they work and live have found much inspiration but relatively little guidance. Only recently has there been rapidly growing interest in the practical application of these ideas at the local level.

In the last few years literature has appeared that begins to support this practical application of ideas. Even more than the paradigms discussed above, this new literature reflects a somewhat bewildering variety of orientations and terminology. The authors include architects, academics, and activists, and the terminology includes everything from "neotraditional town planning"" and "pedestrian pockets" to "reurbanization," "post-industrial suburbs," and "sustainable cities."

To help the reader navigate this new applied literature, I have organized it (see Table 3-1) into four broad categories which reflect the backgrounds, world views, or orientations of the various authors: Designers, Practitioners, Visionaries, and Activists.[13]

TABLE 3-1 *Categories in the Literature*

Designers	Practitioners	Visionaries	Activists
The Costs of Sprawl	Sustainable Urban Development	Sustainable Communities	Green Cities
			Eco-cities
Sustainability By Design	Sustainable Cities	Community Self-Reliance	Eco-communities
	Local Sustainability Initiatives		

The Designers category includes the literature on the costs of sprawl and sustainability by design. These authors are, for the most part, architects, planners, consultants, and related professionals. In general, they are oriented toward sustainable "developments," that is, new subdivisions, within a social structure essentially unchanged from what we have become accustomed to over the last several decades (post-World War II). They are, for the most part, not concerned with global sustainability issues such as atmospheric change or social equity.

At the other end of the spectrum are the authors writing about Green cities, eco-cities, and eco-communities, whom together I call the Activists. These authors are, for the most part, writers and community activists who consider themselves bioregionalists, social ecologists, and various other kinds of environmentalists. Their writings are generally oriented toward community change within the context of a society that is recognizing its anti-ecological ways and embarking on a more sustainable course. The "ecotopian" social structure envisioned by these authors differs significantly from the present industrial-bureaucratic structure, which these writers consider unlikely or unable to change fundamentally toward biophysical and social sustainability.

In between these poles are the Practitioners and the Visionaries. The Practitioners category includes the literature on sustainable urban development, sustainable cities, and local sustainability initiatives (such as Local Agenda 21). These authors represent ecologically informed and inspired practitioners: politicians, local government professionals (e.g., planners and staff of local or provincial/state offices of environmental management, energy efficiency, and so forth), occasional academics and, increasingly, citizens and community organizations. They generally define communities as municipalities, and this literature is often directed toward public sector decision-makers.

The Visionaries category includes the sustainable communities and community self-reliance literature. These authors can be described as agriculturists, economists, architects, planning theorists, and appropriate technologists. They generally define communities of association (e.g., women of color) and of interest (e.g., social justice) as well as of place (e.g., municipalities). This literature is often directed toward professionals, academics, and other citizens concerned with issues such as energy conservation, appropriate technology, and community economic development.

Of these streams of literature, some will clearly be more relevant than others to the circumstances facing any particular community or set of communities. Each category of the eco-cities literature makes important contributions to our understanding of the concept. Table 3-2 compares the orientation, focus, and means of each category.

While there are discernible differences in analysis, emphasis, and strategy between the variations discussed above, the "eco-city" theme is broad enough to encompass any and all of them (see Figure 3-1). [14]

The wording of titles in many of the publications discussed here, such as *Toward Sustainable Communities* (Roseland 1998), *Towards an Eco-City* (Engwicht 1992), and *Building Sustainable Communities* (Nozick 1992) is significant. Eco-cities, or sustainable communities, represent a goal, a direction for community development—not simply a marketing slogan. Indeed, many of these authors would argue that the only modern communities that are remotely sustainable at present are some aboriginal communities that have existed for centuries (although other communities have recently undertaken impressive initiatives toward modern sustainability).

It is at present safe to say that there is no (and perhaps should not be any) single accepted definition of "eco-cities" or of "sustainable communities." [15] Inherent in much of the literature is the recognition that communities must be involved in defining sustainability from a local perspective. The challenge is how to encourage local democracy within a framework of global sustainability.

Example Eco-City Initiatives

Numerous examples of citizen and community initiatives abound which demonstrate that creative, transferable solutions to seemingly intractable social and environmental challenges are being initiated by citizen organizations and municipal officials in cities and towns around the world (e.g., see Roseland 1997, 1998). Some notable initiatives are those of Chattanooga

TABLE 3-2 *Comparisons of the Literature Catagories*

	Orientation	Focus	Means
Designers	Architects, planners, consultants, and related professionals	New developments	Reducing sprawl; design to encourage the revival of public life (e.g., townscapes, streetscapes, malls and squares)
Practitioners	Politicians, local government professionals, citizens and community organizations	Existing settlements, municipalities	Local initiatives to create local sustainable development action strategies
Visionaries	Agriculturists, economists, architects, planning theorists, and appropriate technologists	Communities of association and of interest, as well as of place	Reducing resource waste; energy efficiency, stressing passive solar heating and cooling; encouraging local food production and reliance on local resources; fostering creation of on-site jobs and neighborhood stores to revitalize communities and eliminate wasteful commuting
Activists	Writers and community activists who consider themselves bioregionalists, social ecologists, and various other kinds of environmentalists	Human-scale, sustainable settlements based on ecological balance, community self-reliance, and participatory democracy	Decentralized, grass roots, cooperative development

and the San Francisco Bay Area in the U.S., Ottawa, Hamilton-Wentworth, and Greater Toronto in Canada, and Curitiba in Brazil.

In 1969, Chattanooga, Tennessee, was the most polluted city in the U.S. In 1990 it was recognized as the country's best turn-around story. How did it happen? In 1984, the entire community was invited to envision what they wanted their community to be like by the year 2000. The shared vision of sustainable community development that emerged put affordable housing, public education, transportation alternatives, better urban design, parks and greenways, and neighborhood vitality at the top of the community's agenda. Energetic collaboration of government agencies, manufacturers, and citizens resulted in successful initiatives to clean up the air and revitalize a city in decline. Several eco-industrial parks were established to rebuild the city's

Sustainable Development	**Healthy Communities**	**Community Economic Development**
Sustainable Urban Development		**Appropriate Technology**
	Eco-Cities	
Sustainable Communities, Sustainable Cities		**Social Ecology**
Bioregionalism	**Native World Views**	**Green Movement, Green Cities/Communities**

FIGURE 3-1 Dimensions of the eco-city.

NOTE The "eco-cities" theme does not stand alone, but is situated in a complex array of relevant variations. The figure quite deliberately has no arrows, no lines, and no borders, and should be imagined as a holograph.

economic base, proving that economic development and environmental stewardship can be achieved together. Most importantly, all participants determined to set in motion a comprehensive, interrelated, and strategic process for sustainable community development (Gilbert et al. 1996).

The San Francisco Bay Area citizens organization Urban Ecology organized a five-year participatory process to develop their *Blueprint for A Sustainable Bay Area*. It is based on providing residents with new choices for a prosperous life; protecting, restoring, and integrating nature into people's lives; working toward social, economic, and environmental justice; promoting development and transportation alternatives that connect the region; encouraging resource conservation and reuse; designing with respect for local and historical uniqueness; and enabling residents to nurture a strong sense of place, community, and responsibility (Urban Ecology 1996a). The report's 95 recommendations reflect a wide range of tools and initiatives.

The Health, Family and Environment Committee of the San Francisco Board of Supervisors, with the support of the Mayor, in 1997 unanimously endorsed a *Sustainability Plan* that would guide decisions of all city commissions and departments. Air quality, solid waste, biodiversity, and food and agriculture are among the 10 major topics of the 150-page plan. The plan's transportation suggestions include creating 10 auto-free zones over the next four years, increasing the city parking tax, raising gasoline taxes and bridge tolls, and introducing "congestion pricing" charges for driving at rush hour. Conservative critics have called it "eco-totalitarianism." San Francisco joins Santa Monica, California, and Chattanooga as American cities with extensive environmental plans (GreenClips 76, 1997; Langton 1997).

The City of Ottawa's 1992 Official Plan, developed through an elaborate public consultation process with solid community support, is based on the concept of sustainable urban development. The city's commitment to sustainable development is manifested in the mission statement, guiding principles, and the vision for Ottawa found at the beginning of the plan. Specific policies contained in the plan are designed to reflect this commitment. For example, the housing policies promote affordable housing, infilling and intensification, which reflect the guiding principles of adequate shelter, and conservation and enhancement of the resource base. The transportation policies encourage increased use of public transportation, cycling, and walking, which reflect the guiding principle of increasing non-automobile transportation. In addition, the plan outlines the city's environmental impact assessment process, which is designed to address the cumulative impact of everyday practices and development projects on the environment (ORTEE 1995).

The Sustainable Community Initiative is an ongoing collaborative process in which the Hamilton-Wentworth regional government, 75 kilometers west of Toronto, has been working with thousands of citizens to turn a jointly developed community vision into a reality. Since *Vision 2020* was formally adopted in 1993, the regional government has devised an implementation strategy outlining the major policy shifts needed to achieve the vision, along with over 400 recommendations for specific action and a unique approach for monitoring progress—28 sustainability indicators are compiled into an Annual Report Card, which is the basis of an Annual Sustainable Community Day and forum to assess progress in relation to the goals of *Vision 2020*. The Sustainable Community Initiative has already had a profound impact on the way the local government operates and is leading toward significant improvements to the local environment, including development of a bicycle commuter network, creation of habitat corridors for wildlife protection, and a home-energy and waste-auditing program (ICLEI 1995).

In 1988 a Royal Commission on the Future of the Toronto Waterfront was created to examine matters related to "the use, enjoyment and development" of the area. By the mid-1980s the Toronto waterfront had become an embarrassment. Pollution had closed the beaches; expressways and condominium towers blocked public access, contaminated and abandoned industrial lands degraded the harbor area, and rehabilitation efforts were frustrated by jurisdictional squabbling among federal, provincial, regional, and municipal authorities. The commission, headed by former Toronto Mayor David Crombie, initiated broad public discussion of waterfront issues and quickly saw that the conventional, fragmented, waterfront-specific approach would not work, so they expanded its scope from the waterfront to the watershed. This larger area, which the Commission called the *Greater Toronto bioregion*, is home to four million people as well as innumerable other creatures in complex social as well as ecological relations. Before the Crombie Commission, the notion of bioregional or ecosystem planning was little known outside professional and citizen planning circles and did not seem likely to play a major role in guiding practical efforts to reunite economy, community, and ecology. Now it is a real possibility that has begun to be tested in practice (Gibson et al. 1997).

Curitiba, Brazil, has received international acclaim for its integrated transportation and land-use planning, and for its waste management programs. Both are good examples of sustainable community planning. But how did Curitiba manage to become a positive example for cities in both developed and developing countries? In part, the city's success can be attributed to strong leadership—city officials focused on developing simple, flexible, and affordable solutions that could be realized at the local level and adapted to changing conditions. In addition, "the government promoted a strong sense of public participation. Officials were encouraged to look at problems, talk to the people, discuss the main issues, and only then reach for the pen" (Rabinovitch 1996). Jonas Rabinovitch, a long-time advisor to Curitiba Mayor Jaime Lerner, believes the lesson to be learned from Curitiba is that creativity can substitute for financial resources. Any city, rich or poor, can draw on the skills of its residents to tackle urban environmental problems (Rabinovitch 1996).

Cities are now banding together under the auspices of various regional associations and partnerships concerning approaches and solutions to common urban problems. The CO_2 Reduction Program is a partnership coordinated by the International Council for Local Environmental Initiatives (ICLEI). More than 100 local governments from 27 countries have joined an International Cities for Climate Protection Campaign. Participants pledge to meet and exceed the requirements of the Framework Convention on Climate Change by reducing carbon dioxide emissions by up to 20 percent by 2005. As part of this initiative, ICLEI worked with 14 cities to develop comprehensive local action plans to reduce carbon dioxide emissions (Brugmann 1996).

Conclusion

While many of these examples are impressive in scope or design, most have been adopted piecemeal rather than as part of a broader framework. In other words, the elements of eco-cities are being put in place but not, as yet, the necessary synthesis.

The emerging eco-city vision offers this synthesis. I have argued that a set of what may at first appear to be disconnected ideas about the future of our communities and our planet is, upon closer inspection, a broader framework. This is reflected in the new applied literature of the Designers, Practitioners, Visionaries, and Activists, and it is rooted in a set of ideas that reaches back many years and spans many disciplines. From the examples of eco-city initiatives previously cited and sustainability indicators in Part III of the book, we can see that a cohesive vision for human settlements is beginning to emerge.

Notes

[1]   This chapter is based upon Roseland, M. 1997. Dimensions of the future: An eco-city overview. In M. Roseland, ed., *Eco-City Dimensions: Healthy Communities, Healthy Planet.* Gabriola Island, BC: New Society Publishers; Roseland, M. 1997. Dimensions of the eco-city. *CITIES: The International Journal of Urban Policy and Planning* Vol. 14, 4: 197-202; and upon Roseland, M. 1998. *Toward Sustainable Communities: Resources for Citizens and Their Governments.* Gabriola Island, BC: New Society Publishers.

[2]   Register (1987) cites seven key biological/bioregional principles for ecologically rebuilding cities: 1) diversity is healthy; 2) fairly large areas are required for natural species to develop diversity of population; 3) land has a limit (carrying capacity) to the quantity of biological material it can naturally support in a particular climate; 4) there is a green hierarchy in sustainable community planning (e.g., native and useful plants before lawns and ornamentals); 5) make wastes into new resources; 6) biological pest controls and nutrients are generally preferable to chemicals; and 7) species and habitat protection is a regional problem, not simply an urban one (e.g., fences, forest management, long commutes).

[3]   These include, for example, Ebenezer Howard (1902); Patrick Geddes (1915); Paul and Percival Goodman (1947); Lewis Mumford (1964); Ian McHarg (1969); Christopher Alexander et al. (1977); and Anne Whiston Spirn (1984).

[4]   Space permitting, other paradigms which could also be discussed here include environmental justice (e.g., Bullard 1993), ecological economics (e.g., Jacobs 1991), the steady state (e.g., Daly 1973), ecofeminism (e.g., Plant 1989), deep ecology (e.g., Devall and Sessions 1985), the conserver society (e.g., Science Council of Canada 1977), new physics (e.g., Prigogine and Stengers 1984), and the Gaia hypothesis (e.g., Lovelock 1979).

[5]   An earlier version of parts of this survey appeared in Gardner and Roseland (1989).

[6]   Also known as alternative, renewable, soft, intermediate, radical, liberatory, and human-scale technology.

[7]   This definition is based on the founding report for the Centre, written by David Ross and George McRobie in 1987. McRobie was a colleague of E. F. Schumacher. His *Small Is Possible* (1981) was inspired by Schumacher's *Small Is Beautiful.*

[8]   Social ecology is a term with various meanings in various places, e.g., a branch of urban sociology. The social ecology referred to here, however, is focused primarily around the writings of Murray Bookchin.

[9]   Greens in the U.S. have generally expanded this list to include an explicit emphasis on decentralization (see, e.g., Tokar 1987).

[10]   Tokar (1987) adds freedom, equality, and democracy to the list.

[11]   Elements of bioregionalism can be traced back to the writings of, for example, Kropotkin (1904), Geddes (1915), and Mumford (1964).

[12]   I wish to thank Dr. Julia Gardner for this discussion of native world views.

[13]   For a fuller discussion of this literature, see Roseland (1995).

[14]   A similar argument could be made for some of the other terms discussed here, in particular Sustainable Communities (e.g., Roseland 1998), Sustainable Cities (e.g., Haughton and Hunter 1994), and Local Sustainability Initiatives (e.g., ICLEI 1993).

[15]   Another approach to the challenge of defining sustainable communities is to specify necessary conditions for sustainable communities. For example, I have developed the argument (Roseland 1998) that the efficient use of urban space, minimizing consumption of essential natural capital, multiplying social capital, and mobilizing citizens and their governments are necessary conditions for sustainable community development.

References

Aberley, D. 1993. *Boundaries of Home: Mapping for Local Empowerment.* Gabriola Island, B.C.: New Society Publishers.

Aberley, D., ed. 1994. *Futures By Design: The Practice of Ecological Planning.* Gabriola Island, B.C.: New Society Publishers.

Alexander, C., S. Ishikawa and M. Silverstein. 1977. *A Pattern Language: Towns, Buildings, Construction.* New York: Oxford University Press.

Berg, P., B. Magilavy and S. Zukerman. 1989. *A Green City Program for San Francisco Bay Area Cities and Towns.* San Francisco, CA: Planet Drum Books.

Bookchin, M. 1987. *The Rise of Urbanization and the Decline of Citizenship.* San Francisco: Sierra Club.

Boothroyd, P. 1991. *Community Economic Development: An Introduction for Planners.* Vancouver: UBC Centre for Human Settlements.

Brugmann, J. 1996. "Cities Take Action: Local Environmental Initiatives." In World Resources Institute, United Nations Environment Program, United Nations Development Program, The World Bank, *World Resources 1996-97: The Urban Environment.* New York: Oxford University Press.

Bullard, R., ed. 1993. *Confronting Environmental Racism: Voices from the Grassroots.* Boston: South End Press.

Callicott, J. B. 1982. Traditional American Indian and Western European attitudes toward nature: An overview. *Environmental Ethics* 4:293-318.

Capra, F. and C. Spretnak. 1984. *Green Politics: The Global Promise.* New York: E. P. Dutton.

Community Economic Development Center at Simon Fraser University. 1996. A Working Definition of Community Economic Development. Website http://www.sfu.ca/cedc/ (17 Febr. 1999).

Daly, H. E., ed. 1973. *Toward a Steady-State Economy.* San Francisco: W.H. Freeman and Co.

Darrow, K., K. Keller and R. Pam. 1981. *Appropriate Technology Sourcebook.* Stanford: Volunteers in Asia.

Devall, B. and G. Sessions. 1985. *Deep Ecology: Living as if Nature Mattered.* Layton, UT: Peregrine Smith Books.

Dodge, J. 1981. Living by life: Some bioregional theory and practice. *CoEvolution Quarterly* 32 (Winter): 6-12.

Engwicht, D. 1992. *Towards an Eco-City: Calming the Traffic.* Sydney, Australia: Envirobook.

Engwicht, D. 1993. *Reclaiming Our Cities and Towns: Better Living With Less Traffic.* Gabriola Island, B.C.: New Society Publishers.

Farallones Institute. 1979. *The Integral Urban House: Self-Reliant Living in the City.* San Francisco: Sierra Club Books.

Foster, T. W. 1987. The Taoists and the Amish: Kindred expressions of eco-anarchism. *The Ecologist* 17,1: 9-14.

Gardner, J. and M. Roseland. 1989. Acting locally: Community strategies for equitable sustainable development. *Alternatives: Perspectives on Society, Technology and Environment* (October-November), 36-48.

Geddes, P. 1915. *Cities in Evolution.* London: Williams and Norgate.

Gibson, R. B., D. H. M. Alexander and R. Tomalty. 1997. Putting cities in their place: Ecosystem-based planning for Canadian urban regions. In M. Roseland, ed., *Eco-City Dimensions: Healthy Communities, Healthy Planet.* Gabriola Island, B.C.: New Society Publishers.

Gilbert, R., D. Stevenson, H. Giradet and R. Stren. 1996. *Making Cities Work: The Role of Local Authorities in the Urban Environment.* London: Earthscan.

Goodman, P. and P. Goodman. 1960 (originally published in 1947). *Communitas: Means of Livelihood and Ways of Life.* New York: Vintage.

GreenClips 76. e-mail <GreenClips@aol.com> (July 16, 1997)

Haughton, G. and C. Hunter. 1994. *Sustainable Cities.* Regional Policy and Development Series 7. London: Jessica Kingsley.

Howard, E. 1902. *Garden Cities of To-Morrow.* London: Faber and Faber.

ICLEI, International Council for Local Environmental Initiatives. 1993. *The Local Agenda 21 Initiative: ICLEI Guidelines for Local and National Local Agenda 21 Campaigns.* Toronto: ICLEI.

ICLEI, International Council for Local Environmental Initiatives. 1995. *The Role of Local Authorities in Sustainable Development: 14 Case Studies on the Local Agenda 21 Process.* Nairobi: UN Centre for Human Settlements.

ICLEI/ IDRC/ UNEP, International Council for Local Environmental Initiatives, International Development Research Centre and United Nations Environment Programme. 1996. *The Local Agenda 21 Planning Guide.* Toronto: ICLEI and Ottawa: IDRC.

Jacobs, M. 1991. *The Green Economy: Environment, Sustainable Development and the Politics of the Future.* London: Pluto Press.

Kropotkin, P. 1904. *Fields, Factories and Workshops.* New York: G.P. Putnam's Sons and London: S. Sonnenschein & Co.

Langton, J. 1997. City's plan to kill stray cats raises a howl. *Vancouver Sun,* August 11, p. A8.

Lovelock, J.E. 1979. *Gaia: A New Look at Life on Earth.* Oxford: Oxford University Press.

McHarg, I. 1969. *Design With Nature.* New York: Natural History Press.

McNeeley, J. A. and D. Pitt, eds. 1985. *Culture and Conservation: the Human Dimension in Environmental Planning.* New York: Croom Helm.

McRobie, G. 1981. *Small Is Possible.* London: Cape.

Mollison, B. 1978. *Permaculture One.* Winters, CA: International Tree Crops Institute.

Mollison, B. 1979. *Permaculture Two.* Winters, CA: International Tree Crops Institute.

Mumford, L. 1964. *The Highway and the City.* New York: New American Library.

Newman, P. and J. Kenworthy. 1989. *Cities and Automobile Dependence: An International Sourcebook.* Brookfield, VT: Gower Technical.

Nozick, M. 1992. *No Place Like Home: Building Sustainable Communities.* Ottawa: Canadian Council on Social Development.

ORTEE, Ontario Round Table on the Environment and the Economy. 1995. *Sustainable Communities Resource Package.* Toronto: ORTEE.

Plant, J., ed. 1989. *Healing the Wounds: The Promise of Ecofeminism.* Gabriola Island, BC: New Society Publishers.

Prigogine, I. and I. Stengers. 1984. *Order out of Chaos: Man's New Dialogue with Nature.* Toronto: Bantam.

Rabinovitch, J. 1996. Integrated transportation and land use planning channel Curitiba's growth. In World Resources Institute, United Nations Environment Program, United Nations Development Program, The World Bank, *World Resources 1996-97: The Urban Environment.* New York: Oxford University Press.

RAIN. 1981. *Knowing Home: Studies for a Possible Portland.* Portland: Rain Umbrella Inc.

Register, R. 1987. *Eco-City Berkeley: Building Cities for a Healthy Future.* Berkeley, CA: North Atlantic Books.

Register, R. 1994. Eco-cities: Rebuilding civilization, restoring nature. In D. Aberley, ed., *Futures By Design: The Practice of Ecological Planning.* Gabriola Island, B.C.: New Society Publishers.

Roseland, M. 1995. Sustainable communities: An examination of the literature." In *Sustainable Communities Resource Package.* Toronto: Ontario Round Table on the Environment and the Economy.

Roseland, M. 1997. Dimensions of the eco-city. *CITIES: The International Journal of Urban Policy and Planning* 14,4: 197-202.

Roseland, M., ed. 1997. *Eco-City Dimensions: Healthy Communities, Healthy Planet.* Gabriola Island, BC: New Society Publishers.

Roseland, M. 1998. *Toward Sustainable Communities: Resources for Citizens and Their Governments.* Gabriola Island, BC: New Society Publishers.

Sale, K. 1985. *Dwellers in the Land: The Bioregional Vision.* San Francisco: Sierra Club.

Schumacher, E.F. 1973. *Small Is Beautiful: A Study of Economics as if People Mattered.* New York: Harper & Row.

Science Council of Canada. 1977. *Canada as a Conserver Society: Resource Uncertainties and the Need for New Technologies.* Ottawa: Supply and Services Canada.

Smith, E. J. 1989. A perspective from a Nuu-chah-nulth on planning sustainable communities. In W.E. Rees, ed., *Planning for Sustainable Development: A Resource Book.* Vancouver: UBC Centre for Human Settlements, pp. 127-129.

Spirn, A.W. 1984. *The Granite Garden: Urban Nature and Human Design.* New York: Basic Books.

Surpin, R. and T. Bettridge. 1986. Refocusing community economic development. *Economic Deveopment & Law Center Report*, Spring: 36-42.

Swift, R. 1987. What if the Greens achieved power? *New Internationalist,* (May) reprinted in *Utne Reader* 23, (Sept/Oct), pp. 32-33.

Tokar, B. 1987. *The Green Alternative.* San Pedro: R. & E. Miles.

Urban Ecology. 1990. *Eco-City Conference 1990 Report.* Berkeley, CA: Urban Ecology.

Urban Ecology. 1996a. *Blueprint for A Sustainable Bay Area.* Oakland: Urban Ecology.

Urban Ecology. 1996b. Mission statement and accomplishments. *Urban Ecology. Rebuilding Our Cities in Balance with Nature.* Website http://www.urbanecology.org/mission/index.html (17 Febr. 1999).

Voisey, H., C. Beuermann, L.A. Sverdrup and T. O'Riordan. 1996. The political significance of Local Agenda 21: The early stages of some European experience. *Local Environment* 1, 1: 33-50.

Wackernagel, M. and W. Rees. 1996. *Our Ecological Footprint: Reducing Human Impact on the Earth.* Gabriola Island, BC: New Society Publishers.

WCED, World Commission on Environment and Development. 1987. *Our Common Future.* New York: Oxford University Press.

WHO, World Health Organization, Health and Welfare Canada and the Canadian Public Health Association. 1986. Ottawa charter for health promotion. *Canadian Journal of Public Health* 77: 425-427.

An Example of Eco-City Development: Urban Agriculture

Thomas van Wijngaarden

One seventh of the planet's food supply is grown in cities, and there are 800 million urban farmers in the world (City Farmer 1997). In spite of this fact, the role that urban agriculture can play in sustainable development is not currently recognized to its full extent. Urban agriculture has the potential to considerably contribute to food and income security, pollution prevention, waste reduction, and the improvement of living conditions in urban environments. Urban agriculture has multiple facets, including arable farming (cereals and vegetables), fruit growing, aquaculture, and livestock rearing from a micro scale up to co-operatives with a considerable overall total production. It is most common—and has the largest potential for expansion—in urban areas of developing countries. Urban agriculture is best described in terms of the various production phases; the pre-production phase is characterized by the often readily availability of low-cost resources and inputs, such as land, water, seeds, feeds, and fertilizers. In contrast, the production phase is highly market-oriented, but can be negatively affected by variable factors such as land tenure, theft, and environmental impacts. The post-production phase is highly dependent on the proximity of processing plants, distribution networks, and market facilities that limit transport and packaging costs (UNDP 1996).

Asian cities such as Katmandu, Karachi, Singapore, Hong Kong, and Shanghai produce between 25 and 85 percent of their own fruit consumption (Van der Bliek 1996). It is estimated that Hong Kong, with the highest population density of the world's largest cities, produces two thirds of the poultry, one sixth of the pork, and nearly half of the vegetables that are consumed within the city. Singapore produces 25 percent of its vegetable consumption and has been self-sufficient in meat production at a level of 70 kilograms per capita per year (UNDP 1996). Although urban agriculture is widespread and has a long history in Africa, it has often been undervalued and opposed by generations of government officials. In spite of this lack of support from the government, urban agriculture is thriving in the nongovernment regulated or informal sector. Encouraging urban agriculture can even bring about self-sufficiency. Bamako, Mali, for example, is considered self-sufficient in the production of vegetables and is known to produce half or more of the chickens it consumes. Cities that lack official support still see their urban agriculture flourishing, and new crops and techniques introduced by immigrants are rapidly adapted. For example, Thai mushroom culture, Filipino seaweed production, and Asian vegetable and fruit production techniques were introduced in Ghana, Zanzibar and Côte d'Ivoire. The proportion of families engaged in farming in Dar es Salaam rose from 18 percent in 1967 to 67 percent in 1991 (UNDP 1996). In Lusaka, capital city of Zambia, former policies discouraging urban agriculture were changed due to worsening economic conditions and food supply. Nonetheless, a new policy was adopted in 1999 that forbids maize crops along roadsides, since such crops act as breeding grounds for mosquitoes and hiding places for criminals (UNDP 1996). South American urban agricultural systems are renowned for their high productivity and sustainability; this ancestral knowledge is now being rediscovered. Although based on traditional systems, urban agriculture has also been greatly influenced by several external factors, such as foreign immigrants and international organizations that have introduced new farming techniques For example, in Peru, fish are produced using wastewater following an Asian model. Moreover, some native animals, such as guinea pigs and iguanas have been successfully adapted for raising in urban areas (UNDP 1996).

While urban agriculture in European and North American cities has declined since the late nineteenth century, it started to make a comeback in the 1970s, in the wave of the Green movement. In the twenty-first century, social interests largely predominate over food security concerns, as exemplified by the popularity of ornamental gardens. North American cities such as Vancouver and Chicago have gone beyond just the greening of the urban environment and actively support urban agriculture initiatives. For example, in Chicago, the city environment department constructs gardens atop several city buildings as part of a U.S. Environmental Protection Agency program studying ways to help cool cities and reduce smog. The city of Vancouver works together with City Farmer, a very active non-profit association that promotes urban food production and environmental conservation. This cooperation led to the opening of a new public garden which demonstrates conservation methods such as contouring of the ground, soil conditioning using compost, collection of rain water, and the use of native plants (City Farmer 1999). The renewed interest in urban agriculture also calls for a change in urban spatial planning. Urban planners will have to take into account the need to produce food inside the city, an urban function which has been largely neglected in our land-use plans. The contribution of metropolitan areas to the value (in U.S. dollar terms) of the agricultural production of the United States rose from 30 percent in 1980 to 40 percent in 1990 (UNDP 1996). Although current population trends—in particular the increasing urban population density—reduce the opportunities for urban agriculture, it is important to stress the necessary contribution of urban agriculture to sustainable urban development. The increasing consumption of food, water, energy, raw materials, and minerals in cities has lead to increased pollution and waste production. Urbanization often reduces the green areas in cities, resulting in poorer drainage, changes in micro-climate and increased dust production. Urban agriculture can, on the one hand, reduce input flows to urban environments, mainly by reducing transport and packaging needs. In Britain, food transport represents one fourth of all journeys and twelve percent of fuel consumption. On the other hand, output flows can also be reduced, mainly by recycling organic waste.

Opportunities and incentives for urban agriculture can be captured in the framework of one or several of the three spheres of sustainable development: social, economic, and environmental sustainability. Urban agriculture can assist in combating environmental degradation, promoting ecological restoration, reducing resource consumption, improving health and nutritional status, food and income security, ecological education, local economic development and diversification, stimulating community building, and an overall feeling of well-being.

Opportunities need to be analyzed in the perspective of the need and availability of resources. We can distinguish between:

* Human resources: who will provide the labor, and how much time can he/she spend on it
* Physical resources: land or space, water, light, and nutrients
* Financial resources for the purchase of inputs: hardware, technology, seeds, stocks, supplements, training, etc.

Additional factors that should be studied are economic benefits, environmental considerations, social aspects and individual well-being, and supportive or discouraging policies.

Urban agriculture faces multiple constraints, such as land availability, land tenure insecurity, lack of economic incentives, poor availability of credit and capital, lack of skills, low solar access, pollution, policy or regulatory obstacles, theft, vandalism, efficient production technologies, as well as the need for mitigating measures against odor and noise nuisances to neighbors. In addition, urban agriculture definitely has conflicts of interest. Apparent improvements in urban environments can negatively affect specific socioeconomic groups. For example, if cities in developing countries implemented mass collection of urban waste, poorer households would not be able to afford feed supplements for their goats to compensate for the feed previously scavenged from urban waste (Richardson and Whitney 1995).

The challenges faced by urban agriculture lie first and foremost with the governing agencies and local authorities. Official recognition of urban agriculture provides opportunities for supply of services such as extension services, technology development, organization for marketing, strengthening of communities, and veterinary services to name a few that will assist in safeguarding human health and environmental quality (Waters-Bayer 1995). Recognizing the role of urban agriculture in sustainable development implies acknowledging its contribution to poverty alleviation, gender issues (e.g. in Kenya and other East African countries three-fifths to two-thirds of the primary urban farmers are women), and positive effects on waste management and material flows.

Changes in the mentality of planners, designers, and decision-makers on one hand, and in management practices on the other hand, will open new perspectives for people to use their urban environment and its improvements more sustainably. This can be achieved by allowing the use of public or waste land for urban agriculture, by giving urban farmers rights for grazing their cattle inside the city, by promoting useful trees and crops in public areas, as well as by stimulating community involvement in real estate development and architectural design.

References

City Farmer. 1997. *Creating a Center for Sustainable Urban Agriculture and Food Systems at the University of California Gill Tract in Albany.* Urban Agricultural Notes, City Farmer, Office of Urban Agriculture, Canada.

City Farmer. 1999. Urban agriculture notes. City Farmer. Website http://www.cityfarmer.org (5 May 1999).

UNDP, United Nations Development Program. 1996. *Urban Agriculture. Food, Jobs and Sustainable Cities.* Publication Series for Habitat II, Volume I. UNDP. New York.

Van der Bliek, J. 1996. Landbouw in de Stad: Een contradictio in terminis? *Derde Wereld* 15, 2: 139-150.

Richardson and Whitney, 1995. Goats and garbage in Khartoum, Sudan: A study of the urban ecology of animal keeping. *Human Ecology* 23, 4: 455-475.

Waters-Bayer, A. 1995. Living with livestock in town: Urban animal husbandry and human welfare: Livestock production and diseases. In Proceedings of the 8th International Conference of Institutions of Tropical Veterinary Medicine. 25-29 September 1995. Berlin.

Chapter 4

Sustainable Development at the Local Level: The North-South Context

Dimitri Devuyst

Summary

Although urban population growth rates slowed down in the 1980s and 1990s, the number of people living in urban areas is expected to double between 1990 and 2025 to more than 5 billion people. Some 90 percent of this growth will take place in developing countries, which implies a major challenge for urban planners, local authorities, and people aiming for a more sustainable urban future. In this chapter it is argued that each city has a unique set of problems in dealing with sustainable development and each city requires a specific vision and tailor-made measures with which to solve them. Generally speaking, priorities will be very different in cities of the developed and the developing world. In the Northern Hemisphere, there is a need to reduce both the consumption of resources and the production of wastes. In the Southern Hemisphere, there is still a priority to provide the basic necessities for the population. But even within an individual city, different policies are needed for the wealthier and poorer sections of that city. (The upcoming case study of Durban, South Africa, a Third World city with both affluent and poor areas, illustrates this point.) On the one hand, the least developed areas should focus on the provision of basic human needs, such as decent shelter, basic health care, and education, clean water, energy, and sanitation. On the other hand, the more developed areas need to look into ways of minimizing the use of natural resources. The importance of community development and collaboration with low-income people both in rich and poor countries is stressed. The most important resources for the city of the future are the knowledge, creativity, and skills of its citizens; and it is up to local governments to make sure that the community can maximally benefit from this vast pool of human resources.

Introduction

Global and urban population growth rates will bring about an important expansion of cities in the coming decades. By the year 2000, the urban population is estimated to reach 2.9 billion people and is expected to grow to 5 billion by 2025; 61 percent of the world population will then live in urban areas. Estimates show that by 2015, the world will contain around 560 cities with more than one million people. Urban growth is also demonstrated by the increase of the number of megacities, which are defined as cities with a population exceeding 8 million. In 1950, New York and London were the world's only megacities. By 1990, there were 21 megacities, 16 of them in developing countries. In 2015, there are expected to be 33 megacities, 27 in the developing world. (Carlson 1996; UNCHS 1996b; WRI 1996). By 1990, the urban population in the Southern Hemisphere was approximately 1.4 billion. This was already more than the total population of East and Western Europe, Russia, North America, and Japan combined (IIED 1995).

However, these figures need to be put in their proper context. As the International Institute for Environment and Development (IIED 1995) indicates, only a small percent of the Southern Hemisphere total population is found in megacities, and most developing countries have no megacities. Urbanization is closely linked to economic growth. In general, the higher the per capita income of a country, the higher the level of urbanization. Low income countries are also among the least urbanized. Most of the urban population in the Southern Hemisphere is found in urban centers with less than one million inhabitants. In other words, as IIED (1995) proposes, we should not merely focus on managing the problems of the giant metropolis, because the common perception of the uncontrolled growth of large numbers of megacities is at odds with reality.

Urban areas throughout the world, whatever their size, are characterized by problems of the sustainability of their development. Generally speaking, cities in wealthy countries have more problems of ecological sustainability due to their high per capita use of natural resources and production of waste, while cities in poorer countries face problems to provide their populations with basic necessities, such as access to adequate shelter, safe and sufficient water, and a healthy environment. The particular problems that emerge vary with the region and developmental level of the area and depend on many factors which interact in complex ways. Atkinson and Allen (no date) indicate that the following major issues can influence the degree and types of urban problems:

➤ Geography and climate—Cities in colder climates require heating in winter which results in serious air pollution during winter months; cities in flat coastal plains are more subject to serious flooding; and desert cities need to pay more attention to securing water supplies

➤ Urban culture—In some areas a rich urban tradition already exists while other regions are urbanizing for the first time in such a way that living habits are ill-adapted to urban requirements

➤ Social and cultural factors—These factors can have an important impact on the way that cities are built and lived in. For example, the use of automobiles as a means of transportation is part of the urban lifestyle for many citizens of cities such as Los Angeles and Bangkok, resulting in a large amount of space taken up by roads and parking spaces, as well as significant amounts of air pollution

➤ Urban economy—Depending on the types of industries present in a city, various types of pollution may occur; the state of the economy may stimulate uncontrolled urban expansion in some areas, but abandonment in others

➤ Size and structure—The size and structure of cities and the speed and orientation of their growth and change has an effect on the capacity to plan and manage them

➤ Carrying capacity—The carrying capacity of the urban subregion indicates the importance of the natural resource base of the region. A region with abundant natural resources will be able to support more people than one with a limited resource base. A city, for example, cannot use more water than what is regionally available

In 1996, urban problems were considered serious enough by the United Nations to trigger the organization of the United Nations Conference on Human Settlements (Habitat II) in Istanbul. It presented the following issues as dominant program principles and strategies for the development of human settlements: democracy, human rights, participation, sustainability, decentralization of government, empowerment of women, and public-private partnership (Carlson 1996). The conference encouraged all governments to take the responsibility to provide cities with the basic infrastructure and services such as: sanitation, health services, education, property rights, credit, affordable child care, clean water and air, a decent environment, and other necessities for survival (Abzug 1996). However, some critics say that Habitat II (UNCHS 1996a) made little progress in suggesting how the concept of sustainable development could be operationally applied in urban areas, even though the term was mentioned repeatedly (Cohen 1996). My commentary will give an indication of the variety of problems faced by urban areas, focusing on the contrasts that are related to the level of economic development. The South African city of Durban is examined in a case study.

Developed vs. Undeveloped Worlds Within One City

South African society is characterized by a pattern of very poor and very rich areas, a result of many years of an apartheid regime. The separated coexistence of residents of mansions and sprawling shantytowns makes these places an ideal "laboratory" for research on sustainable urban development. The city of Durban, located on the east coast of the country, is an example of this kind of world, one which is divided by vast disparities in wealth, resources, culture, and development. In 1996 the Durban Metro Council released its study "Durban's Tomorrow Today. Sustainable Development in the Durban Metropolitan Area" (Hindson et al. 1996). The report is the first phase of the city's Local Agenda 21 process. As apartheid was abolished, Durban was the first South African city to develop a Local Agenda 21 during this time of important social change. To be able to develop more sustainably, Durban's Local Agenda 21 encompasses aspects of sustainable development typical for western cities, as well as aspects that are more common in the sustainable development of cities in developing countries. In this section Durban will be presented as a case study to discuss the difference between the problems and development patterns of affluent communities and poorer areas. The basis for the discussion is the report by Hindson et al. (1996).

Overall, Hindson et al. (1996) come to the conclusion that Durban's affluent communities need to adopt new lifestyles which maintain quality of life while using fewer material resources, while Durban's poorer communities will need to modify their aspirations to achieve an improved quality of life by a step-by-step advancement toward realizable and sustainable goals. The following are some of the most remarkable findings of the study (Hindson et al. 1996).

As a result of planning under the apartheid regime, the Durban Metropolitan Area (DMA) has an urban form that is fragmented, racially divided, and one where the vast majority of the poor are located in the urban periphery and the affluent in the core. Recently a residential de-concentration took place, as people moved out of overcrowded townships onto vacant land, placing additional pressure on a rapidly depleting natural resource base. Although offering opportunities to people otherwise trapped in rural poverty or confined in overcrowded townships, unimproved informal suburban settlements provide unhealthy and unsafe living conditions.

Housing conditions reflect the graded social structure of the DMA and range from extreme affluence in some White suburbs to overcrowded, poorly constructed and serviced estates in historically colored and Indian areas, to townships and shack areas in varying states and levels of utility services.

Affluent housing is associated with extensive land use, exotic gardens, high energy and water consumption, and cars and commuting. Such settlements are unsustainable, but nevertheless provide a model for the rest of society.

The case of Durban clearly shows that both richer and poorer population groups have to deal with sustainability problems. In South Africa an afford-ability-expectations dilemma exists among the poorer population groups; in other words, they strive for the same standard of (unsustainable) living as those in the richer communities. It is clear that sustainability implies simultaneous adjustments in expectations and consumption patterns among both the rich and the poor. Both in housing and water/sanitation matters, the affordability-expectations dilemma becomes obvious. People in Black communities expect housing and levels of service similar to those in formerly White, colored, and Indian areas. However, costs place these beyond the reach of most of the poor without extensive state subsidization. Housing and residential development programs provide the greatest single opportunity to transform the urban form in a more sustainable way and provide major opportunities for community involvement.

In DMA 40 percent of potential consumers have access to seriously inadequate water supplies and 38 percent of households have minimal or no access to sanitation. It is estimated that some 500,000 water connections and 380,000 sanitation connections are needed to eliminate the backlog over the next ten years.

A double standard also exists in the management of waste. Affluent residential communities and industry produce waste on a scale similar to that of major western industrial cities. Waste management in black townships degenerated during the 1980s along with the general breakdown in local authorities and service delivery. In informal areas which have been colonized by the poor without authorization or planning by the government, there has been little or no infrastructure put into place and therefore little or no basic services have been implemented, such as waste removal. The accumulation of solid waste in these areas now poses a serious and growing threat to health and quality of life. It also causes infrastructural damage such as the blockage of stormwater drains and sewers, leading to a flow of effluent on the surface. Waste reduction in the form of minimization and recycling is not widely practiced in the DMA.

In the DMA, due to the residential concentration of the poor in the periphery of the city, this segment of the population must make long journeys to work in the city center. Population growth and urbanization are associated with rapidly increasing traffic flows. Automobile ownership is rapidly increasing, and likely to double over the next 15 years. DMA has an extended public transportation system, but there has been a sharp decline in rail use and increased use of combi taxis rather than buses. The use of public transport is skewed to lower income groups, where ownership and access to private transport is low.

Energy use in South Africa is characterized by very high dependence on coal and coal-fired electricity production, massive investments in producing oil from coal and gas, intensive use of energy in the industrial development sector, and energy scarcity among low income households. Domestic consumption of electricity varies greatly across the socioeconomic spectrum. Wealthier households are heavy consumers, whereas poorer homes use mainly paraffin, candles, and rechargeable batteries. There is also local fuel wood collection. People that live in dwellings without electricity are exposed to health risks associated with the fuels they use. Domestic health risks have the greatest direct impact on women and children, who are confined for the longest periods to indoor domestic environments.

Economically, Durban is characterized by a powerful corporate sector and a relatively undeveloped and under-resourced small and micro business sector. Economic activity is concentrated in core metropolitan areas, where jobs are inaccessible to the majority of the poor. Unemployment is a problem everywhere in the city, but especially in poor settlements in the periphery, followed by townships and informal settlements.

The racial and class structure of the DMA, which manifests itself with extremes of affluence and poverty, also leads to high levels of conflict and violence. The consequences of violence in the periphery are the breakdown of local government, administration and services, and the severe burdening—and in many areas near-collapse—of the educational and health sectors. This violence has resulted in a large scale displacement of people, heightened insecurity, and a more difficult struggle for survival, especially for the most vulnerable members of society: women, children, and the elderly. Moreover, violence leads to the breakdown of family cohesion and parental control, increases the rate of crime and the abuse of women and children, as well as increases psychological problems associated with acute social dislocation and trauma. Violence has shattered value systems that bound families and communities together during the period of repression, and has created a climate of despair.

Informal settlements are the most vulnerable to a range of health problems and the burden of care is disproportionately placed on women. Predictably, health problems among the richer and poorer population groups are distinctly different: health problems related to poverty include diseases such as measles, tuberculosis, diarrhea, acute respiratory illnesses, and nutritional problems, whereas a western, more affluent lifestyle include such diseases as heart disease, cancer, stress, and age-related diseases.

There is also a spatial pattern in education that corresponds to the socio-spatial structure of the DMA. White areas have the best educational facilities, whereas informal settlements in the periphery and rural and semi-rural areas have very poor facilities.

The study conducted by Hindson et al. (1996) also included discussions with focus groups. These groups consisted of citizens from all different communities in the DMA. In the focus group sessions, participants were asked: "What, if anything, would you be willing to sacrifice to ensure that your children and grandchildren inherit a residential area with a resource base capable of meeting their needs?" It is remarkable to see the difference in the answers from the citizens of the richer and poorer areas. Participants from the more affluent areas questioned whether there was any room in society to make sacrifices, taking into account the many demands that society makes on individuals. Participants from the poorer areas were shocked that they were asked to make sacrifices, given the little they had of any resources and the injustices they had experienced in the past from the apartheid regime. In the end, group discussions usually spontaneously led to an agreement that the only sacrifices that individuals could make were time, effort, and care. People can, for instance, get more involved in the community and devote more time and attention to their children's needs and education. Moreover, it was felt that there was room to make for material sacrifices of some kind, such as giving up alcohol, tobacco, and drugs.

Hindson et al. (1996) also sought to probe perceptions on the desired future shape—in ten years time—of the DMA in the different residential communities. There was a notable degree of similarity across the different residential areas (both rich and poor) on what Durban should ideally look like in the future. The following are some of the recurring themes:

- The DMA should be a harmonious and integrated city, with increased residential integration, overcoming past racial antagonisms and building racial harmony
- The DMA should be a safe and secure city, without crime and violence.
- The DMA should be an economically successful city with full employment, with people earning reasonable incomes
- The DMA should be a city with good housing and services for all. People should have access to services of a reasonable standard, such as water, sewerage, electricity, and transportation

❧ The DMA should provide high quality educational and health facilities for the entire population

❧ The DMA should be a clean and green city

❧ The DMA should be a city where people govern themselves

❧ The DMA should be a city of community activity, culture, sport, recreation, and fun

Sustainable Urban Development in Northern and Southern Hemispheres

From the Durban case study, it becomes clear that less developed urban areas have different sustainable development needs than more developed areas. It shows the need for different measures to attain a more sustainable state, depending on the development level of the city in general or the affluence of the various neighborhoods within the city.

The basic needs in poorer areas of cities or cities in less developed countries include:

❧ Affordable and decent housing which helps to ensure well-being, security, skills development, empowerment, health, and urban functional efficiency

❧ An easily accessible supply of clean water

❧ A sanitation system (both for solid waste and waste water) which meets hygienic and environmental standards

❧ Improved energy services, which provide basic access to renewable energy sources. The most promising solutions for the energy problem are solar energy and energy from waste

❧ Education and health services with special attention to the most vulnerable and marginalized groups (i.e., women and children)

❧ More jobs and economic development near poor residential areas, including training the local population to set up small businesses

❧ Promotion of community development, including programs which foster safety and stop violence

The more wealthy areas of cities and urban areas in industrialized regions should first focus on:

❧ Reducing the consumption of energy and natural resources

❧ Reducing the consumption of material goods

❧ Reducing the production of waste

✤ Reducing the need to travel individually by car and encourage the development of pedestrian and bicycle friendly cities

✤ Increasing attention for the local community by reversing the trend of protecting wealthy communities as if they were fortresses where the citizens "hide" from problems outside the walls of their compounds

Since every city is a mix of neighborhood types, sustainability issues in the various areas of the city are linked and need to be tackled in a holistic manner. The following are some examples of measures that are beneficial for both affluent and poorer areas simultaneously:

✤ Both wealthy and poorer areas will benefit from an integration of land-use, transportation, and environmental components in spatial planning.

✤ Open spaces or natural areas should be protected against unrestrained development in all neighborhoods of the city

✤ There is a need to dismantle extreme residential separation in cities. The benefits of a greater social mix in deprived areas include the stimulation of the local economy by the increased purchasing power of the higher-income groups and the presence of role models for youth in such neighborhoods. Much attention must be given to getting rid of unsustainable traditions and transforming institutions and systems, leading to an improved integration of natural, social, economic, and governance components of society. This entails building new relationships between those who were previously in conflict with one other, and installing new value systems and lifestyles that are compatible with long-term, sustainable development and that meet the needs of all those living now and future generations

Local Community Development and Sustainable Development

To become more sustainable, social inequality in urban areas needs to be urgently eradicated. This can be realized more successfully if the local community is involved. Therefore, local community development is an important component of sustainable development. Local community development is a process of organization, facilitation, and action that allows people to create a community in which they want to live through a conscious process of self-determination. Local community development is thus a process in which the ideals of sustainable development can be implemented by both allowing and encouraging people to act as catalysts for sustainable social change at the community level (Maser 1997).

It is possible to imagine an alternative to the current functioning of local government, where the central role of government is to support individual and community initiatives within all urban centers and to ensure that all urban dwellers have access to essential infrastructure and services. It is very important that low-income groups are supported and encouraged to organize, that they receive technical assistance from outside agencies, and that collaboration takes place between community organizations and local governments in the installation of infrastructure and services (Hardoy et al. 1992a and 1992b).

To fight social exclusion and discrimination, and to promote human rights and peace in urban areas, Sachs-Jeantet (1994) proposes that the following measures to be taken:

* Fighting the fragmented city and struggle against social exclusion with policies to alleviate urban poverty, promote social integration, and generate employment. Unemployment is often the trigger of exclusion; however, exclusion is not only economic and social, but also political, cultural, and symbolic. It is important to satisfy the claim for dignity of marginalized populations

* Building partnerships for change between the civil society, the state, and the market in the context of "mixed economies," with special emphasis on participatory management and increased citizen involvement

* Empowering of local communities through enablement strategies for urban self-reliance by making available the resources and techniques that cannot be mobilized locally, in particular, funding community initiatives. Community involvement is indispensable for social cohesion

* Strengthening of local capacity to cope with the rapidly changing environment, and making changes in priorities

* Implementing a holistic and multisectoral approach to urban regeneration, targeted at neighborhoods as the essential building block of urban change

It is important that local community habits are examined for their compatibility with goals of sustainable development. In case that such community traditions are likely to help the sustainable nature of development, their role in society should be taken seriously. An example of such a community tradition is the *Bayanihan* in the Philippines, as described by Cubelo (1998).

Bayanihan is a Filipino word referring to a spirit of communal unity and cooperation. Although *Bayanihan* can manifest itself in many forms, it is most clear in the tradition of neighbors helping a family to move by getting enough volunteers to carry the whole house and moving it to its new location. This relocation is linked to a party, making the move day a truly festive community event.

The *Bayanihan* is an old Filipino neighborhood tradition of cooperation, sharing, and interaction. It is very much evident in the local neighborhoods, mainly in rural areas. Its activity has been a source of vitality and integration among community people.

People in this community help each other build houses, community infrastructure, and initiate local activities and festivities like the *Barrio Fiesta* (neighborhood party), *sayawan* (local dances) or local beauty contests (for fundraising), planting and greening or beautification projects, nutrition programs, and summer sport leagues among youth.

As an example, the *Bayanihan* staging of the annual Barangay Fiesta in honor of the local patron or saint has been a source of local spirituality and bonding, and a renewal of family tradition and cohesion where family members unite and celebrate this particular festivity together. These local festivities revitalize streets and neighborhoods by the constant flux of people and preparatory activities days before the actual celebration—where every household in the community welcomes everyone to their community and shares the bounties offered in each house. This also provides an exuberant nightlife for community people through nighttime presentations (10 days before the date of the fiesta) in their community parks sponsored by each organization and various sectoral groups (e.g. Barangay officials, Purok representatives, women's groups, business groups, and youth organizations). The organization of this yearly event enhances community planning skills, fundraising and management skills, and trains people in effectively handling group interests and dynamics.

In communities where there are not enough resources for public infrastructures, the cooperation and sharing of special talents (carpentry, masonry, artistry) or other abilities has helped to build houses, churches, artistic *Purok* landmarks, and public parks in the neighborhood. The strong community attachment can also result in, for example, an effective 24-hour policing of the neighborhood against burglary, crime, and street violence. Today, in slum and squatter areas, the cohesiveness of the local population as a result of *Bayanihan* has prevented evictions by the government of people from their mostly illegally constructed dwellings.

In effect, the *bayanihan* is an important aspect of a community life that helps to maintain the characteristics of an integrated and exuberant neighborhood. The constant social interaction has forged a strong decision-making body which plays an important role in protecting household and community interests. The *Bayanihan* is a manifestation of inherent Filipino traits of *pakikiisa* (integration), *pagtutulungan* (cooperation), and *pakikisama* (sharing).

Developing Individual Strategies for the Sustainable Development of Urban Areas

The Filipino example of *Bayanihan* only survives today in very few areas of the country and, sadly enough, is almost completely dying out in urban neighborhoods. Emphasizing its importance reminds us of the prime role of communities and neighborhoods in the development process. In retrospect, *Bayanihan* played an important role in pioneering neighborhood political action and community development. *Bayanihan* at the neighborhood level has a great potential of bringing together the metropolis, residence, and workplace. It can only be realized when *Bayanihan* is fully re-introduced in metropolitan planning and is brought back into the heart of community urban life.

Cherishing the *Bayanihan* tradition could be an example of a new and more sustainable way of managing urban problems, since the most important resource for the future city is the knowledge, ingenuity, and organizational capacity of citizens themselves. Hardoy et al. (1992a) indicate that in the near future in most developing countries, economic circumstances are unlikely to change for the better in terms of the provision of basic needs by governments. Therefore, the future cities will, to a large degree, be built and shaped by people with low incomes. What will keep urban areas running is the dynamism shown by the people—in a strong community, people will organize illegal occupation of land, will erect housing there, will organize a defense against attempts of eviction, and will mobilize others to negotiate for secure tenure and for some public provisions of infrastructure and services.

While solving the problems of the poor will mainly be a task in cities of the Southern Hemisphere, it should be clear that every city—including North America and Europe—will have to deal with neighborhoods in crisis, in which problems of dereliction, inadequate housing, long-term unemployment, crime, social exclusion, pollution, and lack of natural green spaces or bodies of water are concentrated. In other words, fighting social exclusion and stimulating local community development is a necessity in industrialized, prosperous countries.

Obviously, no two cities are the same and each one has a unique set of problems. Therefore, it is impossible to prescribe one standard solution for the sustainable development of cities. However, a number of sustainable development scenarios can be a source of inspiration for cities when they develop their own set of answers for a sustainable future. Because of the uniqueness of the sustainability issues in each city, it is essential that each local authority develops its own long-term vision for the future and explores various sustainable development strategies. Strategies for urban areas in Europe and North America could include new ways of urban planning, such as making more compact urban neighborhoods that would consequently be less dependent on the use of large amounts of natural resources, reducing the need to travel by car, and stimulating inhabitants to lead a more sustainable lifestyle. Strategies for cities in developing regions may have to focus more on the sustainable provision of basic human needs, such as decent housing, water and energy supply, sanitation systems, education, and health care services. Both in the Northern and Southern Hemispheres, local community development strategies are needed.

Sustainability assessment can help in the implementation of sustainable development strategies, and can be applied in any city, at any level of development, and within any strategy of sustainable development. Therefore, sustainability assessment should be seen as an instrument that gives an account of the need and degree to which the principles of the local sustainable development strategy or vision should be incorporated in all kinds of local authority decisions.

Working Toward a Global Sustainable Society

Strategies for achieving social equity, social integration, and social stability are essential underpinnings of sustainable development (UNCHS 1996a). Planning for sustainable development therefore needs to take into consideration the underlying economic, social, and political causes of poverty and deprivation.

Poverty, exclusion, and inequality have taken on such important dimensions throughout the world that they appear to be the main reasons for instability today (Bessis 1995). The gap between rich and poor, powerful and powerless remains wider than ever (Lubelski and Carmen 1999). An increasing number of academics and practitioners are convinced that the problems are of such magnitude that they necessitate drastic remedies that offer a new structural and long-term problem-solving dimension to development aid policies. In other words, it is clear that, although local community development can play an important role in sustainable urban development, this will not solve existing global problems. There is an urgent need for a critical evaluation by world leaders of the causes of continuing social problems that face humanity and the necessity of considering major reform in the way that the global society functions today.

Sustainable development is a multidimensional concept which can only be realized through an approach "where the social is in control, the ecological is an accepted constraint, and the economic is reduced to its instrumental role" (Bessis 1995). This means that we need to stress a new value system, in which economic efficiency would be measured according to its capability of satisfying the social needs of all human beings on this planet, not just the wants of a few.

In the end, what is needed to solve current problems of unsustainable development is the gradual move towards a global sustainable society. For people living a western lifestyle today, living in a truly sustainable society is hard to imagine because it requires a radical change of behavior and leaving behind a certain intrinsic lifestyle.

An example of what a sustainable society could look like is the so-called conserver society, developed by University of New South Wales academic Ted Trainer. The essence of Trainer's (1985) conserver society is the small-scale, cooperative, and highly self-sufficient local economy, in which people can live well but modestly on far lower rates of resource consumption than we have now. The conserver society must provide adequate and satisfactory material living standards at the lowest reasonable levels of resource consumption without having any desire to increase levels of consumption over time. This should be accompanied by a constant attempt to improve the quality of our services, technology, educational systems, hospitals, and cultural activities.

The following are some characteristics of a conserver society (Trainer 1995):

* Cutting back on unnecessary consumption, more recycling, designing things to last and to be repaired (instead of replaced)
* Development of as much self-sufficiency as possible, at the national level, at the household level, and especially at the neighborhood or local and regional levels
* Introduction of many small businesses, local family businesses, and cooperatives
* Local production of food with market gardens located throughout suburbs and cities, cutting the cost of food by 70 percent (as a result of reduced transport costs)
* A more communal and cooperative way of life in which people could, for example, volunteer one day a week in community work projects and share materials and tools
* The economy must be a zero-growth economy. Basic economic priorities must be democratically planned according to what is socially desirable, mostly at the local level. Much of the economy could remain as a form of free enterprise carried on by small businesses, as long as their goals were not profit maximization and growth

➤ People would live well without much need for cash incomes, because they would need to buy less. Consequently, people might work only one day a week for money and spend the rest of the week doing a wide variety of interesting and useful activities around the neighborhood, many of which would benefit the community

In the eyes of many people living a western lifestyle, the conserver society might seem unrealizable and even undesirable. To the poor families living in informal settlements in less developed countries it may sound like heaven on Earth. The conserver society is certainly not the only possible way to organize a sustainable society, but seems to meet the requirements of protection and enhancement of the environment, meeting of social needs, and promotion of economic success. Would it function in practice? We do not know, but it is certainly worthwhile to study the idea, examine possible alternatives, incorporate aspects of it in our current societies, or even try it out in reality on a small scale. Again, sustainability assessment can play a catalyzing role in such a process.

The conserver society appears to be a radical change from our current lifestyles and that may be an important contribution to reach a sustainable state on a planet where ten billion human beings are estimated to live by the middle of the next century. In the end curbing population growth remains a top priority. Access to family planning, better education and health care for children, and improving the social status of women are all key factors needed to slow population growth. Educating girls to become strong, independent, and educated women who have access to free birth control anywhere in the world is of utmost importance for the future of humanity.

Conclusion

Differences in the development level of urban areas influence the way in which the concept of "sustainable development" takes shape. Not only is every city unique and characterized by its own problems, but urban environmental, social, and economic problems also differ from one continent and region to another. Moreover, it is important to realize that huge inequalities exist within urban areas.

It is not necessary for city plans to be carbon copies of one another. Cities should instead develop their own holistic views and long-term visions for solving local problems, striving for an equal and satisfying quality of life of the population and simultaneous preservation of the natural ecosystem on which the city survives.

Many problems in urban areas today are a result of social inequality, and we therefore need to examine ways to fight social exclusion and poverty. On the one hand, much can be done through local community development or by empowering the local population. On the other hand, it is absolutely necessary to look for more structural ways of solving poverty, focusing on the sources of unequal development and unbalanced sharing of resources. In the context of this book, we can conclude that sustainability assessment can be applied by any local authority or even by local community groups, independent of their level of development. The result of the assessment itself will, however, vary from place to place and be dependent on the local sustainable development strategy or vision.

References

Abzug, B. S. 1996. The challenge of our times. *The Urban Age* 4, 2. Washington, DC: The World Bank Group.

Atkinson, A. and A. Allen. No date. *The Urban Environment in Development Cooperation. A Background Study.* Brussels: European Commission Directorate General IB/D4, Environment and Tropical Forests Sector.

Bessis, S. 1995. From social exclusion to social cohesion: A policy agenda. *Management of Social Transformations (MOST)—UNESCO Policy Paper 2.* Website hppt://www.unesco.org/most/besseng.htm (21 May 1999).

Carlson, E. 1996. The legacy of Habitat II. *The Urban Age* 4, 2. Washington, DC: The World Bank Group.

Cohen, M. 1996. Habitat II: A critical assessment. *The Urban Age* 4, 2. Washington, DC: The World Bank Group.

Cubelo, E. L. 1998. *The Challenge of Metropolitan Planning in Manila: Integration, Diversity, and Sustainability.* Master thesis. Brussels: Human Ecology Department, Vrije Universiteit Brussel.

Hardoy, J., D. Mitlin and D. Satterthwaite. 1992a. The future city. In J. Holmberg, ed., *Making Development Sustainable.* Washington, DC: Island Press.

Hardoy, J., D. Mitlin and D. Satterthwaite. 1992b. *Environmental Problems in Third World Cities.* London: Earthscan Publications.

Hindson, D., N. King and R. Peart. 1996. Durban's tomorrow today. Sustainable development in the Durban metropolitan area. *The Durban Metro Council.* Website http://www.durban.gov.za/central/urb_devenviro/a21/ srep/cover.htm. (30 Sept. 1998).

IIED, International Institute for Environment and Development. 1995. Urban environment: Out of control? *EC Aid and Sustainable Development Briefing Paper 7.* Website http://www.oneworld.org/euforic/iied/bp7_gb.htm (30 Apr. 1999).

Lubelski, M. and R. Carmen. 1999. The North-South dimension of Local Agenda 21. In S. Buckingham-Hatfield and S. Percy, eds., *Constructing Local Environmental Agendas. People, Places, and Participation.* London: Routledge.

Maser, C. 1997. *Sustainable Community Development. Principles and Concepts.* Delray Beach: St. Lucie Press.

Sachs-Jeantet, C. 1994. Managing of social transformations in cities. A challenge to social sciences. *Management of Social Transformations (MOST) - UNESCO Discussion Paper Series 2.* Website hppt://www.unesco.org/most/sachsen.htm (26 May 1999).

Trainer, T. 1985. *Abandon Affluence.* London: Zed Books.

Trainer, T. 1995. *The Conserver Society. Alternatives for Sustainability.* London: Zed Books.

UNCHS, United Nations Conference on Human Settlements. 1996a. *Report of the United Nations Conference on Human Settlements (Habitat II).* Istanbul: United Nations.

UNCHS, United Nations Centre for Human Settlements (HABITAT). 1996b. *An Urbanizing World. Global Report on Human Settlements 1996.* Oxford: Oxford University Press.

WRI, World Resource Institute. 1996. *World Resources 1996-97. The Urban Environment.* Oxford: Oxford University Press.

Part II

Urban Sustainability Assessment Tools in the Decision-Making Process

Dimitri Devuyst

Sustainability assessment as presented in this book is mainly an instrument of the "impact assessment" type, growing out of the existing systems for environmental impact assessment (EIA). Therefore, Part II of the book first looks more closely into EIA and strategic environmental assessment (SEA) and the role of these instruments in the decision-making process. This leads to the development of methodological approaches to sustainability assessment and their application in practice.

Against the background of major changes in the world and the efforts to address them, one can argue that the traditional environmental impact assessment instrument is ready for change. Impact assessment systems that aim to lead our societies to a more sustainable state can no longer report on the environment in isolation. Rather, environmental issues have to be viewed as parts of larger systems, closely coupled to socio-economic developments and political and institutional structures. The current trend, which is also reflected in the sustainability assessment tools presented in this book, is thus toward integrated analyses, including the evaluation of alternative policy options to provide an improved knowledge base for action, political accountability, and public participation. This part of the book aims to examine EIA and SEA in a sustainable development context and discusses future developments in the impact assessment field, especially in relation to the integration of sustainable development principles in impact assessment. Special attention is given to existing practices and applications of sustainability assessment already in place.

Integrated assessment is a major theme of Chapter 5, in which Dimitri Devuyst begins with a description of the concepts of EIA and SEA. The introduction of sustainable development principles in impact assessment systems is also discussed. Moreover, SEA practice is examined with special attention to the introduction of impact assessment systems at the local level. Research shows that the introduction of impact assessment in local planning, policy-making, and decision-making is not obvious and that local authorities will have to be supported, motivated, and trained for such systems to function. Proposals are made to improve SEA as a sustainability assessment instrument. Mostly, this will require efforts in relation to the integration of different processes and issues. SEA should be integrated into planning and decision-making processes and should be linked to other instruments for environmental management. This leads to a proposal for a broad Sustainability Management System that includes components of impact assessment, auditing, monitoring, life-cycle analysis, and environmental care systems.

In Chapter 6 Lone Koernoev looks more closely into the relationship between impact assessment systems and the decision-making process. Impact assessment has the ambitious goal of improving—and making more rational—the decision-making process. It should, however, be clear that decision-making processes are not always rational and can often be influenced by confusion and complexity. Although we might have a rational procedure for impact assessment, linked to a rational procedure for decision-making, we are not always guaranteed a rational final decision; other factors such as incomplete information and that fact that people do not always behave according to theoretical models. Cognitive limitations, mental capacity of humans, and political limitations also influence the final outcome.

A definition of sustainability assessment is formulated in Chapter 7 by Dimitri Devuyst. Existing sustainability assessment initiatives and their introduction at the local level are examined. Belgian, Dutch, and U.S. examples show how sustainability assessment can function in practice. Moreover, an operational methodological framework for sustainability assessment developed by the author is presented. The methodological framework is called ASSIPAC (Assessing the Sustainability of Societal Initiatives and Proposals for Change) and gives an overview of issues that should be examined in a sustainability assessment report. By following this framework, a useful document for discussing sustainability aspects of local authority initiatives is created. The method also aims to stimulate evaluators to think creatively and come up with innovative measures to steer initiatives in a more sustainable direction. Finally, the chapter presents the following five existing methods that should be further examined for possible integration in sustainability assessment: a) system analysis; b) ecocycles and eco-balancing; c) action planning; d) force field analysis; and e) problem-in-context.

Performing a sustainability assessment following the ASSIPAC method should lead to a document that contributes substantially to a meaningful discussion on the sustainability of a certain initiative. Two examples of this are examined in Chapter 8 by Dimitri Devuyst. The first case is the sustainability assessment of two Belgian funds for reviving Flemish urban neighborhoods and city centers. Advantages and disadvantages of the Social Impulse Fund and the Mercurius Fund are discussed in the framework of sustainable development principles. The second case study looks into the compulsory introduction of company transportation management plans in private companies and what this means for the sustainability of the Belgian society. Moreover, the ASSIPAC methodology used in this sustainability assessment is critically evaluated.

Norway is one of the most advanced countries in experimenting with Municipal Sustainability Assessment. The methodology developed in Norway is called "direction analysis" and is presented in Chapter 9 by Carlo Aall. So far, three Norwegian local and regional authorities have participated in developing and testing a scheme for direction analysis. A key element in the model is a suggested list of general municipal sustainability topics. These should be translated into local or regional sustainability goals by the different authorities and provide a basis for picking out sustainability indicators.

Part II of this book shows that sustainability assessment is currently still in an experimental phase and that important research efforts will be needed to further develop and test sustainability assessment methodologies. A practical "learn from experience" approach is proposed for the development of such methodologies. Local authorities and scientific institutions should form partnerships in trying out and improving sustainability assessment systems. To ensure an efficient sustainability assessment system, the necessary political and economic conditions must be met. In other words, sustainability assessment will only function when local leadership is convinced of its usefulness. Moreover, sustainability assessment cannot survive in isolation, but should be part of a strong policy for sustainable development at the local level. Comparable to the role of EIA in the U.S. NEPA legislation, sustainability assessment can function as an "action-forcing mechanism," making sure that all government initiatives take the existing sustainable development policy seriously.

Chapter 5

Linking Impact Assessment with Sustainable Development and the Introduction of Strategic Environmental Assessment

Dimitri Devuyst

Summary

This chapter is an introduction to the concepts of Environmental Impact Assessment (EIA) and Strategic Environmental Assessment (SEA). It provides an overview of good practice in SEA, discusses SEA application at the local level, and examines the links between SEA and sustainable development. Aspects of integration are also examined in-depth. Procedural and methodological issues of integration are discussed in relation to planning processes, decision-making, sustainable development strategies, and Local Agenda 21 processes. Moreover, the integration of environmental, social, and economic issues in SEA studies and the integration of SEA with other instruments for environmental management is discussed.

Introduction to Environmental Impact Assessment and Strategic Environmental Assessment

Impact assessment can be broadly defined as the prediction or estimation of the consequences of a current or proposed action (Vanclay and Bronstein 1995). Environmental Impact Assessment (EIA) was first developed in the US as part of the National Environmental Policy Act (NEPA), signed into law in 1970. NEPA not only provided the US with an environmental policy, but also introduced a new "action-forcing" device, namely the publication of Environmental Impact Statements (EISs) describing in detail the environmental impacts likely to arise from federal actions (Wood 1995). NEPA (CEQ 1986) states that:

> The primary purpose of an environmental impact statement is to serve as an action-forcing device to ensure that the policies and goals

defined in the Act are infused into the ongoing programs and actions of the Federal Government. It shall provide full and fair discussion of significant environmental impacts and shall inform decision-makers and the public of the reasonable alternatives which would avoid or minimize adverse impacts or enhance the quality of the human environment. Agencies shall focus on significant environmental issues and alternatives and shall reduce paperwork and the accumulation of extraneous background data. Statements shall be concise, clear, and to the point, and shall be supported by evidence that the agency has made the necessary environmental analysis. An environmental impact statement is more than a disclosure document. It shall be used by Federal officials in conjunction with other relevant material to plan actions and make decisions.

The following objectives of EIA of the California Environmental Quality Act (CEQA) are very relevant and can be considered basic principles for EIA processes in general. The CEQA aims (Bass et al. 1996):

✦ To disclose the significant environmental effects of proposed activities to decision-makers and the public

✦ To identify ways to avoid or reduce environmental damage

✦ To prevent environmental damage by requiring implementation of feasible alternatives or mitigation measures

✦ To disclose reasons for agency approval of projects with significant environmental effects to the public

✦ To foster interagency coordination in the review of projects

✦ To enhance public participation in the planning process

Wood (1995) emphasizes that EIA is not a procedure that prevents actions with significant environmental impacts from being implemented. However, it is a procedure that seeks the authorization of actions with full knowledge of any environmental consequences.

For impact assessment experts, it is encouraging to see that there is currently a trend to apply the basic features of EIA (such as those indicated in Table 5-1) to other assessment frameworks than just "the environment" in its narrowest sense.

Since the introduction of EIA in 1970, a discussion has been taking place on what is meant by the term "environment" in EIA. Should EIA also deal with social, economic, health, safety, and sustainability aspects of an activity? These are areas that do not, in the strictest sense of the term, make up

TABLE 5-1 *Some Basic Features of EIA*

> ✦ EIA should encourage decision-makers to take into account significant impacts of proposals during decision-making
>
> ✦ EIA should aim to implement a policy through the application of a legally binding procedure (EIA as an "action forcing" mechanism)
>
> ✦ EIA should lead to the making of informed decisions, based on the most relevant, accurate, and scientifically correct information
>
> ✦ EIA should open up the decision-making process for inspection and public scrutiny

the traditional components (such as water, air, soil, fauna, and flora) of an environmental study. Another question concerns whether one impact study that deals with a wider range of aspects is preferable to individual reports for environmental, social, economic, and other aspects.

In Belgium, for example, action has been taken to introduce new forms of impact assessment, such as child impact assessment and emancipation impact assessment. Child impact assessment is described in a Decree of the Ministry of the Flemish Community of 15 July 1997 (Belgisch Staatsblad 1997) and aims to assess the impacts of government policies on children and determine whether the UN Convention on the Rights of the Child is respected.

Another example is gender impact assessment that is used in The Netherlands (Verloo and Roggeband 1996), Canada, and New Zealand. In Belgium, the Flemish Minister for Equal Opportunities has developed a manual for public officials, encouraging them to assess the impact of their proposed policies and activities on women. This has been called *emancipation impact assessment* (Meier 1997). Further examples are fiscal impact assessment, demographic impact assessment, health impact assessment, and climate impact assessment (Vanclay and Bronstein 1995).

The introduction of these different forms of impact assessment indicates a mainstreaming and acceptance of the fact that it is sensible to examine the impacts of proposals before decision-making, and shows that the basic principles of EIA can be applied to other areas than purely environmental issues. This is easier said than done as shown in evaluations of EIA systems. Devuyst (1994) studied 17 texts reporting on EIA experience and evaluation initiatives. These texts deal with U.S., Canadian, Dutch, Belgian, Italian, French, Japanese, Indian, Malaysian, and Brazilian impact assessment experiences. The following recurring problems can be noted:

> ✦ Not all activities with significant environmental impacts are subject to EIA
>
> ✦ EIA of plans, programs, and policies is not often organized

→ EIA does not always deal with crucial issues during the decision-making process, and important information is often missing in environmental impact statements

→ In certain EIA systems alternatives and mitigation measures do not have to be considered or are not taken seriously

→ Public participation is insufficient

→ The quality control or review process is open to bias or is insufficient

→ The project initiator sometimes has the dual role of the competent authority and the expert drafting the impact statements and is financially dependent on the initiator, which means that the objectivity of the studies and the credibility and neutrality of the process may be questioned

→ The EIA system lacks standardized requirements for monitoring

The following recommendations are repeatedly stated in many EIA evaluation studies (Devuyst 1994):

→ There is a need for improved training of all parties involved in the EIA process

→ EIA should be secured in legislation and structure should be provided in the EIA process

→ The selection of activities subject to EIA should be organized in a systematic and comprehensive way

→ EIA should be organized as soon as possible in the planning phase, before final decisions are made

→ EIA should be expanded to include plans, programs, and policies

→ There is a need for an advanced scoping phase, which would result in the identification of the most important issues to be discussed in the EIA

→ The scope of subjects to be considered in environmental impact statements should be broadened and not limited to only environmental aspects

→ Examination of alternatives and mitigation measures are an important part of the EIA process and should not be neglected

→ EIA should be extended to actions abroad and to trans-border projects

→ Public participation and circulation of information should be improved

→ Review and/or quality control of impact reports should be strengthened

→ Monitoring of EIA should be required

It becomes more and more evident today that if relatively small projects are subject to EIA, other actions such as policies, plans, and programs that have a much wider scope, should also be screened for adverse environmental impacts. Therefore, current research in the field of EIA focuses to a large extent on strategic environmental assessment (SEA). SEA or EIA for policies, plans, and programs is not a new idea. In 1970 the National Environmental Policy Act (NEPA) required that "all agencies of the Federal Government... include in... major Federal actions significantly affecting the quality of the human environment, a detailed statement...on the environmental impact" (Sigal and Webb 1989). "Federal actions" also include policies, plans, and programs. Sigal and Webb (1989) state, however, that the NEPA environmental review has seldom been applied to the development of broad national policy by any U.S. federal agency. Therivel et al. (1992) define SEA as:

> The formalized, systematic and comprehensive process of evaluating the environmental effects of a policy, plan or program and its alternatives, including the preparation of a written report on the findings of that evaluation, and using the findings in publicly accountable decision-making.

The term *strategic environmental assessment* is widely used to refer to the EIA of policies, plans, and programs, although other terms are also used in specific countries or by specific institutions. Table 5-2 gives an overview of the variety of terms that exist within the context of SEA

Sadler and Verheem (1996) stress the important role of SEA in the sustainable development debate. Identifying the effect of policy options on the attainment of sustainable development is considered a key issue for SEA. This means, however, that there is a need for substantial adjustments to contemporary SEA practice. The next section will deal with current SEA practice and procedural and methodological issues will be discussed further in this chapter.

Strategic Environmental Assessment Practice Today

Only a few countries, regions, and international institutions currently have official requirements for SEA. The United States, Canada, New Zealand, Denmark, The Netherlands, Hong Kong, California, Western Australia, and the World Bank are among those which have at least some form of SEA provision. This does not mean, however, that SEA is applied on a regular basis in these places. Often, the provisions are only recommendations or guidelines resulting in occasional or experimental SEA applications.

Introducing SEA tends to be easier at the plans and programs level than at the policy level. The main reasons include: the ambiguity of policies, the

TABLE 5-2 Terms Used in the Framework of SEA

Term	Meaning
areawide EIA	EIA for the development of a region or an urban area (US Department of Housing and Urban Development)
supplemental EIA	Additional study to examine the cumulative impacts for a number of Environmental Impact Statements (EISs) of projects which are located next to each other and in which no mention was made of the other projects in the individual EISs (California)
class EIA	Coordinating EIA for a number of small to medium-sized activities which are much alike and have similar effects (Ontario, Canada)
cumulative EIA	EIA which focuses on the cumulative effects of a certain set of activities
programmatic EIA	EIA for legislative proposals, proposals for rules and agreements, developments and the management of natural resources (US NEPA)
sectoral EA	Sector-wide environmental analysis before investment priorities have been determined by the World Bank
regional EA	A tool to help development planners of the World Bank to design investment strategies, programs and projects that are environmentally sustainable for a region as a whole

SOURCE Janssens et al. (1996)

reluctance of policy-makers to disclose their intentions, and occasionally an unclear definition of what actually should be considered as "policy" (Eggenberger 1998).

The Commission of the European Communities is currently seriously considering the introduction of SEA. A proposal for a directive on the environmental assessment of certain plans and programs was adopted by the Commission in December 1996 (CEC 1997). This proposal aims to strengthen the existing environmental assessment system for projects by extending it to the land-use planning system in areas such as energy, waste, mineral extraction, and transport. At present, it is not clear when the proposal will reach the legally binding stage of a directive and what exactly will be its content. Discussions on SEA have been going on for many years in the EU and much resistance has been put forward by lobbyists. Once the directive has been finalized, it will have an important effect, since all member states of the European Union will be required to introduce SEA into their legislation.

The proposal for a directive has been discussed in the European Parliament in its Committee on the Environment, Public Health, and Consumer Protection (Gahrton 1998). Proposals for more stringent SEA provisions and additions to the proposed directive are being presented in the European Parliament and include the following important aspects (Gahrton 1998):

* The directive must cover policies as well as plans and programs
* The objective of sustainable development should be spelled out more clearly
* It should be made clear that SEA must be carried out at an early stage in the process
* The number of different sectors considered should be increased
* The conditions governing public access and involvement should be set out in greater detail
* The reasons for not carrying out SEA must be published
* SEAs area of analysis should be spelled out more clearly
* It should be stipulated that alternatives must be analyzed including zero options

The Presidency conclusions of the 1998 Cardiff European Council also show support for the future European initiative on SEA at the policy level in the following statement: "The European Council endorses the principle that major policy proposals by the Commission should be accompanied by its appraisal of their environmental impact. It notes the Commission's efforts to integrate environmental concerns in all Community policies and the need to evaluate this in individual decisions, including on Agenda 2000" (Cardiff European Council 1998).

In 1996, Sadler and Verheem (1996) prepared a report which aimed to show the current status of SEA around the world and examine challenges and future developments. Tables 5-3 and 5-4 show a number of major results of Sadler and Verheem's study (1996) in relation to the principles of SEA and a generic framework for good practice of SEA.

An important term within the SEA framework is *tiering*. This term points to the hierarchy of policies, plans, programs, and projects. As policies, plans, programs, and projects are linked and tiered, so will the SEAs related to them. Higher-level SEAs set the context for lower-level SEAs, which in turn set the context for project EIAs (Therivel and Partidario 1996). Tiering is recommended as a logical approach in which to focus and streamline SEA and EIA. Once in place, tiering ensures that the environmental implications, issues, and impacts of development decision-making can be addressed at the appropriate level(s) and with the degree of effort necessary for informed choice. SEA and EIA should be consistent with and reinforce each other, with the former providing a frame of reference for the latter (Sadler and Verheem 1996).

TABLE 5-3 *Principles of SEA*

Initiating agencies are accountable for assessing the environmental effects of new or amended policies, plans and programs.

The assessment process should be applied as early as possible in the proposal design.

The scope of assessment must be commensurate with the proposal's potential impact or consequence for the environment.

Objectives and terms of reference should be clearly defined.

Alternatives to, as well as the environmental effects of, a proposal should be considered.

Other factors, including socio-economic considerations, should be included as necessary and appropriate.

The evaluation of significance and determination of acceptability should be made against a policy framework of environmental objectives and standards.

Provision should be made for public involvement, consistent with the degree of potential concern and controversy of the proposal.

Public reporting of assessment and decisions needs to take place (unless explicit, stated limitations on confidentiality are given).

Independent oversight of process implementation, agency compliance and government-wide performance should be organized.

SEA should result in incorporation of environmental factors in policy-making.

Tiered to other SEAs, project EIAs and/or monitoring for proposals that initiate further actions.

SOURCE Sadler and Verheem (1996)

In order for SEA to work, Partidario (1996) indicates the need for openness and flexibility in decision-making, whereby political will to use SEA is accompanied by an appropriate administrative mechanism to implement it. Partidario also stresses the need to identify a policy framework, that is, a sustainable policy and strategy that will provide a) objectives, principles, and strategies for sustainable development and b) sustainability benchmarks. There is also a need to develop the necessary regulatory framework that will ensure that SEA principles, methods, and procedures are consistently applied in a clear and effective way.

The results of the SEA Workshop of the 18th Annual Meeting of the International Association for Impact Assessment provide an interesting insight into the specific characteristics of SEA and the direction this instrument should take in the future (Eggenberger 1998). There are too many differences in legal status, initiatives studied, and methodologies applied for there to be one "ideal" SEA system. In order to develop harmony in the diversity of SEA, it was deemed important to clearly define what SEA should be doing or to clearly state the objectives of SEA. This implies defining of a clear set of sustainability targets and agreement on a set of non-dogmatic principles regarding the content and procedures of SEA. SEA is part of an ongoing and tiered process which can not be introduced all at once, but requires a gradual introduction, occurring within the context of parallel processes such as change in the mindset of government officials and in communication and information systems. Integration is considered very important in SEA. It involves the explicit introduction of environmental and sustainability considerations in higher decision-making levels, meaning that environmental and sustainability concerns should not only be integrated in specific projects, but also in plans, programs, policies, and other decisions. Integration means not just adding a *pro forma* environmental or sustainability chapter to the decision-making documents, but to seriously take into account environmental and sustainability issues from the start.

TABLE 5-4 *A Generic Framework for Good Practice of SEA*

Apply a simple screening procedure to initiate SEA or exempt proposals from further consideration.

Use scoping to identify important issues, draft terms of reference where necessary for SEA, determine the approach to be followed, and establish other alternatives for consideration. This stage should be "objectives-led", clarifying the relationship between the priorities met by the policy proposal and the likely effects or implications for environmental protection goals, standards and strategies.

Specify, evaluate and compare alternatives, including the no action option. The aim is to clarify the trade-offs at stake, showing what is gained or lost, and point, where possible, to the best practicable environmental option.

Conduct a policy appraisal or impact analysis to the extent necessary to examine environmental issues and cumulative effects, compare the alternatives, and identify any necessary mitigation or offset measures for residual concerns.

Report the finding of the SEA, with supporting advice and recommendations to decision-makers in clear and concise language.

Review the quality of the SEA to ensure the information is sufficient, and relevant to the requirements of decision-making. Depending on the process, this activity can range from a quick check to an independent review.

Establish necessary follow up provisions for monitoring effects, checking that environmental conditionalities are being implemented, and, where necessary, tracking arrangements for project EIAs. For policies, plans and programs that initiate projects, tiering EIA to the SEA can significantly improve process effectiveness and efficiency.

SOURCE Sadler and Verheem (1996)

The following important recommendations were made by the participants of this SEA Workshop (Eggenberger 1998): a) SEA should be regarded as a learning and communicative process and should be developed in such a way that it creates useful information (which also indicates value judgments) for both decision-makers and citizens affected by the policy or plan proposal that is being examined; b) SEA should be based on simple and flexible tools and processes that enable checks and balances; and c) SEA case studies should be developed.

A proposal for the introduction of an SEA system in the Flemish Region of Belgium (see following box) also points towards the need for flexibility, integration, and the gradual introduction of the system (van Wijngaarden et al. 1997; Devuyst et al. 1998b). This study shows how recommendations for an efficient SEA system can be translated into a functional SEA system.

Highlights of the Proposed Flemish SEA System (Belgium)

* Flexibility is a built-in feature of the proposed SEA system. Organizations are given relative freedom, at least in the starting period, for adjusting the speed of introduction, the extent of investigation, and the procedural form. The system consists of: a) modules that can be incorporated into existing procedures or b) self-sufficient procedures. It is suggested that the latter option be implemented in services where procedures for the development and assessment of decisions are not well structured, transparent, or sometimes even existing, or where the SEA system will be imposed by law.

* At least the following modules should be integrated into the existing planning procedure: screening; scoping; drafting of an SEA statement; public participation; publication of the proposal and its adjustments.

* A phased implementation is proposed for those departments which do not have planning procedures yet or experience with impact assessment: five different self-sufficient procedures are proposed and classified as "1-star to 5-star procedures" according to increasing complexity. The 1- to 5-star procedures provide options for the depth of investigation and a time frame for gradual implementation. All administrations could immediately start with the 1-star procedure, i.e. the use of the screening workbook. Gradually they may consider stepping up to a higher star procedure, as they gain experience and motivate their staff. When reaching the 3-star procedure, they will comply with the minimal requirements of the proposal for an EC Directive on SEA for certain plans and programs (CEC 1997). The 4-star procedure corresponds to the proposals for SEA made by the "Best Available Practice" study of Janssens et al. (1996).

* The introduction of an SEA unit is proposed. This unit should support and supervise the introduction of the SEA system and should be involved with quality control during the different steps of the SEA procedure. It should be a forum for the exchange of information and a center for training of all involved in SEA.

The following different steps are introduced in the proposed SEA procedures:

* Screening: the use of a screening workbook is the first step of the star procedures. It should serve as a logbook, accompanying the proposed initiative from the first concept to implementation.

✢ Strategic Environmental Discussion (SED): a SED takes place between screening and scoping, based on the idea of the Environmental Overview as described by Brown (1997). In the care of a coordinator, the SED brings together representatives from all stakeholders and experts for brainstorming sessions.

✢ Scoping: after checking the screening documents or the outcome of the SED on the basis of the advice of the SEA unit, a full scale assessment may be found necessary. Once notified by the initiator of the proposed initiative, the SEA unit will set up a steering commission of independent SEA experts. In consultation with them, the initiator will set the scope of investigations, contents of the report, methodologies to use, ways of programming public participation and proposals for a communication plan.

✢ Full scale SEA reporting: a full scale SEA is advised whenever neither the screening documents nor the SED have provided sufficient information on the acceptability of the environmental impacts of the proposed initiative. Particular attention is required in case of transboundary effects. In contrast to the project EIA, the initiator is allowed to draw up the SEA report and may choose to call on external experts. Experts in the SEA system do not need to be accredited because of the wider scope of SEA and the possibility of performing SEA internally.

✢ Public participation: a concrete plan for information and communication should be drawn up. Opportunities for information and for participation can be proposed for each step of the procedure. The norm of at least one round of public participation for SEA, suggested by Janssens et al. (1996) has been retained.

✢ Quality control and evaluation: evaluation and post-evaluation of the SEA process and reports are important sources of feedback on the efficiency and efficacy of the process. Quality control and evaluation are essential to the continuous improvement of the SEA system, and for informing the wider public on its successes and shortcomings. The two major actors are the SEA unit and, through participation, the public. A "Total Quality Management (TQM)" system would be ideal (Cohen and Brand 1993). The authors stress that TQM demands a new way of acting: staff are no longer simply following orders but participate in the organization of the work. Once in place, the TQM is a continuous self-improving process.

✢ Monitoring: the use of indicators is particularly important for monitoring the impacts of the proposed initiative. These indicators will be selected and measured at an early stage of development of the proposed initiative, according to a monitoring plan. It is impossible to carry out a base-line survey which is of the same level of detail as for project EIA. The relevance of indicators increases therefore with the level of abstraction of proposed initiatives.

SOURCE van Wijngaarden et al. (1997); Devuyst et al. (1998b)

Environmental Impact Assessment and Strategic Environmental Assessment at the Local Level

SEA is hardly ever applied at the local level today. It exists mainly at the planning level in a number of large metropolitan areas. In California, city governments are very active in the preparation of SEAs. Two types of SEA are permitted by the California Environmental Quality Act (CEQA) of 1970: the Program Environmental Impact Report (PEIR) and the Master Environmental Impact Report (MEIR). Plans examined by city public authorities in PEIRs and MEIRs submitted in California include: city urban redevelopment plans, city or county comprehensive plans, regional transportation plans, traffic congestion management plans, community plans, and citywide zoning/density plans (Bass and Herson 1996). SEA has also been applied in several cases in the city of Hong Kong, as shown in Table 5-5. Another example is the Brussels Capital Region of Belgium, which is active in SEA because its EIA system requires the environmental examination of land-use plans at the neighborhood level.

TABLE 5-5 *Examples of SEA in Hong Kong*

Project proposal	Aim of project proposal
1. Port and Airport Development Strategy 1989	To formulate a comprehensive airport and port strategy for Hong Kong for the year 2011.
2. Tseung Kwan O New Town Feasibility Study of Opportunities for Further Development 1989	To accommodate a total population of 400,000 in conjunction with a 70 ha industrial estate in Tseung Kwan O New Town.
3. North Lantau Development Plan 1992	Development of two new towns, with a total population of 250,000, over an area of 750 ha and development of the new Chek Lap Kok airport site of 1,000 ha.
4. Railway Development Study 1993	Railway development for freight and passenger transport in Hong Kong.
5. North West New Territories (Yuen Long District) Development Statement Study 1994	Preparation of statutory land use plans to resolve potentially incompatible land uses among fishponds, the Ramsar site, low density developments, and high rise developments.
6. Freight Transport Study 1994	To determine a strategy for dealing with freight transport.
7. Territorial Development Strategy Review 1996	Development of a strategic land use - transport - environment framework for Hong Kong up to the year 2011.

SOURCE Au and Wong (1997)

A survey among Flemish local authorities revealed that EIA even at the project level is hardly established in local communities. Almost 41 percent of Flemish municipalities were never involved in EIA in any way, and 40 percent were only required to provide information in relation to the preparation of an Environmental Impact Statement (EIS). No more than 11 percent of Flemish communities had ever made an EIS themselves. Only 21 percent indicated that they took into consideration the environmental consequences of policy proposals, while 20 percent took into consideration policy plans. Only 27 percent think EIA will become an important instrument to assess the environmental consequences of projects at the local level in the future; 20 percent think EIA will become important in respect to policy planning, and 22 percent think EIA will become important in respect to policy proposals. The local authorities expect that the introduction of EIA at the local level in Flanders would cause major problems: 60 percent complain about the lack of educated staff members, 74 percent claim they do not have the time or the financial means for doing EIA, and 47 percent fear the process will take too much time. A considerable percentage—21 percent of local authorities—bluntly agree with the statement that "EIA is not interesting at the local level." On the other hand, a limited number of communities are very interested in EIA and SEA. It is no coincidence that these are the same communities which are involved in developing a Local Agenda 21. This survey indicates that the introduction of SEA at the local level will not be easy and will need to be accompanied by financial and technical support and by the motivation and training of the staff members and decision-makers of local authorities (Devuyst et al. 1998a). Since these survey results show local authorities are reluctant to introduce EIA and SEA, one can predict that implementing sustainability assessment systems at the local level will be equally difficult.

Linking Impact Assessment to Sustainable Development

The recent level of attention given to the subject of SEA is the result of a call by academics, EIA experts, and concerned citizens, decision-makers and politicians for widening the range of human activities subject to EIA, especially at higher levels of decision-making. Also, it's increasingly apparent that decision-making processes which foster sustainability is definitely needed.

Today, only a few EIA systems are formally required to look into sustainability aspects. Some EIA legislation, such as the California Environmental Quality Act (CEQA), does not have the term "sustainability" or "sustainable development" in their policy statements or statutes because the term was not vernacular at the time of drafting. As Dennis et al. (1998) show in the case of the Californian situation, this does not mean that sustainability principles are not implicitly present. For example, the CEQA recognizes the need to examine ecological limits, the worldwide and long-term character of environmental problems, responsibility for meeting the needs of future generations as well as our own, and commitment to long-term maintenance and restoration of environmental quality (Dennis et al. 1998).

An example of a regulation that has explicit requirements for the examination of sustainability principles is the one introduced in New South Wales, Australia. The Environmental Planning and Assessment Regulation of 1994, made pursuant to the New South Wales Environmental Planning and Assessment Act of 1979, states that it is necessary to study potential biophysical, economic and social impacts, and adherence to the principles of ecologically sustainable development for a wide range of proposals submitted for approval to the competent authorities. Following the guidelines of the Department of Urban Affairs and Planning (1996), continued reference should be made in the EIS to the question: *"Is this proposal ecologically sustainable?"*

Ecologically Sustainable Development (ESD) is defined as "using, conserving and enhancing the community's resources so that ecological processes, on which life depends, are maintained, and the total quality of life, now and in the future, can be increased" (Resource Strategies 1998). In regard to ESD, any EIS (Environmental Impact Statement) must take into consideration the following:

➤ The precautionary principle—namely, that if there are threats of serious or irreversible environmental damage, a lack of full scientific certainty should not be used as a reason for postponing measures to prevent environmental degradation

➤ Intergenerational equity—namely, that the present generation should ensure that the health, diversity, and productivity of the environment is maintained or enhanced for the benefit of future generations

➤ Conservation of biological diversity and ecological integrity

➤ Improved valuation and pricing of environmental resources (Resource Strategies 1998)

The Australian National Strategy for ESD (NSESD) identifies two key features that distinguish an ecologically sustainable approach to development. These are:

1. The need to consider the wider economic, social, and environmental implications of decisions and actions for Australia, the international community, and the biosphere in an integrated way.

2. The need to adopt a long-term rather than a short-term view when making decisions and taking actions.

By following an ecologically sustainable path of development, it is anticipated that there will be a reduction in the occurrence of serious environmental impacts arising from economic activity and hence fulfillment of the goals established in the NSESD (Resource Strategies 1998).

As Pearce (1993) indicates, we should be critical about the results of SEA, since it is only a tool in the hands of a particular decision-making climate. It can be no better or worse than the political and economic conditions that are prescribed for it. Pearce (1993) also fears that SEA cannot create sustainable development alone unless the right accounting systems are in place.

Procedural and Methodological Issues of Integration

In this section, a number of suggestions will be made for EIA and SEA procedural and methodological adaptations, making them more efficient sustainability assessment tools. However, it should be clear from the start that impact assessment is introduced in many different decision-making processes, for a wide range of initiatives, and is used by a variety of institutions and a multitude of people with a wide range of backgrounds all over the world. This means that EIA and SEA procedures and methodology should be flexible and adaptable to the local situation in which they are applied. Impact assessment systems should gradually develop and take form on the basis of experience. This is possible if the systems are subject to a continuous process of self-improvement.

Proposals for improving SEA as a sustainability assessment instrument often refer to the need for additional integration: a) integration of SEA in planning processes; b) integration of SEA in decision-making; c) integration of SEA into different levels of sustainable development strategy and in Local Agenda 21 processes; d) integration of environmental, social, and economic issues in individual SEA studies; e) integration of SEA and other instruments for environmental management. Each of these trends toward more integration will now be discussed.

Integration of SEA in Planning Processes

For impact assessment to be able to play a more important role in sustainable resource management, Smith (1993) proposes a more integrative approach, in which impact assessment is placed in the broader context of environmental planning. The link between impact assessment and planning has kept impact assessment professionals busy for many years. Already in 1978 Jain and Hutchings (1978) published a book that looks into the methodological consequences of using EIA in planning. Since then more studies have been done, but progress is quite limited. Planning processes in which SEA is integrated are still rather the exception than the rule. Therefore, there is a clear need for learning from experience. We should not stick to theoretical approaches to EIA and SEA, but apply it in practice, do pilot studies, set up impact assessment experiments, and build partnerships for SEA between local authorities, planners, EIA practitioners, and academics.

Local authorities develop many different types of plans which have important consequences for the sustainable development of society. Land-use plans should especially be considered because the way in which we divide the land for different purposes influences to a large degree how inhabitants of the area will have to live. The tight links between the ability of individual people to live a more sustainable life and the way planners design our living environment should not be underestimated. Land-use plans that aim to stimulate more sustainable lifestyles need to consider density issues; transportation and mobility aspects; the design of roads and buildings; the place of ecological processes in the urbanized areas; integration of various compatible functions in one area; and interactions between living, working, and recreation. The elaboration of master plans for the long-term development of cities and municipalities can have a tremendous impact on the sustainability of future actions in the area. Moreover, many other plans—such as mobility plans, energy plans, specific site development plans, and nature conservation plans—should refer to and be closely linked with land-use and master plans.

Breheny and Rookwood (1993) and the UK Town and Country Planning Association examined planning initiatives for a sustainable urban environment and came up with a number of changes that would be needed for future sustainability. The following box gives an overview of those changes and also provides a checklist developed by Breheny and Rookwood (1993) for monitoring progress. Regularly consulting such a checklist during the various planning phases would be a first step toward the integration of SEA and sustainability assessment principles in the planning process.

Breheny and Rookwood (1993) Proposal of Changes Needed for Planning the Sustainable City Region and a Checklist for Monitoring Progress

Proposed Changes for a More Sustainable City

Natural Resources:

❖ Increased biological diversity, including positive measures for encouraging wildlife

❖ Big increase in biomass (trees and other green plants) in both town and country

❖ Replacement instead of depletion of groundwater reserves and good quality topsoil

❖ Much greater use and production of renewable materials in place of scarce finite ones

Land-use and Transport:

✦ Shorter journeys to work and for daily needs

✦ Much higher proportion of trips by public transport

✦ More balanced public transport loadings to minimize fuel consumption

✦ Greater local self-sufficiency in non-specialty foods, goods, and services

✦ More concentrated development served principally by public transport

Energy:

✦ Greatly reduced consumption of fossil fuels

✦ Increased production from renewable sources; e.g. sun, wind, tides, and waves

✦ Reduced wastage by better insulation, more use of Combined Heat and Power systems, local power generation

✦ Form and layout of buildings better designed for energy efficiency

Pollution and Waste:

✦ Reduced emission of pollutants, especially from industry, power stations, and transport

✦ Comprehensive measures to improve the quality of air, water, and soil

✦ Reduction in total volume of waste stream

✦ Greater use of "closed cycle" processes

✦ Much greater recovery of waste materials through recycling

Checklist for monitoring progress

Pollution reduced by:

✦ Establishing the environmental capacity of the region for emission of pollutants

✦ Refusing permission for any development that would result in the total volume of emissions exceeding the regional capacity

✦ Setting up inducements and penalties to cut existing emissions

Natural resources conserved by:

✦ Encouraging rehabilitation rather than redevelopment

✦ Stimulating regional production of renewables to replace finite non-renewables

✦ Adopting conservation measures to save topsoil

Total volume of waste stream reduced by measures such as:

✦ Reducing business rates for firms using "closed cycle" processes

✦ Introducing graduated charges for waste collection

Increased recycling of most waste materials including:

✦ Recovery of scarce inorganic materials for reuse

✦ Composting of organic wastes

Reduced energy consumption and increased percentage from renewables by:

✢ Program for raising energy efficiency of all buildings to at least minimum sustainability standards

✢ Increased use of solar gain

✢ Greater use of combined heat and power systems

✢ Development of wind farms and wave power

Major increases in biomass, both urban and rural, by:

✢ More community forests and other rural tree planting

✢ Protection of existing urban open space and creation of new open space in areas of deficiency

✢ Additional urban tree planting and other green vegetation

✢ Gardens on flat rooftops

✢ More green areas in new development projects

Regional water supplies augmented and consumption reduced by:

✢ Tree planting to maximize rainwater retention in watersheds

✢ Metering consumers with graduated charges favoring low consumption

✢ Applying "closed cycle" methods to water use

✢ Separating "gray" water for filtering and return to groundwater reserves

✢ Reducing urban run-off by use of more permeable paving, providing natural channels and lagoons in place of closed drains

Urban decentralization and dispersal reduced by:

✢ Greening and decongesting inner cities

✢ Making inner city housing more attractive by eliminating excessive densities, designing for "defensible space"

✢ Increasing average densities in city suburbs and small towns

✢ Using more concentrated forms for new development

Commuting distances reduced by:

✢ More local production to meet local needs

✢ Local employment to match local skills

✢ More telecommunication-based home-working, especially in rural areas

✢ More mixed development

✢ More housing in major employment centers

✢ More complementary development in adjoining small towns to reduce reliance on distant large cities

✢ Building balanced new communities

Public transport made more attractive and economic by:

✢ Concentrating more mixed-use development at public transport nodes

✤ Coordinating land-use and public transport to achieve more balanced commuter flows

✤ Creating more dedicated public transport routes

✤ Improving frequency and reliability of services

✤ Raising densities to complement improved public transport

Road traffic reduced by:

✤ Locating new development so as to reduce travel demand

✤ Using opportunities to reshape urban areas to reduce private motorized trips

✤ Refusing permission for new car-based out-of-town retailing and business parks

✤ Pricing road use on congested routes

✤ Reducing car parking provision and increasing charges where public transport available

✤ More pedestrian-priority areas

SOURCE Breheny and Rookwood (1993)

Integration of SEA in Decision-Making

Research has indicated that decision-makers do not and cannot operate under a scientific management rationality, but operate in a very complicated political world (Friesema 1978). Many things take place in the decision-making process, and there can be a high level of confusion and complexity. Sometimes decision-makers simply do not really know what to do--or how to do it (Koernoev 1998). The goal of impact assessment to influence and improve the decision-making process is therefore very ambitious. Koernoev elaborates more on the links between decision-making, planning, and impact assessment later on in this volume.

Clark (1997) proposes the ultimate integration of SEA in decision-making. He states that because SEA should lead to a shorter, simpler, more open process in comparison with EIA, perhaps SEA itself should have no procedural requirements. Although it is clear that SEA should be simple and flexible, one can still defend the need for at least a minimal procedure. It should, for example, be possible for all those interested to see a record of how sustainability aspects were examined during decision-making. Not having such a record would mean not having any proof of SEA activity or indications of results. It would, however, be interesting for integration purposes if the decision-makers themselves or individuals preparing the decision-making documents undertake the SEA and not a separate group of SEA professionals. This will only be possible when these individuals are trained in SEA practice.

Bryan et al. (1998) propose an "Integrated Decision-Making" (IDM) approach to impact assessment. The four key components of IDM include stakeholder/issue identification and involvement, a mediation and negotiation process among stakeholders integrally tied to projection of effects, projection of environmental and social effects of proposals, and action-planning to meet goals and objectives once a preferred alternative has been identified.

Integration of SEA into Different Levels of Sustainable Development Strategy and Local Agenda 21 Processes

International, national, regional, and local strategies for sustainable development should be attuned to each other, becoming more detailed and more operational as they go to the lower levels. At each level SEA can be used to assess whether initiatives taken at that level foster sustainability. In other words, at each level the sustainable development strategy can be used as the assessment framework on which judgment can be based.

At the local level it will be useful to integrate an instrument such as SEA into the Local Agenda 21 process. Both SEA and Local Agenda 21 processes will reinforce each other. For example, on the one hand, the local community vision on sustainable development that should be developed in a Local Agenda 21 can be used in the SEA process as a basis for judgment. SEA can, on the other hand, be utilized in a Local Agenda 21 to stand guard to the sustainable nature of decisions made. Sadler and Verheem (1996) propose that SEA be used to scope toward sustainability; that is, ensuring that policies, plans, and programs are in accordance with national "green plans" or consistent with the commitments made in endorsing Local Agenda 21. At the level of local authorities, SEA could be used—in the spirit of NEPA—as an enforcement device that stands guard to the implementation of the goals of a Local Agenda 21.

The introduction of both SEA and a Local Agenda 21 will require authorities to open up to a process of change. Both processes require a new style of transparent governance and the participation of the public. It is encouraging to see that research in the Flemish Region of Belgium has indicated that those municipalities that are most interested in introducing an SEA system are also those which are most advanced in the development of a Local Agenda 21 (Devuyst et al. 1998a).

Integration of Environmental, Social, and Economic Issues in Individual SEA Studies

The difference between all kinds of "discipline specific" impact assessments (such as EIA, social impact assessment, technology assessment, health impact assessment, economic and fiscal impact assessment, and demographic impact assessment) and sustainability assessment is that in the assessment of sustainability, aggregating environmental, social, and economic impacts into integrated appraisals is part of the overall process. Here we are clearly faced with many methodological problems and much more research will be needed to overcome them. As Clark (1997) states: "EIA practitioners have not yet mastered strategic analysis like carrying capacity and sustainability thresholds." Not only are there problems in relation to the methodology of integrating environmental, social, and economic impacts, there are also new challenges in predicting the magnitude of sustainability impacts.

Bellamy and Dale (1998) propose the use of a systems approach in evaluating the impact of regional planning for sustainable and equitable resource use. General systems theory has the potential to provide impact assessment with powerful conceptual tools for analyzing complex uncertain impact chains, and it will be worthwhile to look into this theory for use in SEA in the future. The four underlying principles that are fundamental elements of a systems-based methodology are (Bellamy and Dale 1998):

1. To provide a framework for the integration of social, economic, environmental, policy, and technological perspectives on evaluation of planning processes and support mechanisms.

2. To recognize the underlying assumptions and hypotheses that underpin the planning approach, as well as any related research activity, process, or product development.

3. To base performance assessment on the establishment of practical, valid, and equitable assessment criteria by which change can be monitored and evaluated.

4. To ground the impact assessment within the environmental, policy/institutional, economic, sociocultural, and scientific/technological contexts of planning processes.

In developing of methodology for predicting sustainability impacts, we should take into account the typical environment in which impact studies are prepared. This includes minimum budgets and limited time to carry out the studies. On the one hand, initiators will not accept methods which would raise the budget too much or excessively delay the decision, and on the other hand the public should expect methods which lead to studies which are scientifically defendable, of high quality, and useful in planning, policy-making, and the decision-making process. Researchers should not only look into the development of quantitative methodology. Due to the current lack of scientific knowledge and the vague nature of many issues of sustainable development, it is also worthwhile to advance qualitative prediction and assessment techniques which result in information which can improve public discussions and decisions on sustainability issues.

Integration of SEA With Other Instruments for Environmental Management

In the framework of the development of a sustainability assessment system, it would be a useful experiment to go beyond the scope of impact assessment and integrate different forms of instruments for environmental management (i.e. impact assessment, environmental auditing, and life cycle assessment) into a broad sustainability management system. Developing further the idea of van Wijngaarden (1998), it would be possible to combine the various members of the impact assessment family and other instruments for environmental management with the philosophy of, for example, an environmental management system such as ISO 14,000. This standard is a "systems" standard, meaning that it does not tell organizations how to run their businesses but rather how to define management processes to be followed to control the impact an organization will have on the environment (von Zharen 1996). The following box gives an overview of the major features of an environmental management system. Next to ISO 14,000, other well-known environmental management systems such as the British Standard BS7,750 or the European Eco-Management and Audit Scheme (EMAS) would be equally useful as a source of inspiration for the development of a sustainability management system. Certainly at the local level EMAS may be useful because its application in local government has already been studied (see HMSO 1993). Integrating instruments for environmental management in an overall environmental management system would create a setting for the prediction, monitoring, control, and management of impacts, embedded in an overall framework of sustainable development policy and planning. Extending the scope of investigation from only environmental aspects to all facets of sustainable development would create a truly sustainability management system. It would take into consideration all important issues, minimize effects, and have a continuously self-improving character.

Characteristics of an Environmental Management System

An environmental management system is defined by the British Standards Institution (BSI 1994) as "the organizational structure, responsibilities, practices, procedures, processes, and resources for determining and implementing environmental policy."

Gilbert (1993) summarizes the basic stages of an organization's environmental management system as follows:

* A policy statement indicating commitment to environmental improvement and conservation and protection of natural resources.

* A set of plans and programs to implement policy within and outside the organizational culture.

❖ The measurement, audit, and review of the environmental management performance of the organization against the policy, plans, and
programs.

❖ The provision of education and training to increase understanding of
environmental issues within the organization.

❖ The publication of information on the environmental performance of
the organization.

The steps in the Eco-Management and Audit Scheme (EMAS):

1. A **Policy** stating the organization's overall environmental aims and
commitment to continuous improvement (beyond compliance with
minimum legislative requirements).

2. A **Review** of the environmental impacts of the activities being considered (and their regulatory and policy context).

3. A **Program** of activities to achieve defined objectives which translate
the policy's aims into specific quantified goals for improvement (in
the light of the Review).

4. A **Management System** which defines responsibilities, procedures,
and tools for implementing the program.

5. Periodic **Audits** to assess whether the Program is being followed and
whether the management system is adequate, and any changes
needed.

6. A **Statement** of environmental performance, which must be published.

7. Impartial, external **verification** of the quality and completeness of
the process, leading to formal validation of the public statement and
the right to use a special "statement" and "graphic" to publicize participation in the scheme.

SOURCE   De Weerdt (1998) and HMSO (1993)

A vital component of a functioning environmental management system is
monitoring and acting upon the results of the monitoring initiatives. In
response to the monitoring, all the elements of the system have to be
changed and updated. Monitoring has also been considered a crucial element in an efficiently functioning impact assessment system. Already in
1985, Greene et al. (1985) pleaded for changing the EIA process from a lin-
ear structure to a cyclic structure in which all participants can learn from
experience. Important in such a cyclic system is that a continuity be provided between the implementation of the initiative and the impact assessment process. It also ensures the feedback of experience from one impact
assessment study to the next. It shows that impact assessment can be linked
to environmental auditing by monitoring and evaluating the results of the
impact assessment.

Part III of this book will show that monitoring is also a significant step in
sustainable development. Monitoring progress made toward a more sustainable society can be done through sustainable development indicators. Indi-

cators may become especially meaningful in a sustainability assessment because of the difficulty in accomplishing a detailed prediction of impacts on the sustainability of society. Based on a regular monitoring of sustainable development progress, problem areas can be identified, inviting sustainability assessors to be extra careful in these fields.

Conclusion

Today, the application of SEA at the local level is limited to a few forward-thinking regions and cities. In all likelihood, introducing SEA in government decision-making will encounter resistance and will only be successful if decision-makers open up to a process of change. First of all, government officials must accept the need to assess the interdependencies between human activities and the environment as early as possible in the decision-making process. Second, government officials should accept the need for informed decision-making and participation. It should be clear that SEA cannot be forced upon people or governments, but will only be effective when its usefulness is clear to all those involved.

Improving SEA as a sustainability assessment instrument will mostly require efforts in relation to the integration of different processes and issues. On the one hand, SEA should be more integrated into planning and decision-making processes. On the other hand, SEA should be linked to other instruments for environmental management and form an integrated management system.

The introduction of SEA at the local level is not obvious. Local authorities need to be supported, motivated, and trained in its application. It is clear that the development of a Local Agenda 21 by a local community can be an important catalyst for local authorities to start doing SEA. Encouraging local communities to develop a Local Agenda 21 is therefore, the most direct way of stimulating the future establishment of SEA and sustainability assessment at the local level.

References

Au, E. W. K. and J. Wong. 1997. *Examples of Strategic Environmental Assessment in Hong Kong*. Hong Kong: Environmental Protection Department, Hong Kong Government.

Bass, R. and A. Herson. 1996. Strategic environmental assessments in the United States: Policy and practice under the National Environmental Policy Act and the California Environmental Policy Act. In *Proceedings of the 16th Annual Meeting of the International Association for Impact Assessment*. 17-23 June 1996. Estoril: International Association for Impact Assessment.

Bass, R. E., A.I. Herson and K.M. Bogdan. 1996. *CEQA Deskbook. A Step-by-Step Guide on How to Comply With the California Environmental Quality Act*. Point Arena, CA: Solano Press Books.

Bellamy, J.A. and A.P. Dale. 1998. Evaluating the impact of regional planning for sustainable and equitable resource use: Establishing effective evaluation criteria. In *Proceedings of the 18th Annual Meeting of the International Association for Impact Assessment.* 19-24 April 1998. Christchurch: International Association for Impact Assessment.

Belgisch Staatsblad. 1997. Decreet houdende instelling van het kindeffectrapport en de toetsing van het regeringsbeleid aan de naleving van de rechten van het kind van 15 juli 1997. *Belgisch Staatsblad* 7 October 1997. Brussels: Bestuur van het Belgisch Staatsblad.

Breheny, M. and R. Rookwood. 1993. Planning the sustainable city region. In A. Blowers, ed., *Planning for a Sustainable Environment. A Report by the Town and Country Planning Association.* London: Earthscan.

Bryan, C. H., W. H. Jones and A. T. Wolf. 1998. "Integrated decision-making approach to impact assessment." In *Proceedings of the 18th Annual Meeting of the International Association for Impact Assessment.* 22-24 April 1998. Christchurch: International Association for Impact Assessment.

Brown, A. L. 1997. The environmental overview in development project formulation. *Impact Assessment* 15, 1: 73-88. East Lansing: International Association for Impact Assessment.

BSI, British Standards Institution. 1994. *British Standard for Environmental Management Systems: BS7750BSI.* London: British Standards Institution.

Cardiff European Council. 1998. *Presidency Conclusions.* Document SN 150/98, Cardiff, 15-16 June 1998.

CEC; Commission of the European Communities. 1997. Proposal for a council directive on the assessment of the effects of certain plans and programs on the environment. *Official Journal of the European Communities No C 129/14,* 25 April 1997. Brussels.

CEQ, Council on Environmental Quality. 1986. *Regulations for Implementing the Procedural Provisions of the National Environmental Policy Act.* Reprint 40 CFR Parts 1500-1508. Washington DC: Council on Environmental Quality.

Clark, R. 1997. Making EIA count in decision-making. In *Proceedings of the 17th Annual Meeting of the International Association for Impact Assessment.* 28-31 May 1997. New Orleans: International Association for Impact Assessment.

Cohen, S. and R. Brand. 1993. *Total Quality Management in Government. A Practical Guide for the Real World.* San Francisco: Jossey-Bass Publishers.

Dennis, N. B., R. A. Grassetti, and R. Odland. 1998. An analytical framework for addressing sustainable development in environmental impact assessment in California. In *Proceedings of the 18th Annual Meeting of the International Association for Impact Assessment.* 22-24 April 1998. Christchurch: International Association for Impact Assessment.

Department of Urban Affairs and Planning. 1996. *EIS Guidelines.* Sydney: Natural Resources and Environmental Policy Branch, New South Wales, Australia.

Devuyst, D. 1994. *Instruments for the Evaluation of Environmental Impact Assessment.* Ph.D. thesis. Brussels: Human Ecology Department, Vrije Universiteit Brussel.

Devuyst, D., T. van Wijngaarden and L. Hens. 1998a. The introduction of strategic environmental assessment at provincial and municipal level in the Flemish region (Belgium): Examining attitudes of persons involved, pProposals for an SEA system and future challenges. In *Proceedings of the 18th Annual Meeting of the International Association for Impact Assessment.* 19-24 April 1998. Christchurch: International Association for Impact Assessment.

Devuyst, D., T. van Wijngaarden and L. Hens. 1998b. The introduction of strategic environmental assessment in Flanders (Belgium): Examining attitudes of persons involved, proposals for an SEA system, and future challenges. In *Proceedings of the European Symposium on Environmental Assessment of Plans and Programs.* 10-11 September 1998. Angers: Ecole Supérieure d'Agriculture d'Angers.

De Weerdt, H. 1998. Environmental auditing and environmental management systems. In B. Nath, L. Hens, P. Compton and D. Devuyst, eds., *Environmental Management in Practice, Volume 1. Instruments for Environmental Management.* London and New York: Routledge.

Eggenberger, M. 1998. Report of the workshop on SEA, planning and decision-making—problems, challenges and recommendations. In *Proceedings of the 18th Annual Meeting of the International Association for Impact Assessment.* 22-24 April 1998. Christchurch: International Association for Impact Assessment.

Friesema, H. P. 1978. Environmental impact statements and long-range environmental management. In R. K. Jain and B. L. Hutchings, eds., *Environmental Impact Analysis. Emerging Issues in Planning.* Urbana: University of Illinois Press.

Gahrton, P. 1998. *Draft Report on the Proposal for a Council Directive on the Assessment of the Effects of Certain Plans and Programs on the Environment.* Brussels: Committee on the Environment, Public Health and Consumer Protection, European Parliament.

Gilbert, M. 1993. *Achieving Environmental Management Standards. A Step-by-Step Guide to Meeting BS7750.* London: Pitman Publishing.

Greene, G. , J. W. MacLaren and B. Sadler. 1985. Workshop summary. In B. Sadler, ed., *Audit and Evaluation in Environmental Assessment and Management: Canadian and International Experience:* 301-316. Banff: Environment Canada and The Banff Center School of Management.

HMSO. 1993. *A Guide to the Eco-Management and Audit Scheme for UK Local Government. A Manual for Environmental Management in Local Government.* London: HMSO.

Jain, R. K. and B. L. Hutchings. 1978. *Environmental Impact Analysis. Emerging Issues in Planning.* Urbana: University of Illinois Press.

Janssens, P., D. Devuyst and L. Hens. 1996. *Studie betreffende de inventarisatie en kritische beoordeling van de bestaande methodologie die op wereldschaal gehanteerd wordt in de m.e.r. betreffende beleidsvoornemens, plannen en programma's (BPP's), inbegrepen de ontwikkeling van een organisatieschema van het wetenschappelijk onderzoek ter ondersteuning van de m.e.r. betreffende BPP's.* Brussels: Human Ecology Department, Vrije Universiteit Brussel and Ministry of the Flemish Community.

Koernoev, L. 1998. Strategic environmental assessment and the limits to rationality in decision-making processes. In *Proceedings of the 18th Annual Meeting of the International Association for Impact Assessment.* 19-24 April 1998. Christchurch: International Association for Impact Assessment.

Meier, P. 1997. *Het Emancipatie-Effecten Rapport (EER). Een instrument voor een gelijke-kansenbeleid.* Brussels: Vrije Universiteit Brussel and Ministry of the Flemish Community.

Partidario, M. R. 1996. Strategic environmental assessment: Key issues emerging from recent practice. *Environmental Impact Assessment Review* 16, 1: 31-55. New York: Elsevier Science.

Pearce, D. 1993. *Blueprint 3. Measuring Sustainable Development.* Centre for Social and Economic Research on the Global Environment. London: Earthscan.

Resource Strategies. 1998. *Environmental Impact Statement of the Cowal Gold Project.* New South Wales.

Sadler, B. and R. Verheem. 1996. *Strategic Environmental Assessment. Status, Challenges and Future Directions.* Den Haag: Ministry of Housing, Spatial Planning and the Environment of the Netherlands; International Study of the Effectiveness of Environmental Assessment and The EIA-Commission of the Netherlands.

Sigal, L. L. and W. Webb. 1989. The programmatic environmental impact statement: Its purpose and use. *The Environmental Professional* 11: 14-24. National Association of Environmental Professionals. Malden: Blackwell Science.

Smith, L.G. 1993. *Impact Assessment and Sustainable Resource Management.* Essex: Longman Scientific & Technical.

Therivel, R., E. Wilson, S. Thompson, D. Heaney and D. Pritchard. 1992. *Strategic Environmental Assessment.* London: Earthscan.

Therivel, R. and M. R. Partidario. 1996. *The Practice of Strategic Environmental Assessment.* London: Earthscan.

Vanclay, F. and D. A. Bronstein. 1995. *Environmental and Social Impact Assessment.* Chichester: John Wiley & Sons.

van Wijngaarden, T., D. Devuyst, R. Van Asbroeck and L. Hens. 1997. *Volume 1. Onderzoeksresultaten. Strategische milieueffectrapportage in Vlaanderen. Ontwikkeling van een gebruiksvriendelijke methodologie ten behoeve van de milieueffectrapportage betreffende beleidsvoornemens, plannen en programma's rekening houdend met de juridische en wetenschappelijke randvoorwaarden.* Brussels: EIA Centre, Human Ecology Dept., Vrije Universiteit Brussel and Ministry of the Flemish Community.

van Wijngaarden, T. 1998. Integrating Impact Assessment and Indicators for Sustainable Development: Towards an Integrated Management System. Unpublished draft text. Brussels: Human Ecology Department, Vrije Universiteit Brussel.

Verloo, M. and C. Roggeband. 1996. Gender impact assessment: The development of a new instrument in the Netherlands. *Impact Assessment* 14, 1: 3-20. East Lansing: International Association for Impact Assessment.

von Zharen, W. M. 1996. *ISO 14000. Understanding the Environmental Standards.* Rockville, MD: Government Institutes Inc.

Wood, C. 1995. *Environmental Impact Assessment. A Comparative Review.* Essex: Addison Wesley Longman Limited.

Chapter 6

Strategic Environmental Assessment and the Decision-Making Process

Lone Koernoev

> If the earthy world of decision-making occasionally leads us to the
> divine world of Ibsen, Kierkegaard, and Senancour, let us be grateful.
> But let us not stay in that world too long. It is hard enough to make
> sense of the simple things without discovering they are really not as
> simple as they look.
>
> (James G. March 1994: 272)

Summary

Strategic Environmental Assessment (SEA) is a work tool used for the consideration of ecological dimensions of policy. Integration of SEA into the planning and policy-making processes is widely recognized as a fundamental component to this instrument (Therivel and Partidario 1996; Partidario 1996; Sadler 1994). To integrate SEA it is necessary to have knowledge on processes taking place within an organization and knowledge on how decisions are made.

This chapter focuses on the subject of rationality in decision-making processes and the implications for the integration of SEA. The main questions in this chapter are: How does decision-making take place? Is the intended rationality the dominant way of reaching a decision? What are the consequences for the development of SEA and the tools to undertake environmental assessment? Is the consequence that we reject the rational approach totally? Or is the solution complementary approaches like learning and communication? The aim is to give a modest contribution to the understanding of decision-making. By the use of organizational theory, the chapter will describe and explain some aspects of decision-making and seeks, in particular, to raise an awareness and discussion of the cognitive and political constraints in regard to rationality. On the basis of a discussion of the relations between SEA, the policy process, and the cognitive bias in relation to decision-making, the paper concludes by demonstrating the need for the development of SEA by incorporating other approaches, such as the learning and communicative processes, in addition to the rational approach.

Strategic Environmental Assessment

Sustainable development is a concept that has created much dispute and discussion. Today, it is accepted that the environmental effects from development have to be considered in the early decision-making phases in order to prevent or reduce environmental damage. Environmental Impact Assessment (EIA) was introduced in 1969 and is now implemented in most industrialized nations. With EIA it is possible to predict and assess the consequences a particular project would have on the environment, such as the environmental impact of building a highway or a bridge. Today efforts are directed toward developing and implementing principles for Strategic Environmental Assessment (SEA). The basic principle for SEA is the integration of environmental considerations, alongside economic and social considerations, into formulation and evaluation of new policies, plans, and programs. Several relationships exist between other policy tools and SEA. Policy tools like technology assessment, land-use planning, Green economics, international environmental law and policy (such as the Biodiversity and Climate Change Convention) correspond to and reinforce SEA (Sadler and Verheem 1996).

Why is it then that SEA is receiving more and more focus? The scope of decisions that can be taken at the strategic level is wide, whereas the potential to prevent an activity's effect on the environment at the project level (EIA) can be very limited. At the project level the environmental assessment is focused primarily on *how* development can take place with the fewest environmental impacts: "At this stage, the prior questions of *whether, where* and *what* type of development should take place are either decided or largely pre-empted by earlier policy-making processes" (Sadler and Verheem 1996: 31). In other words, the formulation of alternatives can be very limited. Furthermore, EIA does not address cumulative and large-scale effects. Several development projects, each with acceptable environmental impacts, can together have unacceptable impacts. SEA is widely considered a better way of preventing damage to the environment. Here decision-makers have to consider broad political alternatives and mitigation at an early stage, when a policy, plan, or program is not yet finalized. Besides the potential environmental benefits from SEA, there are also democratic and participatory issues to consider, which have been argued by various authors (Therivel and Partidario 1996; Wood 1995; CEC 1996; Sadler and Verheem 1996). SEA is used in political systems that are traditionally based upon closed and limited participatory traditions. In reality, policy-making takes place predominantly behind closed doors, and involves a relatively small number of people who set the agenda (Sadler and Verheem 1996; Therivel et al. 1992).

The quantity of research on SEA and SEA literature is increasing considerably internationally. Not surprisingly, the first works covered the assessment of SEA and the principles of SEA (see, for example: Therivel et al. 1992; Therivel 1993; Wood and Djeddour 1992; UNECE 1992; World Bank 1991). These are now followed by collections of examples and experiences (see, for example: Elling 1997; Therivel 1997; Therivel and Partidario 1996; Sadler and Verheem 1996; Partidario 1996). In principle, SEA has gained acceptance among professionals, without any agreement in relation to the scope and the principles for SEA, but with an acceptance that integration of environmental assessment at the strategic level depends on contextual relations.

Strategic Environmental Assessment, Rationality, and Decision-Making

Imperfect and irrational decision-making certainly influences SEA and how successfully SEA is implemented in the decision-making process. However, it can be very difficult to examine whether SEA has had an effect on the decision-making or not. The knowledge of how SEA is linked in reality to the policy and planning process (PPP) is limited, but it is recognized as very important: "The links between SEA and the PPP-making process are difficult to clearly identify and explain, but are crucial to the effectiveness of SEA." (Therivel and Partidario 1996: 183). One should be aware that even if SEA does not result in any specific or definable results, the process can still increase the general awareness of the relationship between the environment and actions.

But what, then, is meant by "*rationality*"? Through examination of literature about decision-making and planning—where the term "rationality" occurs—it becomes obvious that "rationality" has many interpretations. And talking about rationality in decision-making processes needs to be distinguished between the rationality of the *outcome* of the process (substantive rationality) as opposed to the rationality of the *process* itself (procedural rationality). In other words, a rational procedure does not automatically lead to a rational choice because decision-making often involves incomplete or inaccurate information, and people do not always behave as assumed in the models. In this chapter, we will deal with rationality in its narrower definition: procedural rationality. Because the process by which a decision is reached is the focus of SEA, I will use March's definition of a rational decision: "...a decision is described as 'rational' if it is made by a process that follows standard procedures for choosing among alternatives in terms of expectations about future consequences" (March 1994: 224). Few would argue that the aim of SEA is to make sure that the best decision—environmentally-speaking—is taken. But a more precise aim is that environmental assessment is to be used to produce knowledge about environmental effects of different alternatives. This knowledge can then enter into a larger basis for decision-making, in which economic, social, and political relations also play a role. In a rational approach to SEA, decision-making follows a series of analytic steps (see Figure 6-1).

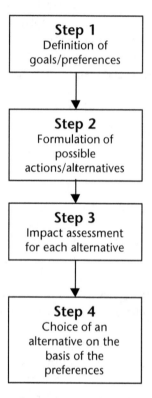

FIGURE 6-1 *A rational approach as to how SEA and decision-making take place. This procedure for choice implies that a systematic analysis of the consequences of alternatives are carried out and that an alternative is chosen on the basis of defined preferences.*

The basis for the discussions in this chapter is: *Is procedural rationality the best way to integrate environmental considerations in decision-making and thereby reach a sensible decision?* It is beyond the focus of this chapter to assess whether the substantive rationality of the result of a decision-making process is rational or not. But let's talk about what is wrong with the rational approach (that is, procedural rationality). I will point out two difficulties:

1. The approach is normative. The procedures are based upon an attitude toward *how* decision-makers should reach a decision instead of *what* they actually do in the decision-making process.

2. The representation of the decision-making process is based upon the idea that preferences (and therefore alternatives) do not change through the process and also that (everyone's?) preferences are looked at simultaneously.

Decision-making processes do not always follow a rational procedure. The study of rationality in decision-making is not new, and the picture of imperfect and irrational decision-making in a complex world is drawn by many and is supported by many studies (March 1965 1987 1988 1994; Katz and Kahn 1966; Simon and March 1958; Lindblom 1959; Scott 1987; Breheny and Hooper 1985). Many things take place in the decision-making process, and there can be a high level of confusion and complexity. Views and preferences change according to the alliances between participants in the decision-making process. In some cases, participants might not know what to do—or how to do it. Knowledge about consequences might not exist. This characterization of complex decision-making can characterize decisions at the strategic level, where the decision-maker has to consider uncertain outcomes, act in a political system, face many new situations and problems, and therefore search for new ways to find solutions.

It is very important to pay attention to the complexity of the decision-making process. So when we develop methods and procedures for SEA, we require a balance between our goal to get organizations to take more sustainable decisions in a more rational way as well as an acceptance and understanding of the complexities of human behavior.

Another important issue is the variation of policy-making and planning processes in organizations that depend on external factors such as the political system, culture, and the like. This means that it can be extremely difficult to search for general models and a clear pattern that can characterize the processes. This was also one of the conclusions in a survey of agency decision-making in all Commonwealth and state spheres of Australian government done by Bailey and Renton (1997: 322): "The most important and informative result obtained from the survey was that, ironically, no trends or pattern with respect to agency policy-making, or consideration of environmental issues during policy-making, were identified."

Future case studies will add to our knowledge and give insight as to how we can link the policy-making and planning processes with the considerations of environmental issues. As a supplement to important practical experience, we can learn from the experience gained within organizational theory and policy science, which both have a long theoretical and empirical history. This chapter draws on one corner of organizational theory—decision-making theory—which is concerned with understanding, explaining, and predicting behavior in organizations.

Understanding Decision-Making and the Limits to Rationality

To understand organizational behavior and hence, decision-making processes, three levels of analysis must be considered: The processes taking place within *individuals* (e.g., attitude and attributes), *groups* (e.g., communication) and *organization* (e.g., the structure) (Scott 1987; Greenberg and Baron 1993).

Although there is a common focus of research in relation to how organizations work, there are major differences in the questions investigators are interested in. The focus of this chapter is on individual processes in order to explain individual behavior within an organization and how this influences the decision-making process and consequently, the integration and use of SEA. To make the link between the policy-making, planning, and SEA, we must start with an understanding of the individual and take account of the context in which the decision-maker and organization functions. In spite of their complexity, organizational decisions of policy-making character are still made by individuals (Katz and Kahn 1966).

Decision-making is one of the most ordinary processes. We make decisions that are both simple (such as deciding which paper the environmental statement is to be printed on) and complex, with far-reaching consequences (such as changing a national strategic decision toward the recycling of paper, glass, metal, garbage, and so forth, to decrease landfill).

The process of problem solution is grouped by Simon (1976) into three overall stages: (1) finding occasions calling for a decision, (2) inventing, developing, and analyzing possible alternative courses of action and (3) selecting a particular alternative from those available. These stages in the process of reaching a decision are affected by several limitations to rationality, which can be divided into two categories: (1) cognitive limitations and (2) political limitations; see Figure 6-2 (March and Olsen 1976; March 1988; Cyert and March 1963).

It is also important to recognize that not all policy decisions involve all three stages. Decisions under immediate pressure may result in an immediate solution with little analysis of the problems, no search for alternatives, and low effort to weigh the consequences (Katz and Kahn 1966). An example of a poor assessment of a problem and its consequences, made under strong pressure, can be found in the actions of the Ministry of the Environment in Denmark some years ago. Decision-makers were suddenly faced

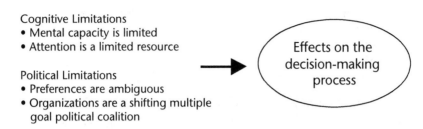

Cognitive Limitations
• Mental capacity is limited
• Attention is a limited resource

Political Limitations
• Preferences are ambiguous
• Organizations are a shifting multiple
  goal political coalition

Effects on the decision-making process

FIGURE 6-2 *Limitations to rationality in decision-making.*

with strong oxygen depletion in Danish waters, which received intensive media focus. In the space of only four months, the greatest ecological project in Danish history was set up without making an impact analysis to illustrate the effects of the "Action Plan for the Aquatic Environment" on the environment, nor in the country's economy. The plan was not able to reduce the nitrogen emission by 50 percent, which was the goal. An estimation is that the plan has reduced the emission by about 15 percent (Schroeder 1990). It will cost much more than the 12 billion Danish kronen to halve the environmental impact of nitrogen. The plan had almost no impact on the content of oxygen in the open sea, which was the initial focus (Schroeder 1990). But how is it possible to be so far from the mark, and why did no one investigate the effect of the plan before implementation? The "Action Plan for the Aquatic Environment" was not a plan: it was a political decision. It is an example of how interested organizations and the media can become an effective unit and how politicians react to such situations. The immediate pressure resulted in an immediate solution that failed to both analyze the nature of the problem or solve it.

From the Rational to the Administrative Model

On what basis, then, do we make decisions? The rational model has its roots in traditional economic theory, which assumes complete rationality. According to these models, "the economic decision-maker" maximizes profit by a systematic search for the best solution to a problem. The basis is complete and perfect information for the decision-maker. What makes the rational economic method special is that it requires that decision-makers consider all alternatives and consequences to make the optimal decision. It is obvious that the theory of the rational-economic model is hardly applicable to reality. The need for an administrative theory has its roots in the knowledge about the practical limits to human rationality.

In an attempt to describe more realistic decision-making, Simon proposes the administrative model, where the principle of *bounded rationality* is expressed (Simon 1957: 198):

> The capacity of the human mind for formulating and solving complex problems is very small compared with the size of the problems whose solution is required for objectively rational behavior in the real world—or even for a reasonable approximation to such objective rationality.

According to the rational model, the decision-maker seeks to maximize and make the optimal choice, while decision-making according to the administrative model seeks to make *satisfying choices*. Simon expresses it as such: "the key to the simplification of the choice process...is the replacement of the goal of maximizing with the goal of satisfying, or finding a course of action that is 'good enough'." (Simon 1957: 204) March provides a description of limited rationality in the process of reaching a decision (March 1994: 9):

> Studies of decision-making in the real world suggest that not all alternatives are known, that not all consequences are considered, and that not all preferences are evoked at the same time. Instead of considering all alternatives, decision-makers typically appear to consider only a few and to look at them sequentially rather than simultaneously. Decision-makers do not consider all consequences of their alternatives. They focus on some and ignore others. Relevant information about consequences is not sought, and available information is often not used. Instead of having a complete, consistent set of preferences, decision-makers seem to have incomplete and inconsistent goals, not all of which are considered at the same time...Instead of calculating the 'best possible' action, they search for an action which is "good enough."

A summary of the differences between the rational-economic model and the administrative model is shown in Table 6-1.

TABLE 6-1 *Characterization of the Rational-Economic Model and the Administrative Model*

	The Rational-Economic Model	The Administrative Model
Rationality of decision-maker	Perfect rationality	Bounded rationality
Available information	Complete access	Limited access
Selection of alternatives	Optimal choice	Satisfying choice
Type of model	Normative	Descriptive

It is very important to be aware that mental capacity and attention span are limited. First of all, the decision-maker does not know all alternatives, consequences, and preferences. Alternatives have to be searched, consequences have to be examined, and preferences have to be found and developed. In theories of limited rationality, attention is seen as a scarce resource. The search for information will always have costs, and here the aspect of the search for *satisfaction* becomes relevant.

The allocation of attention cannot be explained simply by looking at the structural constraints in the organization (who is allowed or required to attend to which decision processes and when), but there are also behavioral variations (March and Olsen 1976). The first alternative considered tends to be close to the immediate situation; marginal changes are examined and the number of alternatives being considered are restricted (March 1994; Lindblom 1959). At the same time decision-makers: "…see what they expect to see and overlook unexpected things" and "…the world is interpreted and understood today in the way it was interpreted and understood yesterday." (March 1994: 11). These issues have to do with the fact that decision-makers are involved in making many decisions and face many outside pressures and have external time constraints, they are capable of dealing with only a limited volume of information, or as expressed by Simon (1976: 79):

> It is impossible for the behavior of a single, isolated individual to reach any high degree of rationality. The number of alternatives he must explore is too great, the information he would need to evaluate them so vast that even an approximation to objective rationality is hard to conceive.

There is a link between (a) attention and (b) search and development for alternatives, consequences, and preferences. The available *attention to search and develop is limited,* which therefore affects the search for information and consequently, the decision itself (March 1994). When we are operating in a very unstable situation, the effort to reach a complete rational decision is unlikely. The number of alternatives are limited both by the capacity of the decision-maker, but also by the available resources. Another aspect is that decision-makers do not have complete information about resulting consequences upon which to determine the best alternative. Secondly, there is the aspect of timing. It is not realistic to assume that everyone can attend to everything all of the time. There is individual variation in the amount of involvement that, among other things, is a result of other time-consuming issues that requires attention for participants in the decision-making process. The distribution of attention can also be a consequence of focus in society at a specific moment: "This fact is reflected by the dynamism of the political agenda for change. Acid rain was focused upon 15 years ago, today it is the problems of CO_2, and tomorrow it may be something that is not yet perceived as an issue requiring attention." (Koernoev 1997: 178)

Politics and Decision-Making

The concept of "bounded rationality" deals with the cognitive aspects of decision-making, while political behavior is omitted. Many have pointed to the implications of power games and political behavior in decision-making processes, where individuals or groups with different preferences seek to influence the decision. Decision-making will be influenced by different preferences and individual interests, which can contradict and most likely change over time. Another limitation to rationality is therefore the *ambiguity of preferences*. (March 1981, 1994; March and Olsen 1989):

> ✦ Preferences are *not consistent*. Both individuals and organizations can have inconsistent preferences and their wishes can contradict other objectives

> ✦ Preferences are *not stable*. Preferences often change over time

> ✦ Preferences are *not exogenous*. The experience with certain issues and their consequences can cause preferences to evolve. This means that the preferences can develop during the decision-making process and be a result of the process itself

Recognizing the ambiguity of preferences, it is not possible to follow a rational procedure in the sense that: (1) you define your goals/preferences, (2) you define alternatives, (3) you assess the consequences of each alternative, and (4) you choose an alternative on the basis of the preferences. This raises the question: Is procedural rationality the correct way to reach a sensible decision? Or do we need to be more open to a process that supports discussion and the development of preferences instead of more or less assuming what the preferences are?

Cyert and March (1963) recognized that a firm, and organizations in general, can be seen as a multiple goal coalition, where the composition of the firm is not given, but instead negotiated; and where the goals of the firm are bargained instead of given. At the same time, each participant joins the process with its own preferences. Conflicts of interests have the basic premise that an organization is a coalition of individuals and groups, which all follow different goals. This will, therefore, lead to conflicts between, for example, the factions who promote environmentally friendly energy production by windmills versus those who seek the preservation of the visual or aesthetics aspects of the landscape. To complicate matters, environmental considerations are in competition with economical considerations. Aside from the fact that individuals and different groups have different views and preferences in the decision-making process, it can also be difficult for dialogue between the different players. Therefore, even the "exchange" of views and preferences becomes challenging. This was experienced in Sweden in the case of integrating SEA with municipal comprehensive land-use planning (Asplund and Hilding-Rydevik 1996: 140):

> The barriers to integration lie mostly at the level of officials, and are to a certain degree built into the organization. Politicians generally have an unclear picture of how the planning process is carried out. Officials are unaware of the connection between the planning process and how decisions are carried out. The difficulty of having a dialogue between professionals from different disciplines, with differing views of the relation between people and nature, is well known, but is not accepted and therefore not dealt with in the planning process.

Having "the coalition" in mind, the search and the use of information through SEA will not always be innocent. Cyert and March (1963: 75) support this attitude:

> Individuals will treat estimates, information, and communication generally as active parts of their environment. They will tend to consider the decision for which the information is relevant, the probable outcomes of various possible biases in information, and the payoff to them of various possible decision results. They adjust the information they transmit in accordance with their perceptions of the decision situation.

Information is crucial for decisions. Information from SEA is made into an instrument for the strategic actors. Information is controlled and communication is shaped with a consciousness of how it might influence the decision, preferences, and interests (March 1995). The fact that SEA is implicated in politics also means that SEA can contribute to the political debate: "...when you look at SEA as a political-democratic process and not just a technical-scientific one, then it leads to the opposite of what is feared: SEA contributes to a return to politics." (Elling 1997: 171)

The Specific Decision Situation and Bounded Rationality

Given the cognitive and political characteristics, decision-making processes and SEA are bound to be affected. We are in a situation where SEA is related to the policy process and under the influence of the cognitive bias. In addition, SEA also has to interact with the existing planning processes that are in place.

In order to add to our understanding of decision-making, we also have to explore the problem of rationality dependent on the decision situation. To see the limitations dependent on cognitive conditions is too simple. When Simon wrote about bounded rationality, it was also a condition he was aware of: "Individual choice takes place in an environment of 'givens'—premises that are accepted by the subject as bases for his choice; and behavior is adaptive only within the limits set by these 'givens'."(Simon 1976: 79) This is shown as the decision situation in Figure 6-3.

The Decision Situation

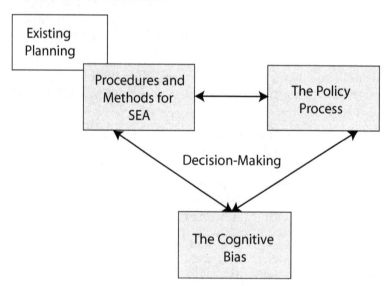

FIGURE 6-3 *The relations between SEA (and the existing planning), the policy process, and the cognitive bias, which take place in the decision.*

Forester criticizes the concept of "bounded rationality" because it only builds on cognitive bias. By reformulating Simon's "bounded rationality" to "the bounding of rationality," Forester (1993: 74) asks; "What binds rational actions? How does such binding take place? Can some of the connection of action be prevented, whereas some cannot? How might social structure influence the connection of action?" Forester speaks in terms of situated rational actions. By doing this he opens up a broader discussion of rationality in decision-making because he adds the focus on the specific and unique decision situation the decision-maker is faced with.

Looking at the cognitive limitations, we get an awareness of the limitations of the human mind, which *seem to be:* 1) inevitable and 2) not socially systematic, but instead nonstructural (Forester 1993). But as Forester (1993: 74) points out:

> First, might not some practical bounds, instead of being inevitable, be artificial and unnecessary, mere artifacts of personality or power relations or custom—any of which might be changed, thus rebounding the action situation at hand? Second, if bounds can be socially constructed by virtue of organizational design or political structure, might not some bounds be socially systematic, derived from existing patterns of social structure rather than from wholly random or ad hoc sources?

By also looking at the decision situation, we find conditions that are not inevitable and structural. In other words, they can be changeable conditions. Forester's concern with the relativity of rationality can also be found in Faludi (1996). Faludi's opinion is to focus attention on *how* decision situations are defined. March is also working with the specific decision situation and looks at behavior in an organization not only as a result of consequence-driven choice, but also a consequence of standardized working procedures, professional standards, cultural norms, and institutional structures (March 1995). So instead of assessing alternatives in relation to the value of their consequences, individuals (and organizations) make decisions and follow rules that will "be appropriate" in the specific situation. This is what March calls *rule-based acting*. The decision situation therefore influences both *how* the decision-maker reaches a decision and *which* decision is taken. The historical constitution, the institutional condition, and the pressure placed upon actors influencing their actions in particular political ways limits rational actions and needs to be considered in defining the decision situation.

A case of how existing institutional structures affects decision-making and environmental assessment was observed during a study of groundwater protection in regional planning in a Danish County in the period of 1993-1997 (Koernoev 1999). The organizational setting is a structure with a number of departments. Each department is working with and preparing a well-defined part of the overall regional plan. The structure proved to have important consequences for decision-making. The departments were each working on the basis of their own "rationalities," and the input from each department was consequently based on their understanding of what could be a problem and what is not a problem and what are possible solutions. In addition, the rationality was defined by the working norms and procedures in each department. The results were: a) the definition of problems were narrow and b) the search for solutions and consequences were kept within each department's own experience and knowledge. The rationalities in the departments were not necessarily consistent, but because of the organizational structure, it did not become very visible (e.g., conflicts between intensive agriculture and the protection of extraction areas). The structure did not support the confrontation between rationalities, in the sense that department division blurred the conflicts. The most important conclusion to be drawn is how *local "rationalities" can limit the learning process*. The problem definition, the search for alternative solutions, the environmental assessment, and finally, the choice of solutions (formulation of policies for groundwater protection, selection of areas for future water extraction, etc.) took place within a limited time frame. This example points to the need to consider how institutional structures can either support or discourage the learning process.

But how is this view to be integrated with SEA and to what extent should we take the cognitive and political constraint of decision-making into account? A large portion of the SEA research up until now has been about how decisions ought to be. We aim for procedures that will make the decision-maker's actions more rational (in the sense of following procedure) and choosing the alternatives that are best when compared with sustainability. This is, for example, reflected in published articles dealing with environmental assessment during the last few years. From January 1994 until January 1998, 63 articles dealing with EA (EIA and SEA) were published in the scientific journal *Environmental Impact Assessment Review*. Out of these, only six articles were focused mainly on the link between environmental assessment and the policy process.

Conclusion: Toward Appropriate Methodology

In conclusion, we cannot reduce rationality just to a technical question: SEA takes place in a political and social world, and we have to be aware of the human aspects and limitations. If we accept this, then at first sight procedural rationality and environmental assessment are essentially incompatible.

In the discussion as to how SEA can be integrated within the existing social and political processes and in the search for complementary approaches to rational decision-making, several issues are to be kept in mind. From this chapter, four theses can be drawn and should be considered while developing and integrating SEA methodology into the existing planning and decision-making processes.

Thesis 1: There is a need to see SEA as a learning and communication process.

Methodology that supports communication in an organization and the learning process:

+ Allows for the development of preferences during the process
+ Appreciates a political discussion
+ Deals with the fact that dialogue is important but difficult
+ Accepts that knowledge is developed during the whole decision process

We know, that: a) preferences develop during the decision-making process and will often be incomplete and inconsistent, b) not all alternatives are being considered and the decision-maker looks at them sequentially rather than simultaneously and c) it can be difficult to have dialogue between professionals from different disciplines and between politicians and officials.

Thesis 2: There is a need for (sustainability) targets.

The decision-maker chooses an alternative that is good enough (instead of maximizing) by seeing whether the alternative exceeds some criterion or targets. Targets simplify a complex world because the decision-maker has to choose an alternative that is "just good enough."

Thesis 3: SEA has to be developed in a way that it produces recognizable and valuable information for the decision-maker.

Decision-makers can only cope with limited information. Interpretation and understanding happen with roots in yesterday's interpretation and understanding. The chosen solutions are therefore not far away from earlier experience.

Thesis 4: Use empirical study of planning and policy analysis to produce insightful knowledge and accounts of the context in which SEA has to be integrated.

SEA and decision-making take place in "a decision situation" defined by the historical and structural context, which influence how a decision is reached and which decision is taken.

There are different ways to handle the human aspects and limitations based upon the way decision-makers make decisions and how information is understood. However, there is an inconsistency as to how individuals use information and how our theories say information is used. To adapt SEA to the observed characteristics of human nature and policy-making processes means that we need more knowledge about how people are using information and how they make decisions. Lessons can be learned from organizational theory and policy science combined with empirical studies of the use of SEA.

Considering the general limitations discussed in this chapter, it would be very useful for the integration of environmental considerations at the strategic level to recognize and appreciate that not all decisions are being carried out in the same way. It is important in relation to SEA to appreciate that rational procedures have their strengths, but if we put almost all the weight on the rational approach, we more or less reject other procedures for choice; like, for example, the use of intuition or the use of tradition and belief when decisions are made.

References

Asplund, E. and T. Hilding-Rydevik. 1996. SEA: Integration with municipal comprehensive land-use planning in Sweden. In R. Therivel and M. R. Paridario, eds., *The Practice of Strategic Environmental Assessment*. London: Earthscan Publications Ltd.

Bailey, J. and S. Renton. 1997. Redesigning EIA to fit the future: SEA and the policy process. *Impact Assessment* 15, 4: 319-334. Fargo: International Association for Impact Assessment.

Breheny, M. and A. Hooper. 1985. *Rationality In Planning. Critical Essays on the Role of Rationality in Urban & Regional Planning*. Great Britain: Page Bros.

Commission of the European Communities. 1996. *Proposal for a Council Directive on the Assessment of the Effects of Certain Plans and Programmes on the Environment*. COM(96) 511 final.

Cyert, R. M and J. G. March. 1963. *A Behavioural Theory of the Firm*. Englewood Cliffs, N.J.: Prentice Hall.

Elling, B. 1997. Strategic environmental assessment of national policies: The Danish experience of a full concept assessment. *Project Appraisal* 12, 3: 161-172. Surrey: Beech Tree Publishing.

Forester, J. 1993. *Critical Theory, Public Policy, and Planning Practice. Towards a Critical Pragmatism*. New York: State University of New York Press.

Faludi, A. 1996. Rationality, critical rationalism, and planning doctrine. In S. J. Mandelbaum, L. Mazza and R.W. Burchell, eds., *Exploration in Planning Theory*. Rutgers: The State University of New Jersey New Brunswick.

Greenberg, J. and R. A. Baron. 1993. *Behaviour in Organizations*, Fourth Edition. New York: Simon and Schuster.

Katz, D. and R. L. Kahn. 1966. *The Social Psychology of Organizations*. Chichester: John Wiley and Sons.

Koernoev, L. 1997. Strategic environmental assessment: Sustainability and democratization. *European Environment* 7, 6: 175-180. Shipley: ERP Environment and Chichester: John Wiley and Sons.

Koernoev, L. 1999. *Strategic Environmental Assessment - a Study of Rationality in Planning and Decision-making*, Draft. Denmark: Aalborg University.

Leavitt, H. J. 1965. Applied organizational change in industry: Structural, technological and humanistic approaches. In J. March, ed., *Handbook of Organizations*. Chicago: Rand McNally.

Lindblom, C. 1959. The science of muddling through. *Public Administration Review* 19: 79-88.

March, J. G. 1965. *Handbook of Organizations*. Chicago: Rand McNally.

March, J. G. 1981. Decisions in organizations and theories of choice. In A.H. de Ven and W.F. Joyce, eds., *Perspectives on Organization Design and Behaviour*. New York: John Wiley.

March, J. G. 1987. Ambiguity and accounting: The elusive link between information and decision-making. *Accounting Organization and Society* 12, 2: 153-168.

March, J. G. 1988. *Decisions and Organizations.* Oxford: Blackwell.

March, J. G. 1994 . *A Primer of Decision-Making.* New York: The Free Press.

March, J. G. 1995. *Fornuft og Forandring. Ledelse i en verden beriget med uklarhed.* Denmark: Samfundslitteratur.

March, J. G. and J. P. Olsen. 1976. *Ambiguity and Choice in Organizations.* Oslo: Scandinavian University Press.

March, J. G. and J. P. Olsen. 1989. *Rediscovering Institutions—The Organizational Basis of Politics.* New York: The Free Press.

Partidario, M. R. 1996. Strategic environmental assessment: Key issues emerging from recent practice. *Environmental Impact Assessment Review* 16: 31-55.

Sadler, B. 1994. Environmental assessment and development policy-making. In R. Goodland and V. Edmundson, eds., *Environmental Assessment and Development.* An International Association for Impact Assessment - World Bank Symposium. Washington: The World Bank.

Sadler, B. 1996. *International Study of the Effectiveness of Environmental Assessment.* Final Report. CIDA and International Association for Impact Assessment.

Sadler, B. and R. Verheem. 1996. *Strategic Environmental Assessment. Status, Challenges and Future Directions.* Den Haag: Ministry of Housing, Spatial Planning and the Environment.

Schroeder, H. 1990. *Et økologisk råd. Om vandmiljøplanen, ilten i havet og økologiske helhedsbetragtninger.* Denmark: Rhodos.

Scott, R. W. 1987. *Organizations. Rational, Natural, and Open Systems.* Englewood Cliffs, N.J.: Prentice Hall.

Simon, H. A. 1957. *Administrative Behaviour* (1st ed., 1945). New York: Macmillan.

Simon, H. A. 1976. *Administrative Behaviour* (1st ed., 1945). New York: Macmillan.

Simon, H. A. and J. G. March. 1958. *Organizations.* Chichester: John Wiley and Sons.

Therivel, R, E. Wilson, S. Thompson, D. Heaney and D. Pritchard. 1992. *Strategic Environmental Assessment.* London: Earthscan Publications.

Therivel, R. 1993. Systems of strategic environmental assessment." *Environmental Impact Assessment Review* 13: 145-168.

Therivel, R. and Partidario, M. R. 1996. *The Practice of Strategic Environmental Assessment.* London: Earthscan Publications.

Therivel, R. 1997. Strategic environmental assessment in central Europe. *Project Appraisal* 12, 3: 151-160.

UNECE, United Nations Economic Commission for Europe. 1992. *Application of Environmental Impact Assessment Principles to Policies, Plans and Programmes.* Environmental Series Number 5. Geneva: UNECE.

Wood, C. and M. Djeddour. 1992. Strategic environmental assessment: EA of policies, plans and programmes. *Impact Assessment Bulletin* 10, 1: 3-22.

Wood, C. 1995. *Environmental impact Assessment: A Comparative Review.* Harlow: Longman.

World Bank. 1991. *Environmental Assessment Sourcebook,* 3 Vols. Technical Paper 139. Washington: World Bank.

Chapter 7

Sustainability Assessment at the Local Level
Dimitri Devuyst

Summary

A definition of sustainability assessment as well as a general methodological frame-work for sustainability assessment, the ASSIPAC method, are provided in this chapter. A number of specific instruments for sustainability assessment for use at the local level are presented, and examples are given of Belgian, Dutch, and U.S. sustainability assessment systems. Methods to be used in a sustainability assessment, such as systems analysis, ecocycles, and eco-balancing, action planning, force field analysis, and problem-in-context are discussed.

Defining Sustainability Assessment

As indicated in Chapter 5, the need for EIA at higher levels of planning and policy-making has given rise to questions concerning an appropriate assessment or reference framework. This discussion is linked to the question of how sustainable development principles can be introduced into the decision-making process.

Dalal-Clayton (1992) states that EIA has suffered a bad image among many of those involved in promoting sustainable development because the tool has been used inappropriately or on limited components of the sustainable development system. Those who have employed EIA have rarely achieved the successful integration of environmental, social, and economic issues.

That EIA is an important instrument in attaining sustainable development is, however, confirmed by the results of the United Nations Conference on Environment and Development (UNCED). The Rio Declaration on Environment and Development states in its Principle 17 that "Environmental impact assessment, as a national instrument, shall be undertaken for proposed activities that are likely to have a significant adverse impact on the environment and are subject to a decision of a competent national authority (UNCED 1992)."

Agenda 21 also calls for the strengthening of EIA systems, the introduction of appropriate EIA procedures, the implementation of EIA before relevant decisions, the further development and promotion of the widest possible use of EIA, and participation and training of people in relation to EIA (UNCED 1992).

As a framework, sustainability analysis should be applicable across the spectrum from the project to the policy level (Dalal-Clayton 1992). Rees (1988) suggests four initial steps in re-creating EIA to serve a viable role in sustainable development:

✦ Extend the scope of activities subject to EIA to cover the full range of ecologically and socially relevant public and private sector proposals and actions

✦ Create a variety of institutional frameworks for environmental assessment adapted to the increased diversity of initiatives and activities to be assessed

✦ Develop methods for environmental assessment that reflect the discontinuous temporal and spatial dynamics, and the resilience properties of ecosystems

✦ Implement the preceding as part of a broader planning and decision-making framework that effectively recognizes ecological functions as limiting factors

In fact, this means that we have to develop a sustainability assessment system that is flexible enough to take into account many different kinds of activities and that works within the limits of the carrying capacity of the environment in which an activity will take place.

It becomes clear that there are two possible ways to introduce sustainability principles in impact assessment: a) we can introduce sustainability principles in the existing EIA and SEA legislation and guidelines or b) we can develop a separate system for sustainability assessment. Either option might be the most favorable solution, depending on the specific situation of the country or region where the chosen option will be introduced.

Sustainability assessment can be defined as a formal process of identifying, predicting, and evaluating the potential impacts of a wide range of relevant initiatives (such as legislation, regulations, policies, plans, programs, and specific projects) and their alternatives on the sustainable development of society. The process includes a written report on the findings of the sustainability assessment in such a way that it improves the publicly accountable decision-making process.

In practice, sustainability assessment as an instrument that stands on its own is not widely used today. Where it exists, it is used on an experimental basis or as part of a temporary project or research initiative. All of the following examples have an illustrative and experimental nature, and are not yet applied on a daily basis at the local level.

In 1997, Partidario and Moura introduced an experimental concept, called Strategic Sustainability Assessment (SSA). The purpose of SSA is to ensure that issues of sustainability have a specific role to play in the planning and decision-making process. A major problem in reaching this goal is that at the present time, it is very difficult to establish what sustainable development is and how sustainability can be measured. SSA can be understood as an integrative approach that aims to translate sustainability priorities and criteria into measurable indicators. SSA aims to help define objectives, targets for environmental indicators and adequate combinations of such indicators (traditional or sustainability indicators) in an aggregated way. Sustainability targets, achievable thresholds, and critical levels are considered in SSA (Partidario and Moura 1997).

Another example of sustainability assessment was developed in 1993 when the World Conservation Union (IUCN) and the International Development Research Center (IDRC) came together with the common interest of assessing sustainability in developing countries and developed an approach called the "Systemic User-driven Sustainability Assessment" (SUSA). SUSA tries to assess whether the well-being of people and the ecosystem is improved and maintained. To assess progress toward this goal, an assessment framework is introduced consisting of the following five questions (IUCN International Assessment Team 1995):

1. What is the condition of the ecosystem, how is it changing and why?

2. What is the condition of people, how is it changing and why?

3. What are the main interactions between people and the ecosystem?

4. What conclusions can be drawn about progress toward the goal?

5. What needs to be done to make progress toward the goal?

More relevant to the aims of this book is a project for municipal policy analysis to assess sustainable development in three cities and two counties in Norway (Aall and Høyer 1997). This project is described in more detail in Chapter 9.

Another interesting initiative is SPARTACUS, a two-year research project within the Environment and Climate research program of the European Union, carried out between 1996 and 1998. SPARTACUS is an acronym that stands for "System for Planning and Research in Towns and Cities for Urban

Sustainability." It is financed by national authorities and the directorate-general for Science, Research, and Development of the European Commission. Research is carried out by consultants and scientists in Finland, the UK, Italy, Spain, and Germany.

The objectives of SPARTACUS are:

✦ To design a system that can help in analyzing the interactions between land-use, transport, economy, the environment, and social factors and forecasting these for the future

✦ To build strategies for urban sustainability using combinations of land-use, transport, and environmental policy instruments

✦ To simulate and assess the long-term effects of introducing these policy actions in each of the selected pilot cities and to compare and explain any differences in results (Lautso 1998)

The SPARTACUS system follows a simulation approach in which policies are tested and sustainable urban policy alternatives defined. The policy testing process follows the following steps:

1. Sustainability policies to be used in the test are selected.

2. On the basis of models, researchers predict which impacts the policies would have if they were to be introduced in the Finnish Helsinki Metropolitan Area, the Italian city of Naples, and the Spanish city of Bilbao. In practice, SPARTACUS simulates what effect the policies would have on specific environmental, social, economic, land-use, and transport indicators. Models for predicting impacts include MEPLAN, SUSTI, MEPLUS GIS, and USE-IT.

3. The effects of the policies on selected indicators are reviewed and evaluated. The economic efficiency of the policies is also analyzed using a decision support tool.

4. If there are harmful side effects, the policies are adjusted and tested again.

5. If there are positive results in some cities but negative results in others, the reasons are analyzed.

6. Finally, general policy recommendations are formulated and urban sustainability strategies for European cities are presented (Lautso 1998).

ASSIPAC, a Framework for Sustainability Assessment

Lawrence (1997) states that the broad principles of sustainability, such as those given in Table 7-1, can be applied to the assessment of whether, and to what extent, various actions might advance the cause of sustainability. The need to refine the objectives of sustainable development will certainly involve the establishment and use of sustainability indicators, applied within sustainability reporting frameworks that clearly illustrate progress toward or away from sustainability.

TABLE 7-1 *Examples of Sustainability Principles to Be Used in a Sustainability Assessment*

* Approach problems from a sustainability systems perspective; define mutually supportive sustainability goals and objectives

* Take a long-term perspective of human activities and environmental conditions; strive to live off the interest and do not discount the future

* Strive to span jurisdictional, disciplinary, professional, and stakeholder boundaries

* Ensure that values and value differences, including the inherent value of the natural environment, are made explicit

* Keep options open to the extent possible

* Be sensitive to the ecological and health-risk consequences of being wrong; this means erring on the side of caution, even when there is a lack of full scientific certainty, when there are threats of serious or irreversible environmental damage

* Ensure that the means to achieve sustainability ends are themselves sustainable

* Design approaches to suit the context, including placing proposals within the context of community needs and aspirations

* Ensure a full accounting of social and environmental costs

* Those responsible for adverse environmental effects are responsible for necessary remedial actions and for paying the costs of action (polluter-pays principle)

* View global environmental management as the shared responsibility of all

SOURCE Lawrence (1997)

Depending on the subject of the assessment, different and more specific assessment frameworks should be applied. In the following section, assessment systems for policy proposals in an urban setting are presented. Similar sustainability assessment instruments can be developed for other types of initiatives. For example, a project proposal to be introduced in a developing country would need a different review framework and a different focus on sustainability principles than a similar proposal to be introduced in a North American or European city.

A sustainability assessment methodological framework for use in a European, North American, or Australian urban context is currently being developed by the author of this chapter (see also Devuyst 1998) at the EIA Centre, Vrije Universiteit Brussel. It is called the ASSIPAC-method. ASSIPAC is an acronym that stands for "Assessing the Sustainability of Societal Initiatives and Proposing Agendas for Change." Two types of sustainability assessment are considered: the Sustainability Assessment Check and the Sustainability Assessment Study. The Sustainability Assessment Check is a short study, screening the initiative for possible conflict with policies or visions for sustainable development. The Sustainability Assessment Study is a more in-depth examination of sustainability consequences of the initiative. This two-tiered system allows for flexibility. Depending on the nature and development stage of the initiative, the person or team performing the assessment and their experience with sustainability assessment, the time and financial means available, either a check or a dtudy can be done. Both types of sustainability assessment can be linked: a decision can be made on the need for a Sustainability Assessment Study on the basis of the results of the Sustainability Assessment Check. The following steps are used in the ASSIPAC method:

Sustainability Assessment Check

The ASSIPAC Sustainability Assessment Check makes use of a checklist approach. The checklist is presented in Appendix 7-1. The method consists of the following 11 steps:

1. Identification of the person or team carrying out the Sustainability Assessment Check: the assessors are the individuals carrying out the Sustainability Assessment Check. They can be part of the team of the initiator (carrying out a self-assessment) or they can be independent of the initiator (external assessment); they can be individuals working for the government, a scientific institution, an NGO, or a private organization. It is important to disclose the identity of the assessors, making clear their link to the proposed initiative. Assessors should be familiar with the concept of sustainable development and with the basics of scientific research. All actions described in the following steps are carried out by the assessors.

2. Gather information on the initiative studied and on alternatives to the initiative that might have been developed. This is done by completing

sections A and B of the ASSIPAC Sustainability Assessment Checklist. Assessors should familiarize themselves with the initiative and get to know all aspects of it in detail.

3. Identify sources of information and data that will be needed to complete all requirements of the ASSIPAC Sustainability Assessment Checklist.

4. Obtain information on sustainable development policies, visions, or strategies that are in place in the area. Existing sustainability targets or standards should be identified. (See Section C in checklist)

5. If available, the assessors should describe the "best available sustainable development practice" in an international context for the initiative studied, as required in Section D.

6. Information should be collected on the reactions of the public in general and specific stakeholders in relation to the initiative, in fulfillment of Section E of the checklist.

7. Following section F, assessors should examine forces existing within society that hinder the attainment of a more advanced sustainable nature of the initiative, through a force field analysis.

8. Every item of the checklist in Sections G, H, I, J, and K should be systematically considered: for each topic the assessors should first decide on its applicability to the initiative studied. If applicable, the assessors should gather information on the way the topic is dealt with in the initiative. At this stage assessors can add topics to the checklist that they consider important in the discussion on sustainable development.

9. Following Section L of the checklist, assessors should decide if the topics in Sections G, H, I, J, and K have been sufficiently considered in the development of the initiative. For each item in the checklist an assessment symbol can be awarded (see list following #9) The assessment symbols are an aid for the assessors and their use is not compulsory. It is more important that the assessors give a written evaluation for each topic in Sections G, H, I, J, and K. During the assessment the assessors should refer to the existing policies or visions for sustainable development and sustainability targets and standards that are in force in the area. If no such policies, visions, targets, or standards exist, reference should be made to general principles for sustainable development or sustainable development policies, visions, targets, or standards of other jurisdictions. Also, in case a "best available sustainable development practice" exists, it should be used as a point of reference. Deviations

from this "best practice" should be discussed. If alternatives to the initiative have been developed, their sustainable character should be discussed in a comparative way.

A	Topic generally well integrated in the initiative, no important aspects neglected.
B	Topic generally well integrated in the initiative, only minor aspects neglected.
C	Topic integrated in the initiative in a satisfactory way despite omissions and/or inadequacies.
D	Parts are well integrated in the initiative, but must be considered unsatisfactory because of omissions and/or inadequacies.
E	Topic integrated in the initiative in an unsatisfactory way, significant omissions or inadequacies.
F	Topic not at all integrated in the initiative, topic dealt with in a very unsatisfactory way.
NA	Not applicable. The topic is not applicable or irrelevant in the context of this initiative.

10. On the basis of the previous assessment it should become clear which are the weaker and stronger sustainability characteristics of the initiative and its alternatives. Taking into account the positions of the initiator, stakeholders and the public in general, and acknowledging the forces that obstruct the sustainable character of the initiative, proposals of an agenda for change have to be developed. As requested in Section M of the checklist, scenarios should be developed which will increase the sustainable nature of the initiative.

11. Finally, conclusions should be made (Section N of the checklist) and a report or discussion document should be written. This should be done in such a way that high-quality information is available to decision-makers and stakeholders for discussing the initiative and its role in the community in a sustainable development context. Assessors should make sure to report in a scientifically correct way, referring to sources of information and justifying statements made (Section O of the checklist).

Overall, the report created should have a structure that is similar to that of an environmental impact statement, including a description of the initiative, the development of alternatives, and a final assessment of the impacts of the proposal. This framework is different from EIA in that it does not focus on the prediction of impacts, but places emphasis on describing the characteristics of the proposal and its alternatives related to sustainability. An assessment is made possible through comparison with cases that are considered to be examples of good practice and existing sustainable development policies, visions, or strategies. Information should be provided about the forces that hinder development of the initiative and the development of ways out of an unsustainable society. Depending on the initiative studied or other external factors, the different topics mentioned in Appendix 7-1 can be examined in more or less detail. This method aims to stimulate assessors to think creatively and come up with innovative measures to steer initiatives in a more sustainable direction.

Figure 7-1 shows the sequence of actions to be taken during the ASSIPAC Sustainability Assessment Check. Gathering basic information on the initiative and looking into existing sustainable development policies, visions, or strategies can take place simultaneously. Bringing all this information together makes an assessment possible

Test results of the ASSIPAC Sustainability Assessment Check show the importance of the person performing the Sustainability Assessment. The following box discusses the profile of the person doing the assessment and looks into the position of the evaluator.

Profile of the Person Doing the Sustainability Assessment

The role of the evaluator is crucial to the outcome of the Sustainability Assessment and the ASSIPAC-method. Reflecting upon scenarios for sustainable development and agendas for change requires a creative mind and a degree of fantasy. Even if certain proposals to make an initiative more sustainable might sound strange or unrealistic today, it is important to introduce them in the discussion. This might be important even if these uncommon proposals only serve an illustrative or educational purpose. It is also important that evaluators make an independent assessment and are not influenced by lobby groups within society.

It is therefore very important to think about the profile and position of the person doing a Sustainability Assessment. In EIA the expert drafting the environmental impact report is supposed to be objective. Experience with the Sustainability Assessment Check has shown that more inspiring results are obtained if the person doing the Sustainability Assessment is a passionate and stubborn promoter of the concept of sustainable development. Unless the evaluator is 100 percent convinced of the importance and need for a more sustainable future, the Sustainability Assessment will likely be reduced to a bureaucratic fulfilment of official requirements.

Therefore, to make the ASSIPAC-method an effective tool, it should be executed by enthusiastic individuals who are advocates of sustainable development. These individuals could then be part of a "Sustainable Development Flash Team", a team of people who are completely independent from the initiatives they study, specialised in short interventions in government departments where important decisions are about to be made. They should surprise decision-makers and stakeholders with the strength, inventiveness, and creativity of their visits. The "Flash Team" should be like a flash light: surprising, brief, leaving a strong impression, enlightening, and full of energy. While assessing the sustainable character of government initiatives, they should not only state what is positive and negative in light of sustainable development, but they should also make proposals for what a more sustainable initiative would look like and develop scenarios for change towards a more sustainable society. Moreover, they should also take a role of training and motivating policy-makers, planners, and decision-makers on sustainable development, since for many of these practitioners, sustainable development remains an unmanageable concept.

Sustainability Assessment Study

The ASSIPAC Sustainability Assessment Study requires more research capabilities, time, and financial means than the Sustainability Assessment Check. The Sustainability Assessment Study attempts a more quantitative approach and therefore pays more attention to the identification of a baseline and target against which the initiative and its alternatives can be evaluated. Since sustainable development is a very broad concept, setting baseline and target values is not obvious and can only be developed on the basis of a broad process of community participation. Examples of North American local and state authorities show that it is possible to translate the general definition of sustainable development into a specific vision for the future of the area under consideration. The vision can lead to detailed goals, action plans, and sustainability indicators and targets. This has been done in the U.S. in the states of Oregon (Oregon Progress Board 1997) and Minnesota (Minnesota Planning 1998), the U.S. cities of Santa Monica (City of Santa Monica 1997) and San Francisco (City and County of San Francisco 1996), as well as the Canadian region of Hamilton-Wentworth (Hamilton-Wentworth 1997). Since sustainable development can have a different meaning depending on the community in which it is introduced, different places will have different sustainable development visions and related goals and targets. Therefore, it is important in the Sustainability Assessment Study to first set these baseline and target values against which the initiative and its alternatives can be evaluated. If no sustainable development vision, goals, or targets have been already set, the assessors have the delicate task of making a proposal for a sustainable development vision, with goals, baseline and target values linked to it. It

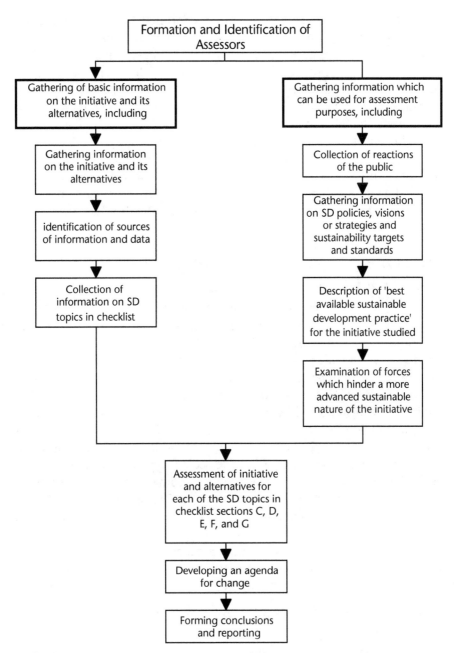

FIGURE 7-1 *Steps in the ASSIPAC Sustainability Assessment Check.*

should be made very clear in the resulting report how these have been developed. Depending on the amount of time and finances available, community workshops can be organized to set the sustainable development point of reference. This point of reference should not always be based on the consensus of a community as a whole, but can also be developed by a specific section within society. If the latter is applicable, the resulting Sustainability Assessment Study should then be clearly identified as a study presenting the ideas and values of that particular segment of the community.

Once the sustainable development vision, goals, and targets have been set, indicators can be linked to it. A prediction can be made of how the indicators will evolve when the initiative is introduced and an assessment can be linked to it. Figure 7-2 shows the most important steps in the Sustainability Assessment Study.

Sustainability Assessment Projects for the Local Level

A number of specific instruments for sustainability assessment in an urban context have been developed. These instruments include a checklist with questions, which is sometimes linked to a scoring system. They can be divided into two major groups: a) those which check whether local authorities are making progress in a sustainable development context and b) those which check whether specific policy proposals developed by local authorities are in line with sustainability goals. The first type of instruments check the current situation in a city and can be considered as auditing instruments. The main question is "Did the community make a step forward in the direction of sustainable development?" The second type of instruments try to assess the sustainability of policy proposals before they are implemented, and are therefore much more in line with EIA. The key question is "Will the proposed policy fit harmoniously into a vision for sustainable development of the local community?"

All of these instruments are still in an experimental phase and unfortunately, no information is currently available on how efficiently they are functioning. The following are four interesting examples:

1. Checklist for Flemish (Belgian) local authorities involved in the Climate Alliance.

2. Instrument measuring the sustainability of local community policies in the Dutch province of Zeeland.

3. Checklist for specific policies proposed in the Dutch city of Tilburg.

4. Checklist as part of the Sustainable City Program of Santa Monica, California.

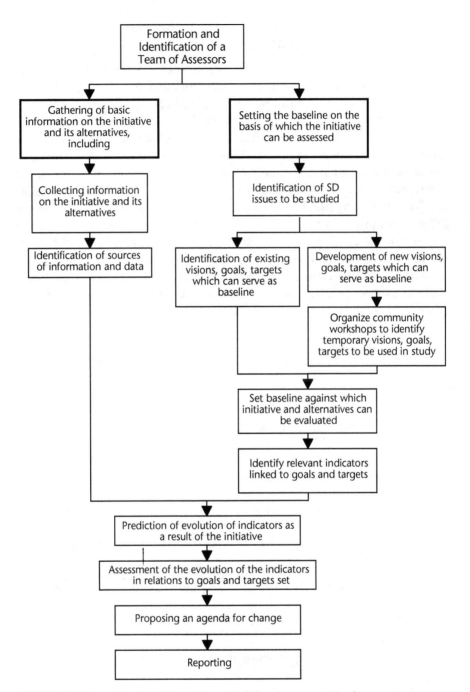

FIGURE 7-2 *Steps in the ASSIPAC Sustainability Assessment Study.*

Checklist for Flemish (Belgian) Local Authorities Involved in the Climate Alliance

In 1990, representatives of European local authorities and delegates of the indigenous peoples of the Amazon Basin founded the "Climate Alliance between European Cities and Indigenous Rainforest Peoples for the Protection of the Earth's Atmosphere." Their goal was to implement local measures against threats to the global climate and the destruction of the rainforests as swiftly as possible. In Europe, this approach is considered one means of developing more sustainable communities. In South America, the Coordinating Body for the Indigenous Peoples' Organizations of the Amazon Basin (COICA) is an official cooperating partner of the Climate Alliance (Klima-Bündnis/Alianza del Clima 1997). The Flemish coordinating body for sustainable development (VODO) developed a simple checklist of 50 questions, making it possible for local authorities to assess how they are performing in relation to the requirements of the Climate Alliance. The checklist has no scoring system and is divided into five themes: 1) energy 2) traffic 3) waste 4) wood and paper and 5) COICA (see box below).

> *Climate Alliance Checklist for Flemish (Belgian) Local Authorities*
> *(Summary and Translation)*
>
> **Energy**
> ✤ Did the local authority calculate CO_2 emission caused by the use of natural gas and electricity?
> ✤ Did the local authority make a CO_2 plan until 2000/2010?
> ✤ Are there specific energy efficiency programs for different target groups?
> ✤ Has the local authority decided against installing electrical heating devices?
> ✤ Have specific measures been developed to cut the use of energy by the local authority itself?
> ✤ Did the local authority send a request to the Flemish Parliament to ask for change of profit mechanisms for intercommunity energy suppliers?
>
> **Traffic**
> *Aspects of land use*
> ✤ Is the local authority assessing the impacts of the location of new buildings on mobility?
> ✤ Did the local authority introduce measures to prevent vacant buildings in town?

➤ Is there a local master plan which aims to reduce the number of cars and develop alternatives?

➤ Has a policy for nature development been introduced?

Infrastructure

➤ Is there a continuous network of footpaths and bicycle paths connecting all neighborhoods?

➤ Are sidewalks wide and without obstacles?

➤ Is the speed of car traffic adapted to other means of transportation?

➤ Are bicycles stalls provided in residential/shopping areas and nearby public transport stops?

➤ Do buses and trams have their own lanes or priorities at intersections?

➤ Is the neighborhood of the train station accessible? Are there efficient ways for making transfers between different means of public transportation?

Parking policy

➤ Does the local authority limit the number of parking spaces?

➤ Is parking time restricted? Are several levels of parking fees introduced?

➤ Is the parking policy controlled?

➤ Is advice given by the local authority on the parking policy of new buildings?

Traffic management plans

➤ Is a coherent approach introduced to solve traffic problems in the region?

➤ Are inhabitants informed of the public transportation system and the provisions for cyclists?

➤ Did the local authority develop a plan to reduce the number of kilometers driven by its staff?

Waste

Exemplary role

➤ Does the local authority take an exemplary role in the field of waste reduction?

➤ Did the local authority develop a preventive policy for acquisitions?

➤ Does the local authority follow a waste prevention policy in all its services?

Policy

➤ Is there a local waste management plan, including measures for waste reduction?

➤ Does the local authority take measures in relation to home delivered advertising papers?

➤ What happens with vegetable, fruit and garden waste? Are there provisions for composting available for the inhabitants?

✢ Is partnership between businesses stimulated to encourage increased waste prevention?

✢ In case new businesses are attracted, is their waste policy taken into account?

Information

✢ Is the waste prevention policy promoted within the services of the local authority?

✢ Is the waste prevention policy promoted towards other organizations and associations?

✢ Is an information campaign organized towards schools?

✢ Are initiatives stimulated, e.g. through a green point system?

CFCs

✢ Are CFCs or HCFCs from refrigerators collected and stored until an environmentally friendly treatment is available?

Re-use

✢ Are second-hand centers introduced or supported?

Wood and paper

Wood

✢ Have guidelines for the purchase of wood been introduced?

✢ Does the local authority provide information towards consumers on tropical wood and wood which was harvested in a non-sustainable way?

✢ Is the re-use of wood products being promoted, e.g. through a second-hand shop?

✢ Does the local authority support reforestation projects?

Paper

✢ Is the consumption of paper by the services of the local authority being reduced?

✢ Is recycled and non-chloro-bleached paper used by the services of the local authority?

✢ Is paper waste being separated?

✢ Is paper waste of the inhabitants being collected?

✢ Is the distribution of advertising brochures discouraged?

✢ Is information provided to consumers about the use and recycling of paper?

COICA

✢ Does the local authority support COICA or its members financially?

✢ Does the local authority inform the public on sustainable development and indigenous peoples?

✢ Does the local authority support COICA politically?

Source: VODO (1997)

Instrument Measuring the Sustainability of Local Policies in Communities of the Dutch Province of Zeeland

The three key questions that are the basis of this instrument are:

1. What happens in Zeeland communities in terms of sustainable development?
2. How do different Zeeland communities compare with one another in terms of sustainable development? and
3. How does local policy in relation to sustainable development evolve?

The instrument focuses on ten policy areas:

1. Waste and emissions,
2. Soil and surface water,
3. Building and living,
4. Energy and climate,
5. Green space planning,
6. International cooperation,
7. Nature, landscape, and agriculture,
8. Nature and environmental communication,
9. Social sustainability, and
10. Traffic and transportation.

For each policy area, approximately ten questions are introduced which deal with plans, measures, consultation and communication, financial means, and results. For each question, a maximum of five to ten points can be earned, depending on the response. Replying to all questions results in a figure between 0 and 100 for each policy area. Different policy areas are awarded a certain weight (based on a survey with all members of the local councils) and a final mark can be calculated (de Vreeze 1998). Table 7-2 gives an example of the type of questions for the policy area "international cooperation."

Checklist for Specific Policies Proposed in the Dutch City of Tilburg

An interesting example of an instrument that checks the sustainability of specific policy proposals is called DOTIS. It is a sustainability check developed for the Dutch city of Tilburg, where it was introduced in 1997. DOTIS is a Dutch language acronym which stands for "Sustainable Development in Tilburg, modern Industrial City." It consists of a set of questions which make it possible to assess if certain sustainability goals and measures for sus-

TABLE 7-2 *Measuring the Sustainability of Local Policies in Dutch Communities of the Zeeland Province—Examples of Questions Within the Policy Area "International Cooperation" (Summary and Translation)*

✦ Has the local authority set goals for international cooperation? Which goals are set in writing? Which goals are measurable?

✦ Has a policy plan for international cooperation been developed? Which specific measures/activities are mentioned in the plan?

✦ Is a budget provided?

✦ Is a public official responsible for international cooperation? For how many hours a week?

✦ Is financial support provided to local groups working in development cooperation?

✦ Is a local working group or a commission on development cooperation introduced in which representatives of local groups also participate?

✦ Is there a link between your city/community and others or the linking of projects (with other local authorities in developing countries, East Europe, West Europe)?

✦ Is specific help provided in case of a link between your city/community and others or the linking of projects (for developing countries or East Europe)?

✦ Which Fair Trade products are bought by the local authority (coffee, tea, sugar, textile, presents, etc.)?

✦ How does the local authority measure the results of the policy on international cooperation?

SOURCE Smit (1997)

tainable development are present in the policy proposals developed by the city government. The effects of the policy proposal on sustainable development are also examined. The field of sustainable development is divided into the following eight major areas of examination:

1. spatial development;

2. economic activity;

3. environmentally conscious performance of households;

4. construction;

5. traffic and transportation;

6. waste management ;

7. energy management; and

8. water management (de Vreeze 1998).

Table 7-3 gives an example of the types of questions presented in DOTIS (Smaal and Wiersinga 1997). The complete list is focused on the issues important to sustainable development in Dutch urban areas and is therefore much more detailed than the general framework presented in the Box on page 188. Goals, measures, and effects have their own scoring systems.

Checklist as Part of the Sustainable City Program of the U.S. City of Santa Monica

Many citizens of the U.S. city of Santa Monica, California, believe that city operations should be the first targeted areas of the practical steps toward sustainability, as part of the sustainable city program. Therefore, the Sustainable City Program's Procurement Working Group developed a checklist to be utilized by all city departments to ensure that broad environmental implications of decisions are considered, and that decision-making occurs in conjunction with its goals. The checklist targets three areas that encompass sustainability issues in the city decision-making process: purchasing, construction and development, and programs and services. Within each area, a decision tree categorizes the type of decisions and subcategories with a list of specific considerations important to the process. For purchasing durable and consumable goods, for example, the list of considerations includes cost, effectiveness, durability, recyclability, material source (new or recycled), resources used during manufacturing, local economic benefits, and existing city purchasing guidelines.

Once complete, the checklist will be supported by two databases. The first will provide city departments with information on environmentally acceptable products, suppliers, consultants, sustainable policy options, best available technologies, and best management practices. The second database will include existing and proposed city regulations and policies, as well as environmental and sustainability considerations (Concern et al. 1998). Although Santa Monica has had good intentions to develop this sustainability assessment approach, widespread use and acceptance of the tool seems to lag behind.

Methods to Be Used in Sustainability Assessment

Examples demonstrate that methods used for sustainability assessment at the local level today are mostly limited to simple checklists. This simple approach should be encouraged because it enables more people to do an assessment themselves, to participate in it, and/or learn more about it. Development of a more advanced methodology for sustainability assessment could, however, complement the checklist approach and result in additional and useful information.

TABLE 7-3 *Examples Found in the DOTIS-System (Summary and Translation)*

	Sustainable urban spatial development	Environmentally conscious performance of households
Goals	Does the policy proposal lead to enhanced spatial coherence and/or quality of different functions; improving environmental quality, nature in the city and public spaces; flexibility/inclusion of future (innovative) forms of traffic infrastructure, waste collection, energy supply, underground constructions; increasing spatial quality?	Does the policy proposal lead to the realization of the following goals of sustainable development: social support and knowledge for the prevention of waste, litter and/or separation of different waste streams; social support and knowledge for limiting the use of energy and water; social support and knowledge of the purchase of environmentally friendly products; social support and knowledge for the increased use of bicycles and public transportation and reduced use of cars; environmentally friendly leisure time of households? Does the policy proposal influence the freedom of action of citizens: does it increase the possibility of citizens to design their immediate surroundings; does it increase choice in consumption, social interaction, possibilities for recreation, and livability of the surroundings?
Measures	Are the following measures included in the policy proposal: using open spaces in the city; building along the outskirts of the city; adding new functions to a monofunctional area; differentiation of types of living quarters, industries and shops; increasing the amount of green spaces; increasing coherence between green spaces; compensating loss of green spaces; moving, cleaning up, zoning and screening off of environmentally harmful activities; improving the identity of the urban space through coherent square structure, pluriform architecture, and strong structural lines; increase social safety through involvement of the citizens, attractiveness of the surroundings, limiting physical vulnerability?	Are the following measures included in the policy proposal: improving the reputation and increasing the use of environmental education centers; training of environmentally sensitive behavior; neighborhood actions: area service centers, youth policy, improving education, day care, policy for senior citizens, play areas, possibilities for leisure activities, neighborhood parties, demonstrations?

TABLE 7-3 *Examples Found in the DOTIS-System (Summary and Translation) (Continued)*

Effects	Will the policy proposal have an effect on: the number of functions in the city/neighborhood; number of kilometers driven by cars and trucks; number of kilometers driven by public transportation; number of kilometers driven by bicycles; number of residences bothered by noise, smell and/or risks; percentage area with soil pollution suspected; area and coherence of green spaces; safety?	Effects: will the policy proposal have an effect on: knowledge and behavior concerning environmentally friendly action; use of environmental education centers; information activities concerning environmentally friendly behavior; quality of separately collected waste (paper, glass, organic waste); livability of neighborhood?

SOURCE Smaal and Wiersinga (1997)

When deciding which methodology to use during sustainability assessment, we need to bear in mind the limitations which all kinds of impact assessments are faced with: a) a limited amount of time to do the study; b) limited financial resources; and c) difficulties in the collection of data. In any case, the methodologies used should deliver results which are useful for deciding which steps to take for the future and should be understandable by as many people as possible, in order to encourage the participation of the public in the sustainability assessment.

The following are five examples of methods that can be especially useful in a sustainability assessment.

Systems Analysis

Planning efforts aimed at long-term solutions and sustainability must somehow analyze and address the systemic aspects of problems. A framework for identifying and analyzing systemic problems, or *systems analysis*, should be provided. Systems analysis provides insight into the functioning of the systems upon which activities depend. It does so by focusing on the interdependent nature of the natural, built, social, and economic systems that support a community or a particular service system. By shifting the focus of assessment to the understanding of systems that are being developed or changed, systems analysis helps planners address the long-term ability of a community to meet its needs (ICLEI 1996).

When using a systems approach, Clayton and Radcliffe (1997) and Checkland (1981) propose to first describe the system under consideration, with the following six steps:

1. Transformation of inputs to outputs: identifying the main flows into, through, and out of the system.

2. Ownership of the system: identifying the decision-makers and stakeholders.

3. Actors in the system: identifying the wider community involved with or influencing the system.

4. Customers of the system: identifying the demands that people make of the system for later comparison with the stated purpose of the system.

5. Environmental constraints on the system: identifying all constraints in the social, economic, and natural environment of the system.

6. Weltanschauung: discussing and clarifying the worldviews, perspectives and perceptions of the participants, owners, actors, and customers.

As the next step, policy options that will lead toward greater sustainability must be identified, and appropriate mechanisms, strategies, and techniques for implementing such policy options must be developed.

A systems approach to sustainability will most likely have to deal with composite problems. The following aspects should be taken into account:

- Interdependence of factors: the important factors in composite problems tend to be directly or indirectly related to one another. In attempts to solve these problems, we need to deal with more than one factor at the same time

- Varying coefficients of interrelations: interrelations between factors can be highly responsive, responsive within certain bounds, or non-responsive Factors may respond immediately or with a delay to change in others

- Positive feedback and cumulative effects. These have the effect of causing impacts to increment beyond proportion to the original impulse

- Negative feedback. This happens when change in one condition is curtailed by change in a dependent condition

- Circular causation. This is where changes in one condition cause change in other conditions and vice versa in a reciprocating evolving dynamic

- Non-equilibrium. Systems change and adapt. Any attempt to interfere with a system and to induce a different outcome is, in effect, an attempt to change the system's adaptive pathway by altering significant control conditions in either the internal structure or the external environment of the system. Interventions are either endogenous, when they derive from the internal dynamic, or exogenous, when they result from an intervention from outside

Clayton and Radcliffe (1997) propose making use of *positional analysis* to clarify choices to be made in the decision-making process. The purpose of any decision-making process is, after taking the relevant facts into account, to arrive at a clear decision that gives the individual or the organization the best chance of achieving the chosen goals. Positional analysis aims to make choices more explicit by way of representing information and clarifying issues and choices. The following steps should be present in a positional analysis:

1. Identifying the relevant conditions, factors, and dimensions when dealing with a compound problem or making a strategic decision.

2. Establishing procedures for quantifying, measuring, ranking or otherwise prioritizing change on each of these dimensions.

3. Measuring or rating the options or scenarios in terms of the change on each dimension.

4. Identifying trade-offs on each of the relevant dimensions.

5. Making a decision on the basis of the overall profile of each option, which may involve assigning weights to each dimension.

The outputs from a positional analysis can be presented in a graphical form, called Sustainability Assessment Maps (SAMs). These consist of a circular-shaped diagram in which each of the important dimensions in a compound problem is represented by an axis. Measurements of change or indications of priorities are then mapped onto these axis. The resulting profile can be used to represent the current situation and possible future scenarios or outcomes. The purpose of the exercise is to make tradeoffs visible, and thereby make assumptions and subconscious decisions part of the explicit decision-making process (Clayton and Radcliffe 1997).

In order to make systems analysis useful in successful planning for sustainability, it should enable participation. Two participatory systems analysis are *service issues mapping* and *networked assessment* (ICLEI 1996):

1. Service issues mapping has been used in the field of development planning over the past decades as a tool for engaging local residents in sharing and discussing information about their local conditions. The mapping process creates a "map" to guide and focus group analysis of issues and problems. Maps can be constructed by a group for a physical area, a neighborhood, family relations, a series of issues, or a natural, manmade, or service system. Service issues mapping is a group brainstorming and analysis technique to help stakeholders identify the full range of issues that must be considered.

2. Networked assessment is an organizational approach used to involve stakeholders in the technical assessment of systemic issues. The basic premise of networked assessment is that the conditions and dynamics of a

system can be better understood if the technical assessment itself is executed by those parties or people who have distinct interests in and day-to-day knowledge of the different components of the system. A network of informed stakeholders is organized to participate, with the help of experts, in the use of a traditional technical assessment method (ICLEI 1996).

Ecocycles and Eco-Balancing

These methods offer residents and municipal officials an understanding of the possible environmental impacts of development alternatives by analyzing the flows of materials and energy in the city. *Eco-balancing* identifies existing or potential imbalances in these flows. *Ecocycles* refers to the fact that all matter on the Earth is used over and over again. In urban systems attention focuses on: 1) the recycling of materials and energy in urban processes and built systems, and the incoming and outgoing flows of materials generated by human activities, and 2) the impacts of urban processes upon natural cycles. An ecocycles or eco-balancing method has been used in Göthenburg, Sweden. Cycles of water, nitrogen, carbon, and mercury have been examined in this city (ICLEI 1996).

Action Planning

The action planning approach mobilizes local resources, and creates "synergies" by combining the efforts of different stakeholders to achieve a common goal. To assure that strategic goals are implemented, an action plan is linked to existing, formal planning processes such as development plans, general plans, and operating and capital budgets. A strategic action plan contains concrete targets for both short- and long-term progress and describes the mechanisms by which the achievement of these targets can be evaluated (ICLEI 1996).

Force Field Analysis

This is an analytical exercise used for priority setting and for selecting and assessing action strategies. The analysis enables a) the identification of specific forces that will either facilitate or hinder achievement of a goal, strategy, or issue; b) the assessment of the relative strength of each force and c) the planning of action strategies to overcome hindering forces and to promote facilitating forces (ICLEI 1996).

Problem-in-Context

de Groot and Stevers (1993) developed a methodological framework, called "problem-in-context," which aims to structure an interdisciplinary search towards environmental problems. A "triple context" of environmental problems is introduced: normative, physical, and social. Analyzing the social context begins with the actors carrying out the problematic activities, their options and motivations, and proceeds to further actors and factors influencing these options and motivations. By this process, one identifies the social options for solving the environmental problem. The social options, with the technical options identified in the physical-science analysis, can feed into the process of designing and evaluating environmental policies and projects (de Groot and Stevers 1993). This method is of interest in a sustainability context because it does not look at the problem in isolation, but tries to find explanations by searching for links and placing the problem in its context. To be valuable in sustainability assessment, the method should be extended to study not only environmental problems and environmental policies, but all kinds of human activities in a sustainable framework.

Conclusion

It is possible to develop impact assessment systems (such as ASSIPAC) which are guided by assessment frameworks based on sustainable development principles. Sustainability assessment should focus on specific aspects, such as: a) context, networks, interactions, connections, partnerships and relationships; b) interdisciplinarity and integration across sectors; c) long-term visions; d) keeping options open, caution, and reversibility; and e) specific environmental, social, economic, planning, and design characteristics. It is proposed that in the absence of well-established sets of sustainability targets and thresholds, the sustainable character of initiatives is assessed with internationally recognized cases of best available practice as points of reference. Moreover, any individual or group of people can develop their own vision for a sustainable future against which an initiative can be assessed. Also, obstacles and forces that hinder a more advanced state of sustainable development of the initiative should be described and scenarios which help us out of an unsustainable society should be developed.

Sustainability assessments, such as those organized by the communities of the Dutch province of Zeeland or the Dutch municipality of Tilburg DOTIS-system, which consist of a simple checklist linked to a scoring system, are very useful, especially for smaller communities that do not have the means to organize a more elaborate assessment system. Instruments such as DOTIS, which follow the EIA approach, and in which the impacts of policy proposals are analyzed, should be used in combination with those following the auditing approach, such as the Zeeland system, in which the already implemented policy of the community is examined.

Specific methodology for sustainability assessment needs to be developed, or existing methods such as systems analysis, eco-balancing, action planning, force field analysis, and problem-in-context should be used. We should, however, bear in mind that the methods should be as simple as possible and easy to use. This will encourage public participation, which is very important in a sustainability context and should result in information that is useful during the decision-making process.

Sustainability assessment should be improved in the future through the introduction of more advanced methods and the development of sustainability indicators and targets. Although these developments are important, they should not stop us from trying to practice sustainability assessment today, on the basis of the existing instruments and systems. Continuous evaluation of this practice will result in a "learning from experience" approach.

References

Aall, C. and K. G. Høyer. 1997. Eco-auditing and sustainable indicators in Norwegian municipalities: Two projects illustrating the development of institutional capacity in Norwegian municipalities. In P. Hardi and T. Zdan, eds., *Assessing Sustainable Development. Principles in Practice.* Winnipeg: International Institute for Sustainable Development.

Checkland, P. 1981. *Systems Thinking, Systems Practice.* Chichester: John Wiley and Sons.

City and County of San Francisco. 1996. *The Sustainability Plan for the City and County of San Francisco.* San Francisco: Department of the Environment.

City of Santa Monica. 1996. *Sustainable City Progress Report. Initial Progress Report on Santa Monica's Sustainable City Program.* Santa Monica: Task Force on the Environment.

Clayton, A. M. H. and N. J. Radcliffe. 1997. *Sustainability: A Systems Approach.* London: Earthscan.

Concern, the Community Sustainable Resource Institute and the Jobs & Environment Campaign. 1998. *Sustainability in Action. Profiles of Community Initiatives Across the Unites States.* Revised and updated 1998 edition.

Dalal-Clayton, B. 1992. Modified EIA and indicators of sustainability: First steps towards sustainability analysis. In *Proceedings of the Twelfth Annual Meeting of the International Association for Impact Assessment.* 19-22 August 1992. Washington DC: International Association for Impact Assessment.

de Groot, W. T. and R. A. M. Stevers. 1993. Problem-in-context: A general framework for problem-oriented environmental sciencerResearch. In B. Nath, L. Hens, P. Compton and D. Devuyst, eds., *Environmental Management. Volume 1. The Compartmental Approach.* Brussels: VUB Press.

de Vreeze, J. 1998. *Duurzame ontwikkeling in de stad. Case-study Leuven.* Master's thesis. Brussels: Human Ecology Department, Vrije Universiteit Brussel.

Devuyst, D. 1998. *Duurzaamheidsbeoordeling.* (Sustainability Assessment) Report of the Environmental Impact Assessment Center. Brussels: Human Ecology Department, Vrije Universiteit Brussel.

Hamilton-Wentworth. 1997. *Hamilton-Wentworth's Vision 2020 Sustainable Community Initiative.* Regional Municipality of Hamilton-Wentworth: Strategic Planning Division, Regional Environment Department.

ICLEI, International Council of Local Environmental Initiatives. 1996. *The Local Agenda 21 Planning Guide. An Introduction to Sustainable Development Planning.* Toronto: ICLEI, IDRC, UNEP.

IUCN International Assessment Team. 1995. Assessing progress toward sustainability: A new approach. In T. C. Trzyna, ed., *A Sustainable World. Defining and Measuring Sustainable Development.* IUCN—The World Conservation Union and the International Center for the Environment and Public Policy, California Institute of Public Affairs. London: Earthscan.

Klima-Bündnis/Alianza del Clima. 1997. *Local Authority Contributions to Climate Protection. Status Report of the Climate Alliance of European Cities.* Frankfurt: Climate Alliance.

Lautso, K. 1998. "The SPARTACUS Approach to Assessing Urban Sustainability." In T. Deelstra and D. Boyd, eds., *Indicators for Sustainable Urban Development. Proceedings of the European Commission (Environment and Climate Programme) Advanced Study Course. 5 - 12 July 1997.* Delft: The International Institute for the Urban Environment.

Lawrence, D. P. 1997. Integrating sustainability and environmental impact assessment. *Environmental Management* 21, 1: 23-42. New York: Springer-Verlag.

Minnesota Planning. 1998. *Minnesota Milestones: 1998 Progress Report.* St. Paul: Minnesota Planning.

Oregon Progress Board. 1997. *Oregon Shines II: Updating Oregon's Strategic Plan.* Salem: Oregon Progress Board.

Partidario, M. R. and F. Moura. 1997. Strategic sustainability assessment—or how to get strategic environmental assessment in the move towards sustainability. In *Proceedings of 17th Annual Meeting of the International Association of Impact Assessment. 28-31 May 1997.* New Orleans: International Association for Impact Assessment.

Rees, W. E. 1988. A role for environmental assessment in achieving sustainable development. *Environmental Impact Assessment Review* 8: 273-291. New York: Elsevier.

Smaal, P. A. and W. A. Wiersinga. 1997. *DOTIS: Duurzame Ontwikkeling in Tilburg, Moderne Industriestad.* Nijmegen: NovioConsult.

Smit, J. G. 1997. *Zeeuwse Meetlat voor Lokaal Duurzaam Beleid. Meting 1997.* Middelburg: VMC.

UNCED, United Nations Conference on Environment and Development. 1992. *Agenda 21: Program for Action for Sustainable Development.* New York: United Nations Conference on Environment and Development.

VODO, Vlaams Overleg Duurzame Ontwikkeling. 1997. *Draaiboek Lokale Agenda 21.* Brussels: Vlaams Overleg Duurzame Ontwikkeling.

Appendix 7-1 The ASSIPAC Sustainability Assessment Checklist

A.	Description of the initiative
A.1.	Title or name of the initiative
A.2.	Type of the initiative
A.3.	General goals of the initiative
A.4.	Specific goals of the initiative
A.5.	Long-term goals of the initiative
A.6.	Phases in the initiative
A.7.	Initiator Including the reputation of the initiator for initiatives in a sustainable development context; how the initiative will be financed; the reputation of the major sponsors in developing initiatives for sustainable development and description of partnerships
A.8.	Geographical description of the region in which the initiative is developed
A.9.	Description of the decision-making process which will be followed
A.10.	Identification of sources of information and data
B.	Description of alternatives for the initiative including a "most sustainable" alternative and other possible alternatives
C.	Description of sustainable development policies, visions or strategies. Describe the sustainable development policies, visions or strategies which are in force in the area. Identify existing sustainability targets and/or standards
D.	Best available practice in an international context for the initiative and its alternatives. Describe on the basis of international literature, interviews with experts, and experience what is the best available practice for sustainable development of the initiative and its alternatives
E.	Discussion of the reactions to the initiative and its alternatives discuss the reactions which were noted during public consultation
F.	Forces which obstruct a more advanced sustainable development of the initiative. Describe which forces hinder a more advanced sustainable development of the initiative
G.	General characteristics of the initiative and its alternatives which could be favourable to sustainable development
G.1.	Integration in strategic visions for sustainable development. Do the initiative and its alternatives fit in the existing visions and policies for sustainable development?
G.2.	Integration and co-ordination with other related initiatives. How do the initiative and its alternatives fit in the network of other initiatives and at what level in the hierarchical structure? Have measures been taken which lead to an optimal integration and co-ordination of the initiative in its context?

G.3.   Integration across different sectors.
Do the initiative and its alternatives take a holistic approach, examining the issues from different sectors and angles? Do the initiative and its alternatives stress "aggregation," "interaction," "connections," and "relationships"?

G.4.   Partnerships across traditional borders within society.
Do the initiative and its alternatives encourage partnerships and co-operation between several groups within society? Did the initiator look actively for co-operation across bureaucratic barriers and traditional borders within society?

G.5.   Empowerment of and co-operation with the local community.
How do the initiative and its alternatives contribute to the empowerment of the local community? Do the initiative and its alternatives lead to a strengthening of the quality, representatives, and resources of the local authority? In what way were the local authority and the local community involved in the formation of the initiative and its alternatives?

G.6.   Keeping options open, caution and reversibility—responding to the needs of future generations.
Do the initiative and its alternatives keep open all possible options for the future? How do the initiative and its alternatives deal with the precautionary principle and the reversibility principle? do the initiative and its alternatives protect the ability of future generations to live a fulfilling life?

G.7.   Budgetary and financial implications.
What will be the implications of the measures for sustainable development on the budget? Are the necessary funds available to realise all sustainability goals of the initiative and its alternatives?

G.8.   Others

H.   Environmental characteristics of the initiative and its alternatives which could be favourable to sustainable development

H.1.   How does the initiative and its alternatives relate to the carrying capacity of the region.
Discuss if and in which way the initiative and its alternatives fall within the carrying capacity of the region and if they have an influence on the carrying capacity

H.2.   Introduction of an environmental care system.
Discuss if and in which way an environmental care system can be incorporated into the initiative and its alternatives

H.3.   Limiting the use of natural resources.
Describe which measures are proposed in the initiative and its alternatives to limit the use of natural resources. What is done for the limited or more efficient use of energy, water, raw materials, and minerals? Is the use of renewable sources encouraged?

H.4.   Limiting the use of materials and the production of waste.
How is the consumption of materials and the production of waste limited? Are measures taken to encourage reuse of materials or recycling, composting, and energy recuperation from waste materials?

H.5.   Protection of biodiversity.
Do the initiative and its alternatives introduce strategies which lead to a protection or increase of biodiversity?

H.6.   Limiting pollution.
Which measures are introduced in the initiative and its alternatives to limit pollution of water, air, soil, and to limit or reduce noise?

H.7.   Restoration and maintenance of ecological cycles.
How do flows of energy and materials generated by the initiative and its alternatives fit in the flows and cycles of the natural ecosystem in which the initiative evolves?

H.8.   Climate change.
How does the initiative and its alternatives deal with the issue of climate change. Does it have an influence on the amount of greenhouse gases released in the atmosphere?

H.9.   Population growth.
Does the initiative and its alternatives have an influence on population growth? Are measures introduced to reach a sustainable population?

H.10.   Others

I.   Social and cultural characteristics of the initiative and its alternatives which could be favourable to sustainable development

I.1.   Empowerment and emancipation of groups within the community.
Do the initiative and its alternatives contribute to the empowerment and emancipation of certain deprived population groups in the community? Are these and other groups encouraged to participate in a discussion about the initiative and decision-making in relation to the initiative?

I.2.   Limitation of social polarisation between groups within society.
Do the initiative and its alternatives lead to a limitation of social differences between population groups within society? Do they lead to more social cohesion within the local community? Do the initiative and its alternatives reduce the gap between poor and rich, North and South, developed and less developed countries?

I.3.   Strengthening local cultural identity and diversity.
Do the initiative and its alternatives value and protect local cultural identity and diversity and the diversity of people within the local community?

I.4.   Protection and improvement of the health of the population.
Do the initiative and its alternatives contribute to the protection and improvement of the health of the population. Are special measures introduced for specific population groups and for a preventive environmental care?

I.5.   Improvement of possibilities for education and training of the local population.
Do the initiative and its alternatives improve the availability of education and training programmes? Is vocational training introduced for the underprivileged, migrants, and unemployed?

I.6.   Improvement of possibilities for local employment.
         Do the initiative and its alternatives improve the availability of jobs for the
         underprivileged, migrants, and unemployed?

I.7.   Increasing possibilities for social, cultural and recreational exchanges between
         members of the local population.
         Do the initiative and its alternatives lead to increased possibilities for social, cul-
         tural, and recreational exchanges between members of the local population?

I.8.   Leading towards a sustainable lifestyle.
         Will the initiative and its alternatives influence the lifestyle of the local popula-
         tion in such a way that it will be less dependent on finite resources and more in
         line with the carrying capacity of the local population?

I.9.   Leading towards strengthened values of a democratic community.
         Do the initiative and its alternatives lead towards strengthened values of a demo-
         cratic community, support diversity, decentralised authority, shared and rotat-
         ing leadership, continuous self-control, and follow the principle that you should
         not expose others to things you would not like to experience yourself

I.10.   Aiming for maximum independence of the local community.
         Do the initiative and its alternatives lead towards a more independent local community?

I.11.   Others

J.   Economic characteristics of the initiative and its alternatives which could be
         favourable to sustainable development

J.1.   Strengthening and diversifying the local economy.
         Do the initiative and its alternatives contribute to strengthening the local econ-
         omy? Are profits made from the initiative reinvested in the local economy? Do
         the initiative and its alternatives lead to a more diversified local economy?

J.2.   Encouraging and supporting private entrepreneurship.
         Do the initiative and its alternatives encourage a flourishing private entrepre-
         neurship, trade, and local industry?

J.3.   Supporting environmentally conscious and ethically responsible trade.
         Do the initiative and its alternatives encourage "fair trade" which is environ-
         mentally and ethically correct?

J.4.   Others

K.   Planning and design characteristics of the initiative and its alternatives which
         could be favourable to sustainable development

K.1.   Promotion of development patterns which reduce the demand for transport.
         Do the initiative and its alternatives encourage an integration of mobility and
         land use planning, reduce the need for car use and promote public transporta-
         tion and non-motorised forms of transportation?

K.2.   Promotion of development patterns which take into account the functions of
         the natural ecosystem.
         Do the initiative and its alternatives lead to integrated land use planning which
         takes into account the functions of the natural ecosystem, vulnerable areas and
         areas which are prone to disasters

K.3. Others

L. Assessment of the sustainable character of the initiative and its alternatives.
Assess the sustainable character of the initiative and its alternatives in a comparative way. Discuss the pro's and cons of the different alternatives. Check how the initiative and its alternatives relate to the best available practice

M. Proposal of an agenda for change. Development of scenarios which lead the way out of an unsustainable society.
Scenarios should be developed which lead the way out of an unsustainable society

N. Conclusion

O. List of references

Annexes

List of scientific methodology used in the different phases of the sustainability assessment

Chapter 8

Sustainability Assessment in Practice: Case Studies Using the ASSIPAC Methodology

Dimitri Devuyst

Summary

In this chapter three examples are given of the application of the ASSIPAC methodology. The first case is a proposal for a Belgian law that would force Belgian companies to develop transportation management plans. Both the negative and positive aspects— from a sustainable point of view—are discussed, using the ASSIPAC Sustainability Assessment Checklist. The forces against the proposal, as well as ideas for an agenda for change are discussed. The second case is a brief illustration of the ASSIPAC Sustainability Assessment Study Approach, using a hypothetical example. The third case deals with existing operational funding systems for reviving Flemish city centers with the Social Impulse and Mercurius Funds. Again, the positive and negative aspects are discussed in relation to sustainable development.

Case Study 1: Compulsory Company Transportation Management Plans in Belgium

The assessment results summarized in this section are the result of a three-day study on the basis of an ASSIPAC Sustainability Assessment Check by Janssens (1999) and Devuyst from the Human Ecology Department, Vrije Universiteit Brussel. They provide a general overview of the sustainability issues in relation to this legislative proposal. Obtaining more detailed information would require a more extensive Sustainability Assessment Study. However, even this limited examination allows us to form an opinion on the sustainable character of the initiative, get an overview of the problems and forces that obstruct the introduction of the initiative, and get insight on proposals for improving the sustainable nature of the initiative.

A legislative proposal introduced by Belgian politician Jos Ansoms for the introduction of company transportation management plans in Belgium is the first initiative we will examine. This law would force all companies with more than 50 employees to develop a plan for the improved organization of traffic generated by its employees travelling between home and work. The main goal of the initiative is to reduce the "car ratio" and "car volume" by 10 percent. The "car ratio" is the ratio (in terms of percentages) between the number of employees travelling by private car over the total number of employees. "Car volume" is the number of kilometers driven annually between work and home by an individual employee. No alternatives to the legislative proposal were introduced by the initiator.

One of the three days of study was needed to collect the necessary background information. The assessment was based upon the following documents: several proposals and drafts of the law, amendments to the law, reports of the discussions of the proposal of the law in the Belgian parliament, and newspaper articles from various Belgian newspapers.

The Belgian authorities did not develop a general policy on sustainable development or sustainable transportation. No targets or standards have been set for the moment. A "best sustainable development practice" for company transportation management plans could not be identified during this three-day study.

No public consultation in relation to the legislative proposal has been organized, but the Sustainability Assessment Check reveals that forces exist that obstruct the introduction of the law. Even though the first proposal of this law was introduced in the Belgian Parliament in 1992, it has taken seven years to come to a draft text. Once the law has been signed, it will take considerable time for the government to introduce implementation measures. It should also be noted that the draft law is already less forceful than the initial proposals. Companies that hire a transportation coordinator, for example, are no longer financially rewarded, as was originally proposed. The following forces are known to obstruct the introduction of the law:

* The employers' organizations: they protest against the compulsory nature of the law. They agree with the goals of the company transportation management plans, but feel that companies should be free to decide if they want one or not. They think the company transportation management plans would place a heavy and bureaucratic burden on companies and would be too costly. They find that the government should spend more money on improving the road network to reduce congestion

* The Belgian State structure: important discussions are taking place in relation to the competence of federal and regional authorities in this matter. For example, proposals for a decree on company transportation management plans are also examined at the Flemish regional govern-

ment level. The state structure, for example, also makes it very difficult to introduce an efficient public transportation system that connects Brussels with its suburbs in Flanders and Wallonia

✦ The employees: people currently driving alone in their car to work fear they will be limited in their freedom of choice in the future. They fear that they will become less flexible in meeting their transportation needs, which is mostly a problem of fulfilling familial duties (for example, an individual who travels alone by car to work, might drop off the kids at school, take the laundry to the dry cleaner and stop by the grandparents' to pick up the children on the way back). They often do not have or perceive the existence of an acceptable alternative to their current mode of transportation. Moreover, they feel that the current system of travel reimbursement would be eradicated, or other penalizing measures would be taken if they continue driving to work alone

✦ The media: certain newspapers highlight the negative aspects of this legislative initiative (For example, printing articles titled *Car Freedom to Be Curtailed*)

The Sustainability Assessment resulted in the following conclusions, which can be considered positive from a sustainable development point of view:

✦ If introduced, the legislative proposal will have a positive impact on the sustainable development of society, especially at the level of sustainable traffic

✦ The initiative will reduce the number of cars travelling at peak hours, reducing the number of traffic jams and resulting in increased accessibility of commercial centers

✦ A decrease in traffic jams will have a positive impact on the livability of the areas which suffer from this problem

✦ A reduction in the number of commuters travelling alone by car will limit the use of gas, will have a positive impact on air quality, and may result in the reduction of noise pollution

✦ Company transportation management plans aim to encourage employees to live more sustainable lifestyles

✦ By encouraging companies to make their sites more accessible to those who do not own cars, the legislative proposal will have an important positive social effect, increasing the equal accessibility of places for all and limiting the social polarization between groups within society

The following can be considered weaknesses of the initiative from a sustainable development point of view:

* There is no integration in strategic visions for sustainable development or sustainable transportation. These visions or strategies have, however, not yet been developed in Belgium

* There is a problem of integration and coordination of similar initiatives taken by the federal and regional authorities

* No attempts have been made to develop a holistic initiative, in which issues from different sectors are integrated

* There are no specific proposals in the initiative to form partnerships among different societal groups, although companies are free to do so in the development of their company transportation management plans

* Local communities are not involved in the proposal, nor are communities in which employees live today, or those in which companies are located

* The initiative does not deal with education or training programs for the underprivileged, migrants, and unemployed

* No attempt has been made to involve local businesses in the proposal. There are no measures to strengthen the local economy or to encourage private entrepreneurship

* The draft text of law proposes a target of a reduction of 10 percent of car ratio and car volume. The fact that there is a clear target is positive. However, no date is set by which this target has to be reached and a reduction of 10 percent can be considered too low from a sustainable development point of view. Moreover, the text also leaves the possibility for the King of Belgium to change this target without any additional set of restrictions

* The draft law follows a traditional system of punishment of companies which refuse to participate. There are no specific proposals for monitoring the effect of the initiative

The final stage of the ASSIPAC Sustainability Assessment Check deals with proposing agendas for change, or in other words, scenarios that lead the society out of an unsustainable state. These agendas should include creative ideas on how to gradually develop a more sustainable society. An agenda for change including company transportation management plans can be developed to include some far-reaching goals of sustainable development and other specific measures. The more detailed this agenda, including quantitative targets and specific dates, the more useful it becomes as a guide for future change. An agenda for change could include the following components:

✦ Over the long-term, we should strive for a sustainable society where humans may live a socially and economically fulfilling life within the boundaries of the carrying capacity of the region. A sustainable society is one in which transportation needs are reduced to a minimum, and when transportation is necessary, people and goods are moved with the most efficient use of renewable resources. This would mean that all trips to and from work can be made by walking, bicycling, and public transportation; non-polluting renewable-energy powered vehicles are used for all trips; and at least 90 percent of commuter trips are made by means other than the private automobile

✦ Much can be done today to reach a more sustainable state through the integration of transportation, land-use, and economic development policies. A policy could, for example, encourage workplaces to be located close together around nodes of public transportation while at the same time concentrating residential areas as much as possible in cities and towns

✦ Economic measures that create more stability and which allow people to maintain their jobs might be an impetus for employees to invest in the local community and live closer to their jobs. Companies could also be asked to initiate partnerships with the local community to develop company housing schemes. As such, employees could be offered a choice of affordable and comfortable homes in the area where they work

Based on the Sustainability Assessment Check, a number of specific recommendations can be made to increase the sustainable character of the draft law. These can be part of an agenda for change, which is developed specifically for this legislative initiative. The introduction of reinforcing measures can take place gradually:

✦ First of all, the initiative would benefit from a more holistic and integrated approach to managing traffic problems. For example, the initiative should be more integrated in a broader policy for sustainable mobility, which tries to solve mobility problems. This should also be linked to land-use policies. An overall Belgian policy on sustainable mobility is still not available. Federal and regional competencies in relation to transportation issues should be better coordinated

✦ The initiative does not attempt to involve the local level authorities or local communities in the development of the company transportation management plans. Making companies and local communities (in the villages where the companies are located) partners in solving traffic problems would be beneficial from a sustainable development point of view

✦ Encouraging social, cultural, and recreational exchanges between the local population and employees of companies located in or near the local community may be a way of convincing employees to invest in the area and live nearby

✦ Companies should be encouraged to set up training schemes for the underprivileged, migrants, and unemployed in relation to the company transportation management plans. It should be determined if company bus drivers, car-pool coordinators, bicycle rack attendants, and so forth, could be introduced, creating new jobs for long-term unemployed individuals

✦ It should be examined if local businesses can be involved in the development of company transportation management plans or if people can be motivated to set up small businesses in combination with the management plans (for example, service related businesses dealing with communal moving of people and goods, bicycle repair shops, etc.)

✦ Companies should not only be encouraged to look into the traffic generated by their employees driving to and from work, but should also consider the transportation of goods and work-related trips by staff members

✦ In addition to focusing on private companies, institutions and organizations such as government institutes, administrative services and schools should be more explicitly asked to develop traffic demand studies and traffic management plans

✦ Much more thorough measures could be taken to convince employees not to use the car on an individual basis to go to work. The legislative proposal introduces the idea of financial incentives for employees who participate in the company transportation management plans. At the same time, however, employees can still receive fiscal benefits from using their car (for every kilometer driven an amount can be deducted from personal taxes). The application of both measures is counterproductive

✦ A reduction of the costs of registration fees and land-taxes collected by the Belgian government when buying a house would make it less expensive to buy and encourage Belgians to get a house closer to their work.

➳ A system of monitoring the effects of the law would be important to be able to tell if the goals are reached and if additional measures need to be taken

➳ It is clear that companies that do not function according to Belgian legislation should be reprimanded. The law may, however, benefit from extra incentives for companies that do an outstanding job

➳ Actions should be taken to resolve problems that are a result of the obstructing forces. These could include publishing studies showing the benefits of company transportation management plans for companies, educating and motivating employees to adopt company transportation management plans, and applying a more coherent government policy in relation to sustainable development and sustainable mobility

Case Study 2: Illustration of an ASSIPAC Sustainability Assessment Study for Company Transportation Management Plans in Commercetown, California

This is an imaginary illustration of the ASSIPAC Sustainability Assessment Study Approach. Suppose that Commercetown, California, would introduce a system of company transportation management plans on a voluntary basis. First, a point of reference for the assessment is set on the basis of the Commercetown vision for sustainable development. Linked to this vision are sustainability indicators and targets. One of the goals of the sustainable development vision for Commercetown is that "more people should make use of public transportation and the use of cars should be reduced." The assessment team decides that the existing indicator "annual ridership on Commercetown metropolitan bus lines" should be part of the Sustainability Assessment Study. A year 2005 target of 20.9 million passengers has been set by the Commercetown authorities, while the 1990 measurement of 19 million passengers is considered the baseline situation. Data from 1993, 1995, 1997, and 1999 show that the number of passengers decreases, with 18 million, 17.8 million, 17.7 million, and 17.5 million passengers, respectively.

A survey carried out as part of the Sustainability Assessment Study might have shown that 15 percent of companies, employing 10,000 people in total, are willing to participate in the introduction of company management transportation plans. The survey may also have shown that as a result of the introduction of the voluntary company transportation management plans, 10 percent of the employees, or 1,000 people, will switch to the municipal bus lines for their daily commute to work. This would result in an additional 480,000 passengers each year (1,000 people, taking a return trip five days a week for 48 weeks a year). Taking into account the estimate of the

Commercetown Transportation Department, which might read 17.4 million bus riders by 2001, we can predict a new 2001 indicators value of almost 17.9 million passengers if the company transportation management plans are introduced. We can therefore state that the introduction of company transportation management plans on a voluntary basis in Commercetown will have a positive influence on reaching the sustainability target, but that to just introduce these plans alone will not result in a major shift towards a sustainable transportation system in Commercetown.

While this example is hypothetical and shows only one small part of a possible Sustainability Assessment Study, it does illustrate how visions, indicators, and targets can be used to predict sustainability impacts.

Case Study 3: The Social Impulse Fund and the Mercurius Fund for Reviving Flemish City Centers

Like many cities in Europe, the historic city centers of Flemish cities face the emigration of Belgian middle class families who prefer to live the car-centered life of the suburbs. A modern house with a garden and abundant surrounding green spaces cannot be found in the congested narrow streets of downtown. Decaying buildings, outdated infrastructure, and crime are even more reasons for many people to leave town. The car-centered lifestyle in the suburbs results in the development of huge shopping malls outside the city centers. Downtown businesses have a hard time competing with the malls, and major parts of city centers are left to more or less marginalized societal groups, such as underprivileged elderly people and immigrants.

The Flemish regional authorities in Belgium aim to solve these problems by introducing an urban policy that aims to make living and shopping in the city center more attractive. Two funding initiatives were introduced to direct financial aid to those areas most in need of redevelopment. The Social Impulse Fund aims to improve the quality of life and environmental quality in the deprived urban areas of Flemish cities and municipalities, and is a cooperative initiative of the Ministry of Internal Affairs, Urban Policy, and Housing and the Ministry of Culture, Family, and Well-being (Ministerie van de Vlaamse Gemeenschap 1996). The Mercurius Fund is a project that aims to revive inner city commercial centers and is an initiative of the Minister of Economic Policy (Ministerie van de Vlaamse Gemeenschap no date).

This preliminary and illustrative Sustainability Assessment Check of the Social Impulse Fund and the Mercurius Fund is based on a one-day examination of the policy documents, legislation and official explanatory documents issued by the Ministry of the Flemish Community, making use of the ASSIPAC checklist methodology developed by Devuyst (1998) and presented in Figure 7-1. The assessment presented here is more a commentary focusing primarily on the general characteristics of the initiatives, in particular in relation to topics of integration across different sectors, integration, and coordination with other related initiatives and cooperation with the local community.

The Social Impulse Fund has a number of characteristics that may contribute to a more sustainable development of Flemish cities, such as a structural basis in regional legislation, a focus on an integration of coherent measures which are traditionally the competence of different ministries or departments, a results-oriented approach, and an aim to have a new style in local governance.

The Social Impulse Fund received a structural basis with the introduction of the Decree of 14 May 1996. This is important in a sustainable development context because the fund, an instrument used to implement urban policy, will not be suddenly halted or radically changed after each election. The Social Impulse Fund is results-oriented, which means that the Flemish regional government signs an agreement with the Flemish cities and municipalities on the basis that certain results must be attained. Flemish regional authorities and local authorities are partners in this agreement. Local authorities and regional authorities decide on evaluation criteria together. Indicators are introduced to measure whether the expected results are reached. Only half of the funds are released in advance, whereas the other half is released after local authorities show proof that the expected results have been reached. The Social Impulse Fund also aims to limit the amount of bureaucracy by trying to keep all administrative tasks to a minimum.

The Social Impulse Fund aims to encourage cities and municipalities to take coherent measures to increase the livability of deprived urbanized areas. Municipalities should propose a package of measures to be introduced and combine activities in family care and well-being, community building, employment, housing, mobility, education and training, economic issues, and land-use. It is stressed that complex problems need an all-inclusive approach and that different ministers need to cooperate in order to find sustainable solutions. The fund itself is a good example of this kind of cooperation, since it is a joint initiative of two ministers.

Moreover, the Flemish regional authorities do not want to impose specific measures, but ask local authorities to develop plans for solving their own problems. As a result, various local measures with the common goal of improving the urban environment receive regional funding. These plans and measures should state how the local authorities propose to solve their livability problems. Since the recovery of the livability of cities cannot be accomplished in a short-term period, the plans developed by the local authorities should at least look three years into the future. Individual measures taken should fit into a broader planning. Another factor is that the Social Impulse Fund requires local authorities to involve the local population. Again, it does not impose certain ways of public participation but requires local authorities to show how the population was involved in the process. Underprivileged social groups should especially be included in the development of such a plan (Vandenberghe et al. 1997).

The Mercurius Fund was introduced within the framework of the Flemish urban policy, and its purpose is to find solutions for city center problems through economic measures. This is being accomplished through the renovation of commercial centers. In comparison with the Social Impulse Fund, the Mercurius Fund is somewhat less innovative, and although a number of goals are in line with sustainable development principles, they are less explicitly stated. The goals of the Mercurius Fund are: for local authorities to develop a vision on how to create a new commercial dynamism in the business centers, to improve the attractiveness of existing commercial centers, to slow down the move of commercial establishments from cities to undeveloped areas outside the cities, and to give deprived areas a new chance for a more vital future. In the context of sustainable development, the Mercurius Fund is positive in the sense that it will reduce growth of commercial centers outside of the city in the scarce open spaces of Flanders and encourage the reuse of more or less abandoned built-up areas. To be able to get funding from the Mercurius Fund, local authorities have to develop a coherent strategic commercial plan in collaboration with local shopkeepers' associations. Like the Social Impulse Fund, the plan has to be results-oriented and the results have to be measurable. Local authorities whose strategic commercial plan is approved by the Flemish regional authorities can make a proposal for funding of a specific project which will make the commercial center more attractive. A selection is made of projects to be funded by the Mercurius Fund on the basis of three main evaluation criteria and seven secondary evaluation criteria. These criteria do not include sustainability principles, except the requirement for more integration of the Mercurius projects, such as the integration of the strategic commercial plan and integration in a set of various measures from different domains (social, economic, environmental).

In the framework of sustainable development, it would have been useful for Flemish authorities to link these funding initiatives to the development of Local Agenda 21, an initiative largely untouched by Flemish local authorities. It is remarkable that the links between the Social Impulse Fund and the Mercurius Fund are very limited, especially since the Mercurius Fund was introduced a year after the Social Impulse Fund, and more steps could have been taken to form a more closely integrated funding system. This is especially striking since one of the aims of the Social Impulse Fund is to also concentrate previously separate funding initiatives. In fact, the Social Impulse Fund brings together the former Special Fund for Societal Well-being, the Special Donation system, and the Flemish Fund for Integration and the Deprived. One wonders why the Minister of the Economy did not join his two colleagues in making a broader "Social and Economic Impulse Fund."

The following additional comments can be made:

* The Social Impulse Fund is more concerned about working in an inclusive and integrated way than the Mercurius Fund, which focuses more narrowly on the commercial vitality of the inner cities

* The Social Impulse Fund requires local authorities to prepare a three-year plan for solving their livability problems. Planning is a first step in a more sustainable direction; however, a three-year term can hardly be called long-term planning

* Neither the Social Impulse Fund nor the Mercurius Fund encourages local authorities to integrate environmental aspects into their planning and projects. The funds would have led our cities in a more sustainable direction if they had also undertaken the goal of reducing environmental impacts. For example, the renovation of buildings in city centers could be organized in such a way that they would make use of renewable energy sources or revived commercial centers could be encouraged to set up common environmental care systems, thereby reducing the production of waste

* The Mercurius Fund is limited in its lack of participation of the local population. Only business associations are involved in the development of a commercial strategic plan. From a sustainable development point of view it would have been more useful to have a strategic commercial plan as part of a broader vision for the future of the community in general. In conclusion, the Social Impulse Fund and the Mercurius Fund can both play an important role in a more sustainable development of Flemish cities.

In conclusion, the Social Impulse Fund is one step ahead of the Mercurius Fund for the following reasons:

* The Social Impulse Fund stresses the need to solve problems of poverty and social exclusion through an integrated approach. Traditional financing of specific short-term projects is replaced by an innovative financing of coherent local plans to combat social problems. In con-

trast, the Mercurius Fund also requires the development of plans, but focusing only on the commercial strategy and leading in the end to the funding of more traditional projects

✦ The Social Impulse Fund shows more initiative for partnership and cooperation than the Mercurius Fund. The Social Impulse Fund is the result of the cooperation between two ministers, whereas the Mercurius Fund remains the exclusive initiative of the Minister of Economic Policy

✦ The Social Impulse Fund stresses the introduction of a new style of governance with more attention to the needs of individual local authorities and requirements for public participation

Conclusion Relating to the Methodology Used

The results of these ASSIPAC Sustainability Assessment Checks are quite general and qualitative, and based on a one- to three-day study of documents. More elaborate and detailed assessments, including the presentation of quantitative results, can be obtained after a much more detailed study. Since the results of this assessment are qualitative in nature, an effort should be made to justify on what basis certain judgments are made. These types of results are open for further discussion and analysis, or can be seen as a stimulus to start a discussion on future improvements of the initiatives.

It is important that a sustainability assessment tool should be flexible and adaptable to the needs of the users. It should be clear that if a sustainability assessment—even a limited one—was implemented before the introduction of the Social Impulse Fund and the Mercurius Fund or the law on company transportation management plans, a number of synergistic alternatives could have been developed. Proposals of agendas for change could have been made as well for improving these fund initiatives to make them more in line with strategies for a more sustainable society.

References

Devuyst, D. 1998. *Duurzaamheidsbeoordeling.* (Sustainability Assessment) Report of the Environmental Impact Assessment Center. Brussels: Human Ecology Department, Vrije Universiteit Brussel.

Janssens, S. 1999. *Duurzaamheidsbeoordeling. Wetsvoorstel tot Bevordering van het Woon-Werkverkeer met Bedrijfsvervoerplannen.* Brussels: Master Programme in Environmental Impact Assessment, Vrije Universiteit Brussel.

Ministerie van de Vlaamse Gemeenschap. 1996. *Het Sociaal Impulsfonds, een Instrument voor het Stedelijk Beleid.* Brochure of the Ministry of the Flemish Community nr. D/1996/3241/208. Brussels: Ministerie van de Vlaamse Gemeenschap.

Ministerie van de Vlaamse Gemeenschap. No Date. *Toelichting voor de Lokale Besturen over het Fonds voor Binnenstedelijke Commerciële Centra.* Unpublished informational report on the Mercurius Fund.

Vandenberghe, J., M. De Coninck and P. De Decker. 1997. Voor Steden en Mensen. Beschouwingen bij Twee Jaar Stedelijk Beleid. In G. Hautekeur, ed., *Naar een Levende Stad.* Brugge: die Keure.

Chapter 9

Direction Analysis: An Example of Municipal Sustainability Assessment in Norway

Carlo Aall

Summary

Three local and regional authorities have participated in a project aimed at developing and testing a scheme for direction analysis. The project which was financed by The Norwegian Association of Local and Regional Authorities (NALRA), was a result of a report made by the NALRA committee on environmental issues in 1993, which recommended the use of direction analysis to elucidate the contents of the goal for sustainable development. The suggested model for direction analysis is based on a sustainability indicator system. But rather than developing a universal indicator list, we have developed a framework for locally adapted sustainability indicators. A key element in the model is a suggested list of general municipal sustainability topics. The suggested topics form the basis for each local or regional authority to "translate" into local or regional sustainability goals, which, in the final instance, will provide the basis for selecting sustainability indicators in every single case of analysis. The following nine municipal sustainability topics form the basis for the model: (1) The satisfaction of basic needs; (2) ecological sustainability; (3) the intrinsic value of nature; (4) the precautionary principle; (5) long-term planning; (6-7) equitable distribution of benefits and burdens globally and over time; (8) cause-oriented environmental policy; and (9) increased emphasis on the participatory principle.

Introduction

Two Norwegian local authorities (Ålesund with 40,000 inhabitants and Stavanger with 100,000 inhabitants) and one county (Akershus—the most populous county in Norway—with 420,000 inhabitants) (see Figure 9-1) have participated in a project aimed at developing and testing a scheme for direction analysis during the period of 1995-1997 (Aall 1998). The "Direction Analysis in Municipal Planning and Policy-making" project was financed by The Norwegian Association of Local and Regional Authorities (NALRA).

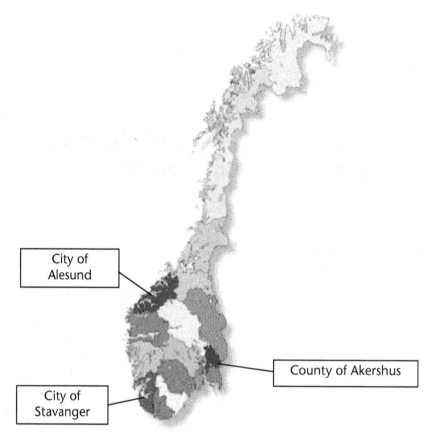

FIGURE 9-1 *Location of regional authorities participating in the Direction Analysis Project.*

The concept of "direction analysis" originates from Sweden. In 1991, the Swedish Association of Local Authorities presented eight questions that were meant to determine whether a plan would contribute in a negative or positive way with respect to sustainability (Månson 1992:15-17, Grundelius 1995):

1. Is energy consumption reduced?

2. Do we change to using renewable energy?

3. Is nature's capacity for building up resources increased?

4. Does biological diversity increase?

5. Do we create closed recycling circles?

6. Do we keep within the limits of what nature and human beings can tolerate?

7. Do we solve problems in a holistic perspective so that we do not create new problems?

8. Do we take the precautionary principle into account?

If all questions are answered by "yes," the subject matter to be assessed should not be in conflict with the goal of sustainable development (Delin and Grundelius 1998). Since 1991, a number of Swedish municipalities have made use of direction analysis in different ways. However, as Delin and Grundelius (1998) point out, the questionnaire seemed successful in triggering a general political debate on the issue of sustainability, whereas the problem of delivering remedial action still remained. Such general debates seemed to result in few realizable measures that change the development proposal in a more sustainable direction. The Swedish approach to solving this problem has been to introduce exergy calculations as a way of making direction analysis more concrete by demonstrating differences in energy efficiency between different options of actions (Wall 1977, Delin and Grundelius 1998). The Norwegian approach presented in this article is somewhat less "arithmomorphic"—to use a phrase from the famous "entropy-economist" Nicholas Georgescu-Roegen (1981). Instead, we have chosen an approach more similar to Strategic Environmental Assessments.In 1993, NALRA appointed a committee on environmental issues. The report from the committee recommends contributing to the development of a methodology for direction analysis (NALRA 1993). High expectations are held of the initiators, interested municipalities, and the Norwegian Ministry of the Environment (MoE) concerning the direction analysis project. On several occasions, direction analysis has been stressed by NALRA as an important tool in the follow-up of the World Commission report and Agenda 21. It has also been referred to as one of five main areas in the MoE's strategy for supporting Local Agenda 21 (MoE 1997b).

We can identify six central guidelines for the development of a workable method for direction analysis: 1) sustainability orientation, 2) strategic orientation, 3) impact-assessment orientation, 4) key-topic orientation, 5) long-term perspective, and 6) precautionary-principle orientation (NALRA 1993, Aall 1998). The report from the NALRA ad hoc committee explicitly expresses that direction analysis must be linked to the goal of sustainable development. Furthermore, it is presumed that direction analysis is to be applied to plans and programs, not to clearly defined projects. However, direction analyses also differ from the popular understanding of strategic environmental impact assessments. They do not comprise all types of environmental consequences, but are limited to a few key topics central to the goal of sustainable development. The report from the ad hoc committee also underlines a multiple generation time perspective. Besides, direction analysis is more oriented toward revealing the uncertainty linked to possible substantial and/or irreversible environmental consequences, thereby linking direction analysis to the precautionary principle.

Sustainability Topics and Indicators for Local and Regional Authorities

Sustainability Indicators

In developing a more concrete and specified methodology for direction analysis, we decided to take advantage of the growing number of sustainability indicator initiatives. The purpose of sustainability indicators is both to measure important features of sustainable development and to act as a catalyst in promoting the same sustainable development. A survey of the international literature on the topic of environment and sustainability indicators shows, however, that the greater part of the literature is concerned with the design of indicator systems and, to a lesser extent, with experiences in putting them to practical use (MacGillivray and Zadek 1995, Høyer and Aall 1997). Until recently, environment and sustainability indicators were developed mostly for use at the international or national levels. However, in the wake of the Rio conference and from the call for Local Agenda 21, local authorities in a number of countries have shown a growing interest in sustainability indicators. In addition, the most prominent attempts to develop and actually *adopt* sustainability indicators are now being found at the local level (MacGillivray and Zadek 1995, MacGillivray 1997, Høyer and Aall 1997). The development and test of a scheme for direction analysis in Norway also seems to fit well in this picture.

In order to communicate information and have a function in a decision-making context, it is not sufficient for indicators to only be accurate; information must also lead to what we might call *resonance* for the intended target group. The target group must understand the information and become motivated to act on the basis of this information. Several authors advocate public participation in designing sustainability indicators, replacing the more expert-based indicator models. Thus, we can distinguish between "cold expert-based" and "warm participation-oriented" indicator systems (MacGillivray and Zadek 1995).

The Universality of Sustainable Development

In determining sustainability indicators, the question of participation introduces the issue of *universality*. Sustainable development is concerned with a number of general norms. Introducing the local level in designing sustainability indicators is in itself no guarantee that the result will be in accord with sustainability norms; it might even have the opposite effect. We are concerned with norms of a global, long-term, and distributional character, remote from the ordinary local perspective of problems. Therefore, in a practical context the challenge is to arrive at indicators that are "just the right temperature," so that they have local resonance through their "warmth" and at the same time are sufficiently in accordance with the goal of sustainable development. In the direction analysis project we have attempted to take this into consideration. Instead of developing a universal indicator list, we have developed a framework for local adaptation of sustainability indicators. A key element in our model is a deduction of a number of universal norms, which in turn forms the basis for deducing general municipal sustainability topics. These norms must form the basis for a local or regional "translation" in the form of adopting sustainability goals which provide the concrete basis for picking out sustainability indicators in every single case of a direction analysis (see Figure 9-2).

The starting point for our sustainable indicator framework model is the goal of sustainable development as outlined by the World Commission and Agenda 21. Whereas the word *sustainable*—with its origin in ecology—has a clear-cut meaning, the term *development* is far more vague and implies much more serious scientific and political disputes over its exact meaning (Høyer 1997). The World Commission (WCED 1987) explicitly states that development is very important and that human beings come first (WCED 1987). At the same time, the World Commission points out that the minimum requirement for sustainable development means that the natural systems sustaining life on earth are not endangered. Several authors have argued that we have to consider the requirement for sustainability as a restriction to the goal of development (Lafferty and Langhelle 1999, Høyer 1997). Crucial questions in determining policies for a sustainable development will then be (Høyer and Aall 1997):

✦ Which needs are basic and which are not?

✦ How many people should have their basic needs met?

✦ What resource quantities exist and what quantities can the environment tolerate in terms of exploitation?

✦ What emission quantities are tolerable to the environment?

✦ How many generations should be taken into consideration in planning?

✦ To what extent can social reorganization and technological development influence the size of the population and the satisfaction of present and future needs, given the restrictions in the supply of resources and in the capacity for deposits?

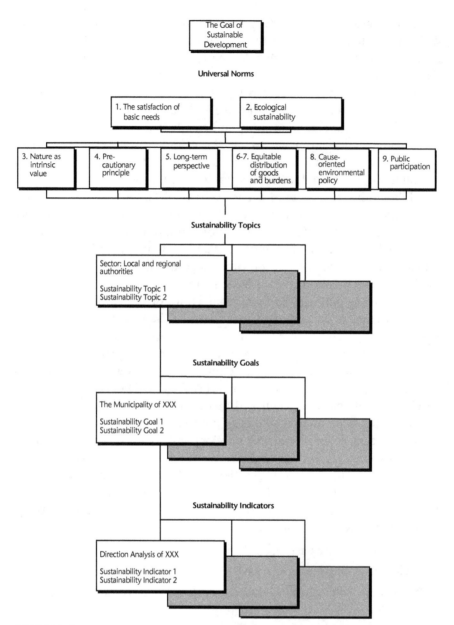

FIGURE 9-2 *Structure of the proposed sustainability indicator model.*

The requirement of satisfying *basic needs*—with emphasis on the needs of the poorer parts of the world—represents the core of the development aspect in the concept of sustainable development. It constitutes a necessary precondition, along with ecological sustainability. The other characteristics have little meaning unless these two preconditions are fulfilled. There is still a significant difference between the goals of development and ecological sustainability. Maintaining ecological sustainability is a *negatively* limiting obligation. It is a question of restricting the extent of human-impacts in nature in order to maintain the necessary ecological sustainability. It is not a primary objective to develop maximum ecological sustainability at the expense of satisfying basic needs. Concerning the development aspect, however, it is a question of a *positively* developing obligation. At present a large part of the world's population does not get its basic needs satisfied. These people must be given priority, even if it implies a reduction of biological diversity. At the same time, the total population figures are too high, and measures must be implemented to reduce the population figures if ecological sustainability is to be maintained on the long-term. The World Commission report emphasizes that a standard of living beyond the necessary minimum to satisfy basic needs is only sustainable if all consumption standards—at present and in the future—are established in line with what is sustainable over a long period.

The concept of sustainability has its origin in ecological science. It was developed to express the conditions that must be present for the ecosystems to sustain themselves in a long-term perspective. In the World Commission report, there are several strong references to the necessity of *ecological sustainability.* This implies a requirement of sustaining biological diversity. Biological diversity must be maintained as a necessary—but not sufficient—condition for sustainable development. A further confirmation of this is found in the Rio Convention on biological diversity. According to the convention, biological diversity is defined as the variability among living organisms, at the species level as well as at the ecosystem level.

In its most elementary form, an understanding of *nature's intrinsic value* implies that biological diversity must be attributed as a value in itself, independent of the instrumental value it has for humankind. The fact that the earth's biological diversity must be protected for its own sake was initially stated in an international political context in the UN Natural Protection Charter from 1982 (Næss 1992). The initiative to promote the charter was taken by a number of poor countries. The UN General Assembly passed it by a majority of 111 to one. The United States cast the single opposing vote. This view was restated in the World Commission report (WCED 1987: 51): "Protection of the environment is not only a development goal. It is also a moral obligation to other living things and for future generations."

In the report, *The Precautionary Principle: Between Research and Politics* by the National Research-Ethical Committee for Natural Science and Technology (Norway), it is argued that the precautionary principle is the most important component in the sustainability goal (Kraft and Storvik 1997:6).

This principle was first defined within the framework of the UN Commission of Sustainable Development (UNCSD) in 1990 at an international conference in Bergen as part of the preparations to the Rio conference in 1992 (MoE 1990). The principle is incorporated in many international environmental treaties, and is also incorporated as an important principle in Norwegian environmental policy:

> In those cases where there is a danger of irreversible environmental consequences, the lack of full scientific proof should not be an argument for failing to implement actions and measures which reduce environmental problems.

> (MoE 1997a)

The precautionary principle is subjected to various interpretations. Two aspects of this principle seem to create confusion: the elements of prevention and uncertainty. Both theoretically and in its practical implications, the precautionary principle is a principle of *preventive* environmental policy. Prevention implies that damages are kept from happening, instead of being repaired retroactively. This does not mean that all forms of preventive environmental policy and standards are at the same time precautionary. However, a standard or a measure implemented on the basis of the precautionary principle can never be non-preventive. If the precautionary principle is interpreted as a general preventive environmental principle only, it could be used in so many different contexts that it might eventually lose its political impact (Cameron and Wade-Gery 1992). *Uncertainty* is a basic premise of the precautionary principle. A lack of full scientific certainty must not be used as a reason for postponing measures to avoid environmental degradation. The existence of such a reasonable doubt is sufficient basis for the implementation of preventive or regulative measures regarding the environmental issues to which the principle applies. Furthermore, there must also be the risk of substantial or irreversible environmental damage. The precautionary principle is *not* applicable when it is certain that environmental damages will occur. Neither is it applicable when the damages are likely to occur, and there is uncertainty with regard to more serious environmental damages. In such cases, decisions must be made on another basis, e.g. related to standards established as a part of preventive environmental politics. Cost-benefit analyses may be one of several instruments in such a context. Sustainable development presupposes a multigenerational perspective. This is a pervasive theme in the World Commission report and relates to both the sustainability and the development aspects of the concept. The theme has been very much in the forefront of the economic considerations of the consequences that can be drawn from sustainable development in terms of criteria for the exploitation of natural resources. Sustainable development gives no acceptance for neo-classical economic ideas of complete substitutability between human-made capital and natural capital. On the contrary, the sustainability goal presupposes a reduction of the total consumption of non-renewable energy and material resources. Consequently, the earth's natural capital is seen as a finite quantity and its use must be spread out over time in order to benefit as many generations of people as possible. This introduces the question of maintaining a long-term perspective. The demand for long-term perspective also implies an obligation to emphasize future environmental impacts at least as much as the present ones. The issue of an *equitable distribution of goods and burdens globally and over time* are closely linked to the question of what is included in the notion of "basic needs." The World Commission emphasizes that basic needs must be satisfied throughout the world and for all future generations. The UN conventions are of some help to us to concretize such an ethical consideration. This is particularly true with the 1948 World Declaration on Human Rights and the 1966 UN International Convention on Human Rights. The convention on human rights suggests minimum requirements for a "good" life by stating that human beings have the right to work, to fair and good working conditions, to social security, to a satisfactory standard of living for themselves and their family, adequate food, clothing and housing, a highest obtainable health condition both in physical and mental terms, the right to an education, and the right to participate in cultural activities. The application of such norms requires further specification. This has been carried out by UN organizations such as the Food and Agriculture Organization (FAO) with its nourishment standards concerning the physiological minimum requirements per person to satisfy the need for calories, proteins, vitamins, certain mineral substances, and so on. Similarly, there is a good reason for maintaining the principle of an equitable distribution of burdens. At present environmental burdens are not equally distributed; the consequences are more serious in some parts of the world than in others. A case in point is the greenhouse effect. Similarly, future generations will be harder hit than the present one. This implies that the remedial actions must take into account the situation for the least favored members of the global society, now and in the future. The World Commission report outlines two major approaches to environmental policy (WCED 1987: 223). The first is characterized as the "standard program," reflecting an attitude to environmental policy, acts, and institutions with the main emphasis on environmental effects. The second approach reflects an attitude focusing on the practices that cause these effects. The World Commission thus distinguishes between an *effect-oriented* and *cause-oriented* environmental policy. The Commission emphasizes that, so far, the first approach has prevailed, whereas it is the second that must be included in an environmental policy for sustainable development. On the one hand, we have so-called "end of pipe" solutions, where attempts are made to reduce the problems after they have emerged. Various types of pollution abatement technologies can exemplify such solutions. On the other hand, we have source-oriented solutions. This requires control and management of what goes in; that is, the processes causing the environmental problems. The World Commission report contains references to *public and popular participation*, but the main emphasis is placed on national and inter-governmental initiatives through international organizations. During the 1992 Rio Conference much emphasis was placed on stimulating various types of participation. The municipalities are given a decisive role in taking initiatives to arrive at an agreement as to what constitutes local challenges in the work toward sustainable development, as well as to distribute the responsibility among important players locally and regionally. This is a role that in many contexts goes beyond the formal role with which local and regional authorities in most Western countries today are assigned through legislation and directives.

A Suggested Model for Direction Analysis

Direction analysis can be defined in the following way:

> A formalized, systematic process for analyzing the direction of development in policy, programs or plans related to operational goals for a sustainable development, the preparation of a written report of the findings, and using the findings in a publicly transparent decision-making process.

> (Aall 1998)

On the basis of universal norms discussed above and experiences from the Direction Analysis project, 15 municipal sustainability topics have been suggested (see Table 9-1). In a Norwegian context the topics indicate a balance between what is practically "possible" and what is scientifically "necessary" in order to assess the "direction" of a suggested plan or policy program. The model presupposes that each municipality must relate its own environmental policy to the suggested sustainability topics, examining whether a decision has been made on sustainability goals for each of the suggested sustainability topics. If any of the topics have not been decided upon, the municipality must first decide on goals within the relevant topics before the direction analysis method can be employed. When the municipality has "translated" the sustainability topics by adopting sustainability goals for all the topics, then the foundation is laid for developing specific sustainability indicators. The choice of indicators is to be made in each individual case whenever there is a need for a direction analysis. Consequently, the concrete situation will lead to the fact that selections of sustainability indicators may vary significantly in each case of direction analysis, but the specific framework in the form of municipal sustainability goals will remain firm for each municipality. Furthermore, the specific framework for direction analysis may vary between the municipalities, whereas the *general* framework in the form of municipal sustainability topics will allow a common methodological platform.

TABLE 9-1 *Proposal of Municipal Sustainability Topics in a Norwegian Context*

I Policy principles	⇒ The precautionary principle
	⇒ Equitable distribution within present generations
	⇒ Equitable distribution between present and future generations
II Policy content	⇒ Protection of biological diversity
	⇒ Reduction of energy consumption
	⇒ Reduction of harmful emissions in the atmosphere
	⇒ Reduction of consumption of material resources
	⇒ Reduction of total mobility
	⇒ Emissions to local recipients within nature's own tolerance level
	⇒ Sustainable exploitation of natural resources
	⇒ Human-made changes of the environment must not harm our health
III Policy processes	⇒ Popular participation
	⇒ Making all sectors accountable
	⇒ International commitment
	⇒ Working to change peoples' attitudes

SOURCE Høyer and Aall (1997)

There is a clear need to distinguish between the objective and the basis of the analysis. The *objective* of the analysis may be a plan, a large project, or a general outline of a specific policy. However, before a local or regional authority can adopt the use of direction analysis, existing statements on environmental goals have to be assessed in relation to the suggested sustainability topics thereby processing the *basis* of the analysis. The following procedure is suggested: 1) examine existing municipal environmental goals and structure these according to the suggested sustainability topics; 2) find the essence of the existing environmental goals within each sustainability topic and pinpoint preferably one or at most three key goals for each topic; 3) in the cases where one or more sustainability topics are poorly covered by existing environmental goals, one out of three alternative steps may be taken (These steps are: a) indicate in the basis of the analysis that no political basis exists for assessing the sustainability topic in question; b) determine key goals exclusively for use in direction analysis which have no managerial function ; and c) await a political decision on the environmental policy concerning the sustainability topic in question); and 4) finally, propose the analysis basis for political sanction to the appropriate body. The aim is to assign a strategic function to direction analysis. Therefore, it is important to justify the use of direction analysis, e.g. in the general municipal master plan and to formulate unambiguous directives for the adoption of direction analysis. The directives should encompass the following: 1) Allocating the administrative and political *responsibility* for conducting the direction analysis; 2) developing *interception criteria* specifying when direction analyses should be made; and 3) describing *follow-up procedures*; that is, how the results from the direction analysis should be used. Interception criteria may relate to the *size* of the initiative (e.g., investment bounds), *character* (for example, type of plan), and possible degree of *conflict*. We can further distinguish between two main agents in a direction analysis: 1) *responsible authority;* that is, the institution with authority to request and alternatively sanction a direction analysis (e.g., the environmental committee) and 2) *responsible operator;* the institution to be audited in the direction analysis. The municipality may choose whether the responsible authority or operator should actually carry out the direction analysis; it may also be done in cooperation between the two. Responsible operators may be required to explicitly specify in the last presentation how they have decided to relate to the conclusions and possible suggestions for improvements in the direction analysis report. When a decision has been made to carry out a direction analysis on a concrete object, the actual *analysis* will comprise the following elements: 1) Precise statement of the purpose of the analysis; 2) choice of indicators; 3) data acquisition; 4) determination of development direction; and 5) assessment of possible improvement actions. We recommend that municipalities keep careful watch of the following aspects when choosing *indicators*: the number of indicators should be small; if possible, there should not be more than one indicator per key goal. The indicators should be quantifiable as much as possible. The data should be easily available. The indicators should be easily understandable. Furthermore, we can distinguish between three types of indicators: 1) *state of the environment*: the development over a given period of time and up until the present situation, e.g. development in energy use for transportation; 2) *short-term plan effect*: alterations as a result of the actual plan; for instance, re-allocation of a specific area in a land-use plan; and 3) *long-term implementation effect*: alterations as a result of implementing the plan, e.g. increased transport as a result of developing a new motorway. Indicators describing status and short-term plan effects are quantifiable in most cases, whereas long-term effects of implementing plans are often more difficult to determine. In many cases, we either have to be content with an estimate of the size of the indicator or recommend a surveillance program in order to determine the value after the plan in question has been passed and implementation attempted. Calculations and *conclusions* will be at two levels: individually for each sustainability topic and collectively for all topics. In addition to determining development direction, it is important to include *suggestions for improvement*, i.e. for possibly correcting the direction. Three different categories for improvement actions exist: 1) *general* improvement actions not specifically related to the analysis object. This may, for instance, imply suggestions for altering existing planning procedures and 2) suggestions for altering the *contents* of the analysis object (i.e., the plan). This might imply smaller or larger adjustments, and in extreme cases, recommendations not to pass the plan; and 3) suggestions for *follow-up actions*, normally actions to check whether the goals and intentions of the plan are actually reached. A direction analysis report may consist of four major chapters (see Table 9-2). In the conclusion, it is important to clearly state the relationship between sustainability topics, sustainability goals, choice and quantification of indicators, determining development direction, and possible suggestions for improvement actions. *Justification* should refer to the formal justification of the direction analysis; for instance, the municipal master plan or similar. *Procedure* should refer to a concrete resolution to carry out the direction analysis or to how interception criteria have been satisfied if this is specified. The formal procedure for the direction analysis from commission to follow-up should be described, as should the integration of the analysis into the process related to the analysis object. Under the heading *Organization of Implementation*, the bodies responsible for implementing and sanctioning the direction analysis should be described. Before the presentation of the results of the direction analysis, the priorities given to sustainability topics, sustainability goals, and choice of indicators should be discussed. The inclusion of the entire analysis basis as an appendix to the report might also be appropriate. The chapter on acquisition and computation of data should discuss how the data have been collected and how the indicator value has been estimated for each sustainability topic and for each indicator. The analysis chapter should explain the reasoning behind the determination of development direction for each sustainability topic and for all of them combined. Further, suggestions for improvement actions must be given.

TABLE 9-2 *A Suggested Template for the Direction Analysis Report*

1. Conclusion

2. Background
   - ❧ Justification
   - ❧ Procedure
   - ❧ Organisation of implementation
   - ❧ Priorities for sustainability topics and key goals
   - ❧ Choice of indicators for the various sustainability topics
3. Acquisition and computation of data

4. Analysis

Appendix: Analysis basis

The Use of Direction Analysis in Practice

The Norwegian Context

Norway is divided into 18 county units and 435 municipal units. The county and the municipality both have separate elected councils. The counties, and particularly the municipalities, vary considerably with regard to population and geographic context. The average municipal size is about 9,900 inhabitants, and one-third of the municipalities have fewer than 3,000 inhabitants. The municipalities and counties are negatively restricted in their activities; that is, they may take on any function that the law does not forbid them to carry out, or that has not been specifically delegated to other institutions. At the same time, however, the municipalities are subject to general legislation and rules of law, unless a special exception has been made. The revenue base consists of several elements, the most important of which are taxes on capital gains and income, property taxes, and municipal fees. In addition, the state provides direct transfers of funds to municipalities and counties through a system of allocations which was established in 1986.

In 1989, provisions concerning environment impact assessments (EIA) were incorporated in the Planning and Building Act. However, these provisions only apply to a small extent to the municipalities because municipal plans seldom trigger demands for EIA. In other cases—that is, whenever private and public developers trigger demands for impact assessments—the municipalities are only involved to the extent that the development plans require municipal approval *after* the impact assessment has been carried out. General provisions for strategic environmental assessments (SEA) have not been established in Norway except in regard to hydroelectric development. Since the late 1980s and early 1990s there has been a major shift toward decentralization in Norwegian environmental policy. Many new environmental issues are located at the local level, and the municipalities are in general given a more important role in Norwegian environmental policy. In 1988, the MoE and NALRA initiated the very important reform Environment in the Municipalities (EIM) through a joint program involving 91 out of 435 municipalities. The EIM-reform implied government funds for the appointment of an environmental officer in each municipality. The reform had three superior goals (MoE 1991): 1) to establish a municipal environment organization; 2) to strengthen municipal environmental planning; and 3) to lay the foundation for delegating more environment tasks to the municipalities. By 1996 nearly 90 percent of the municipalities had employed an environmental officer, whereas 13 out of 19 counties had employed an environmental officer without support from government funds. Seventy-one percent of the municipalities have made a resolution or assert that environmental planning has been started. Since 1993, the percentage of municipalities with adopted action plans has more than doubled, from 20 to 42. Moreover, 40 percent of the action plans are formally based on the Local Planning Act (Lafferty et al. 1998).

The Actual Test of Direction Analysis

Direction analysis was attempted in two municipalities and one county on three different planning processes, and with different starting-points for establishing the basis of analysis (see Table 9-3). The Ålesund project was not concluded because of insufficient funds in the municipality. The

TABLE 9-3 *Basis and Objectives of Analysis in the Trial of Direction Analysis*

Municipality	Objective of analysis	Basis of analysis
County of Akershus	Land use sub-master plan for the region of Romerike	Objectives from a sub-master plan / Regional Agenda 21
City of Stavanger	Review of the land-use master plan	15 key goals derived from a sub-master plan / environmental action plan
Municipality of Ålesund	A thematic sub-master plan on green structures	Objectives from the master plan

projects in the City of Stavanger and the County of Akershus are thematically relatively similar, but Stavanger has to a larger degree followed the recommended set-up for carrying out the direction analysis. For reasons of space, I have therefore concentrated on the project in the City of Stavanger, whereas the Akershus project will only be referred to briefly.

The City of Stavanger: Sustainable Land-Use Planning?

Prior to the direction analysis project in Stavanger, the municipality drew up a basis of analysis through the formulation of a number of key goals (see Table 9-4). The political basis for formulating the key goals was found in the municipal Environment Plan. The basis of analysis was presented for approval to the Committee on Environment.

TABLE 9-4 *Basis of Analysis for Carrying Out the Project on Direction Analysis in the Municipality of Stavanger*

Sustainability topics	Key goals
Protection of biological diversity	The municipality shall endeavour to protect and ensure maximum biological diversity, including the protection of biotopes of nationally and locally endangered species
	Minimise development in agricultural, nature, and recreational areas
Ensuring and planning areas for recreational activities	All inhabitants of Stavanger are to have access to a continuous network of hiking trails within 500 m from their residences
Protection of cultural monuments and cultural environment	(not decided)
Reduced consumption of energy	Energy consumption for transport purposes must be reduced
	Energy consumption in municipal buildings must be reduced
	Energy consumption in industry and households must be reduced
Reduced harmful emissions to the atmosphere	Harmful emissions from transport to the atmosphere must be reduced
Discharge to local recipients must not exceed natures tolerance level	Discharge to sea and watercourses must not lead to overload of nature's own tolerance level
Sustainable harvesting of natural resources	Harvesting of fish and game must be carried out in such a way that maximum biological diversity and stocks are preserved
	Minimise development in agricultural, nature, and recreational areas
Reduced consumption of material resources	The waste volume must be reduced
	The consumption of material resources for the municipality of Stavanger must be reduced

TABLE 9-4 *Basis of Analysis for Carrying Out the Project on Direction Analysis in the Municipality of Stavanger (Continued)*

Sustainability topics	Key goals
Human-made changes of the environment must not be harmful to our health	There must be no local health problems connected with air and water pollution
	By 2005, no one in his own residential or recreational area must be subjected to noise above current norms and guidelines
Popular participation	The population must be actively involved in municipal planning and development
Accountability for all sectors	All municipal sectors have an independent responsibility for the environment
	Environmental protection is a superior objective for the municipality of Stavanger
International commitment	The municipality is to give its contribution to reducing the global environmental problems
Working to change peoples attitudes	The municipality is to organise active information work concerning environmental protection, challenges, and opportunities
The precautionary principle	The municipality must take long-term considerations (50-100 years)
Living conditions and quality of life	Stavanger is to be a city safe for everyone and with a stable environment and good local service. Conditions should be equal regardless of person and residence, and there must not be too much accumulation of health and social problems

Mandatory land-use plans must be reviewed every four years. In Stavanger the direction analysis was included in the formal consultation round for the review of the land-use plan. The analysis was presented as the statement from the municipal environmental committee in the consultation document. Table 9-5 shows the choice of sustainability topics and indicators. The municipality has distinguished between two types of indicators: 1) current status and 2) effects of the proposals in the land-use plan. The analysis is based exclusively on data found in the planning document itself. In other words, there has been no specific collection of data in connection with the direction analysis, which may be considered a weak point in the test. According to the municipality, the reason for this is insufficient working capacity because this was a project, and therefore comes as an addition to ordinary duties.

In the analysis Stavanger concludes that the land-use plan is on the whole in accordance with the key goals (listed in Table 9-4), but adds some minor suggestions for adjustments. Furthermore, the suggestion is made that the

TABLE 9-5 *Choice of Sustainability Indicators for Direction Analysis of the Land-Use Plan in the Municipality of Stavanger*

Sustainability topics/key goals	Sustainability indicators	Status today	Effects of proposals in municipal master plan
Theme 1: Protection of biological diversity			
The municipality works to ensure and conserve maximum biological diversity	Development of areas given municipal protection status	Localities not sufficiently surveyed	New development areas situated outside landscapes given priority in the Environment Plan
Minimize development in agricultural, nature, and recreational areas	m^2 built-up area/ inhabitant	300 m^2	300 m^2
	m^2 new regulated green areas/new residence	-	260 m^2
	m^2 new regulated green areas/residence	-	200 m^2
Theme 2: Ensure and plan areas for recreational use			
All inhabitants of Stavanger have access to a continuous network of hiking trails within 500 m from their residence	Planned areas / ensured areas	(1,600/800) hectares	The land-use plan follows up the goal in the Environment Plan
	Planned hiking trails / developed hiking trails	(180/120) km	The land-use plan follows up objective in the Environment Plan
Theme 4: Reduced energy consumption			
Energy consumption for transport purposes must be reduced	KWh/1,000 inhab./ year		The land-use plan does not cover transport issues –
	Private car km/person	No data available	
	Share of public transport of all travel		
	Share of bicycle rides of all travel		
	Number of electric vehicles		

TABLE 9-5 *Choice of Sustainability Indicators for Direction Analysis of the Land-Use Plan in the Municipality of Stavanger (Continued)*

Sustainability topics/key goals	Sustainability indicators	Status today	Effects of proposals in municipal master plan
Theme 5: Reduce emissions harmful to the atmosphere			
Emissions from transport which are harmful to the atmosphere must be reduced	Number of people exposed to local pollution (NO_2) and dust above recommended limits		The land-use plan does not cover transport issues – No data available
	Reduced emissions of CO_2, NO_x, SO_2, and NM-VOC		

work with land-use planning and transport planning must be integrated to a larger extent. It has not been possible to trace any direct effect of the direction analysis on the land-use planning process. This may be because the direction analysis was carried out so late in the planning process. The fact that this was a project was probably another reason why the results from the test of direction analysis were not given the same weight in the decision-making process as an established routine would have done. Nevertheless, the municipality has resolved to use direction analysis as one of three instruments to ensure the follow-up of the Environment Plan:

> ...(it can be used) in cases where plans/initiatives are of an extent which would justify more thorough assessments of environmental impacts, but where impact assessments according to the Planning and Building Act are not mandatory.

> (City of Stavanger 1997)

In addition to the use of direction analysis, the Environment Plan of Stavanger also proposes the use of environmental auditing and the formulation of an annual environment report as instruments to ensure the follow-up of the Environment Plan.

The County of Akershus: Pinpointing the Unsustainability of Air Traffic

In the county of Akershus environmental goals from the recently adopted Regional Agenda 21 were used as a basis of analysis for the test of direction analysis. Akershus is the first Norwegian County to adopt a plan entitled

Regional Agenda 21. The plan has been adopted as a county sub-master plan pursuant to the Planning and Building Act. The plan seeks to identify the regional responsibility for following up the recommendations of Agenda 21, and it represents an innovation in a Norwegian context, possibly also in a European context (Lafferty 1998). Akershus approved—even before the actual test of direction analysis was finished—the incorporation of direction analyses in its regular planning system:

> …(it can be used) to assess future decisions and plans in relation to the goals in Regional Agenda 21. Result indicators are worked out to follow the development concerning adopted goals in Regional Agenda 21.

> (County of Akershus 1998)

The direction analysis was tested in relation to a county sub-master plan for land-use planning in the region of Romerike (the central part of the county of Akershus). The need for a special land-use plan for Romerike was triggered by the development of the new national airport, Gardermoen. The goal of the land-use plan was to allocate business development in a way that would save as much "unspoiled" nature and arable land as possible, and to strengthen the position of public transportation. The land-use plan provides guidelines for mandatory land-use planning in the municipalities. In Akershus the direction analysis was carried out as part of an informal consultation before the plan was subjected to a formal one, making the test in Akershus happen at a much earlier stage in the planning process than was the case in Stavanger. The direction analysis includes the following topics: 1) protection of biological diversity; 2) reduction of energy consumption and air emissions; 3) sustainable harvesting of natural resources; 4) protection of human health; 5) public participation; and 6) efforts to change peoples' attitudes in a more environmentally friendly direction. The analysis concludes that the land-use plan lays the ground for unsustainable development, because the newly established national airport at Gardermoen will evidently lead to a dramatic increase in business development in this region. In the direction analysis report, the county suggests a checklist for the municipalities in their further land-use planning, so that they can assess the direction of development in their planning processes.

Conclusion

Implications of and Prerequisites for Using Direction Analysis

The idea that "a certain type of question will result in a certain type of answer" applies to most indicator-based analyzing tools, including direction analysis.

Two critical issues will determine whether the utilization of direction analysis by local and regional authorities will make development more sustainable:

1. To what extent local and regional authorities adopt an environmental policy that actually accords with the goal of a sustainable development.

2. To what extent recommendations on remedial actions in the direction analysis report are actually followed up.

Even though municipalities manage to establish an operational analysis basis, there are still some conditions connected with the actual use of the tool:

+ Sufficient commitment from the political and administrative leadership of the municipality

+ Sufficient administrative capacity, especially with respect to data collection and the report phases of the direction analysis

+ Sufficient knowledge both in the administration and among the elected representatives with respect to the true content of a sustainability-oriented environmental policy

Given these prerequisites, local and regional authorities can expect various effects of adopting the method of direction analysis. Using direction analysis may contribute both substantially and instrumentally to sustainable development. The methodology of direction analysis is "known" from the tradition of SEA, whereas the substantial contribution depends on political clarification. The introductory exercise of determining the analysis basis—based on the suggested sustainability topics—may lead to an alteration of the current environmental goals in a more sustainable direction. Direction analysis is by definition an arena for assessing sustainability. Both internal and external processes can be assessed in this way. Direction analysis may also form a basis for making strategic choices at an early point of decision and hence lead to decisions tending in a more sustainable direction.

Are Norwegian Municipalities Ready for the Sustainability Challenge?

The crucial question, however, is not whether municipalities can make use of a specific SEA technique like direction analysis. What matters is the political will to give priority to the goal of sustainable development. The explicit reference in the NALRA ad hoc report on linking direction analysis to the goal of sustainable development implies that the concept of sustainable development in some way differs from the traditional concept of environmental policy. The Norwegian White Paper no. 58 (1996-97) "Environmental Policies for a Sustainable Development" (MoE 1997a) states that sustainable development is to become a general principle for Norwegian environmental policy. The White Paper further states that the concept of sustainability must rest on three equivalent perspectives: 1) ecological; 2) generation; and 3) welfare perspectives. The ecological perspective is concerned with maintaining the ecosystems, whereas the generation and welfare perspectives—often referred to as "social sustainability" or the "distributional perspective"—are concerned with a more equitable distribution of the satisfaction of basic needs between and within generations, respectively. Moreover, it is the distributional perspective that represents the really new and challenging aspect that the concept of sustainability brought into environmental policy. The White Paper of 1997 assigns a far more central role to local authorities with respect to achieving sustainable development than do the two previous White Papers on the World Commission's report (MoE 1989) and Agenda 21 (MoE 1992). However, it is not evident which perspectives the government considers that local authorities use as a basis in achieving the goal of sustainable development. Already in the introduction of the recent White Paper, an important reservation is made where it states that the Government will only, "... clarify and further develop the ecological perspective" (MoE 1997a:1). A report made by the previous director of the Norwegian Pollution Prevention Agency on how Norwegian authorities should address the issue of sustainable consumption and production lays the grounds for such a reservation:

> I have deliberately avoided any discussions about distributional questions. Although they are important, they are politically controversial and extremely complex.

> (Rensvik 1996:1)

The most important political challenge for local and regional authorities in their efforts to develop Local Agenda 21 strategies, or to gain full advantage of sustainability-oriented tools such as direction analysis, is to predict the consequences of introducing the distributional perspective into local environmental policy. This especially relates to what the White Paper no. 58 (1996-97) denotes as the "welfare perspective" on sustainable development (MoE 1997a). Emphasizing the welfare and generation perspective on sustainable development will yield two kinds of effects on municipal environmental policy: new areas of politics related explicitly to distributional issues are brought onto the agenda of environmental policy. Furthermore, and probably more controversial, emphasis on distributional considerations will in many cases intensify political ambition within the environmental policy arena. The recommendations in the World Commission report of a 50 percent reduction of the energy consumption can only be understood on such a basis:

...(the low-energy scenario) stipulates a 50 percent reduction in energy use per capita in industrialized countries, and a 30 percent increase in developed countries. (This will) require enormous structural changes...and it is not likely that more than a few countries will reach this goal in the next say 40 years....The Commission (still) believes there are no other realistic alternatives for the 21st century.

(WCED 1987: 130-131)

The discrepancy between such recommendations and the current environmental policy in Norway is dramatic. In 1998, Norway reached the goal of curbing the growth in—and in the long run stabilizing—energy consumption (MoE 1997). Consequently, it is far from dramatic to maintain that, so far, the distributional aspect of sustainable development has not been prominent in Norwegian environmental policy as well as in most, if not all industrialized countries. The White Paper no.58 (1996-97) gives a contradicting message: on the one hand, the distributional perspective represents something new, and on the other hand, the report reserves itself against adopting this new aspect in national environmental policy. Obviously this gives grounds for asking the question: Which perspectives should local and regional authorities adopt—a limited ecological perspective and/ or an extended distribution perspective? The Regional Agenda 21 plan submitted by the county of Akershus shows that at any rate some local and regional authorities have recognized that this is precisely where the challenge lies. Like the government, the county admits that it only relates to the ecological perspective of sustainability. In contrast to the government, however, the county also hints at the consequences of an extended perspective in which the challenges in Agenda 21 are taken up in earnest. In a case document to the final treatment of the Regional Agenda 21 plan by the county executive committee, the chief county executive officer describes this challenge in the following way:

... the starting-point for the planning process was a resolution to draw up a county plan for environmental policy. The ... focusing of Agenda 21, not only on an ecological, but also on a social and economic sustainable development has thus not been sufficiently considered. This implies, for example, that the environmental problems to a very small extent have been discussed in relation to the issue of global distribution. Besides, the issue of distribution in our own society, as a consequence of the implementation of policy measures suggested in the plan, has not been taken up for debate. The latter problem is particularly relevant in relation to the use of economic policy measures such as taxes on environmentally harmful consumption.

(County of Akershus 1997, case 2/97)

It might be hard to get acceptance for a local and regional "counter-cyclical policy," and such a strategy may easily become a bitter price to pay both politically and economically. Even if it might be desirable from an environmental-political point of view for "local" and "regional" Norway to go for a radical sustainability strategy, it is reasonable to believe that we will end up with a more realistic "mainstream strategy." Another possibility is that in some contexts the differences between a radical local or regional environment policy can be visualized, and a sober assessment made of what can be practically carried out, as shown in the example of direction analysis from the county of Akershus. One of the more important recommendations given by the Direction Analysis project is that NALRA should take an attitude toward the national political aspects of adopting direction analysis. The project suggests that NALRA should take a stand as to what sustainability topics should be used as a reference point for local and regional authorities in a direction analysis. If not, the method might easily lose a legitimate common methodological justification, which will probably be crucial to the credibility of the method. So far, one year after the close of the project, nothing has been done by NALRA, and no local or regional authorities other than Stavanger and Akershus have started using direction analysis.

References

Aall, C. 1998. *Direction Analysis for a Sustainable Policy in Municipal Planning and Policy-Making*. (in Norwegian with English summary). VF-report 1/98. Sogndal: Western Norway Research Institute.

Cameron, J. and W. Wade-Gery. 1992. *Environment and Energy in the Nordic Countries – Energy Scenarios for the Year 2010*. (in Swedish). Report 1992.548. Stockholm: Nordic Council of Ministers.

City of Stavanger. 1997. *Environmental Plan 1997-2009*. (in Norwegian). Stavanger: City of Stavanger.

County of Akershus. 1997. *Regional Agenda 21. Sub-master Plan on Environment*. (in Norwegian). Oslo: County of Akershus.

Delin, S. and E. Grundelius. 1998. *Development and Direction Analysis, Exergy-Calculations in the Municipalities of Lulå and Strängnes*. (in Swedish). Lulå: Lulå Technical University.

Georgescu-Roegen, N. 1981. *The Entropy Law and the Economic Process*. Cambridge: Harvard University Press.

Grundelius, E. 1995. *Preconditions for a Sustainable Development*. (in Swedish). Stockholm: The Swedish Association of Local Authorities.

Høyer, K. G. 1997. Sustainable development. In D. Brune, D.V. Chapman, M.D. Gwynne and J.M.Pacyna, eds., *The Global Environment. Science, Technology and Management* 2: 1185-1208. VCH Weinheim.

Høyer, K. G. and C. Aall. 1997. *Environment and Sustainability Indicators. An International Review.* (in Norwegian with English summary). VF-report 13/97. Sogndal: Western Norway Research Institute.

Kraft, N. and H. Storvik, eds. 1997. *The Precautionary Principle: Between Research and Politics.* (in Norwegian). Oslo: NENT-publication no 10.

Lafferty, W. 1998. Local and Regional Agenda 21 – A politically viable path to sustainable development. *Graz Symposium on Regions - Cornerstones for Sustainable Development. 28-30 October 1998.* Graz.

Lafferty, W., C. Aall and Ø. Seippel. 1998. *From Environmental Policy to Sustainable Development in Norwegian Municipalities – the Transaction from EIM to Local Agenda 21.* (in Norwegian). Oslo: ProSus report 2/98.

Lafferty, W. M. and O. Langhelle, eds. 1999. *Towards Sustainable Development: The Goals of Development - and the Conditions of Sustainability.* Indianapolis: Macmillan/St. Martin's Press.

MacGillivray, A. and Zadek, eds. 1995. Accounting for change. *International Seminar, Toynbee Hall, October 1994.* London: The New Economics Foundation.

MacGillivray, A. 1997. Community indicators resource pack in the UK" In P. Hardi and T. Zdan, eds., *Assessing Sustainable Development Principles in Practice* 45-52. Winnipeg: International Institute for Sustainable Development.

MoE. 1989. *White Paper on Environment and Development. Norway's Follow-up of the World Commission Report.* (in Norwegian). Oslo: Norwegian Ministry of Environment.

MoE. 1990. *Action for a Common Future.* ECE-conference report. Oslo: Norwegian Ministry of Environment.

MoE. 1991. *On Municipal Environmental Policy. White Paper No. 91 (1990-91).* (In Norwegian) Oslo: Ministry of the Environment.

MoE. 1992. *White Paper on Norway's Follow-up on the UN Conference on Environment and Development in Rio de Janeiro.* (in Norwegian). Oslo: Norwegian Ministry of Environment.

MoE. 1997a. *White Paper on Environmental Policy for a Sustainable Development.* (in Norwegian). Oslo: Ministry of the Environment.

MoE. 1997b. *MoE's Strategy for Supporting LA21.* (in Norwegian). Draft 1 December 1997. Oslo: Ministry of the Environment.

Månson, T., ed. 1992. *Eco-cycles: The Basis for a Sustainable Development.* (in Swedish). Report from the Swedish Ministry of Environment, SOU 1992:42. Stockholm: Miljövårdsberedningen.

NALRA, The Norwegian Association of Local and Regional Authorities. 1993. *Think Globally, Act Locally. Priorities for Environmental Protection at the Local Level.* Oslo: The Norwegian Association of Local and Regional Authorities.

Næss, A. 1992. Sustainability. The integral approach. In O.T. Sandlund, ed., *Conservation of Biodiversity for Sustainable Development.* Oslo: Scandinavian University Press.

Rensvik, H. 1996. *Furthering the Aim of Sustainable Development Through Influence on Consumer Behavior Patterns.* SFT Report 96:02. Oslo: National Agency on Pollution Prevention Control.

Wall, G. 1977. *Exergy—a Useful Tool in Resource Management.* (in Swedish). Göteborg: Chalmers Technical University.

WCED, World Commission on Environment and Development. 1987. *Our Common Future.* Oslo: Tiden Norsk Forlag.

Part III

Tools for Setting a Baseline and Measuring Progress in Urban Sustainability Assessment
Dimitri Devuyst

Part III of this book deals with the indicator approach to sustainability assessment. Monitoring, or the repetitive and periodic measurement of certain variables, is an important activity in our attempts to achieve sustainable development. Data resulting from monitoring initiatives can inform policy-makers about whether they are on the right track or have diverted from their planned path towards a more sustainable state. Indicators for sustainable development can be used in monitoring activities to set the baseline situation and measure progress toward sustainability goals and targets determined by local authorities. This Part aims to provide a state-of-the-art guide in the development of a monitoring system for sustainable development, including indicators, targets, and reporting systems. It looks into the application of indicators in sustainability assessment systems and links indicator systems to target setting and sustainability reporting.

Since we cannot actually assess when a community has become truly "sustainable," sustainability indicators cannot tell us how far we have to go. They can, however, let us know if we are moving in the right direction and are therefore useful in a sustainability assessment process. In Chapter 10 Thomas van Wijngaarden gives a general overview of indicators for sustainable development. Different types of indicators and their essential features are discussed, and various frameworks for the development of sustainable development indicators (SDIs) are presented. Although it would be ideal to have standard sets of SDIs applicable all over the world, huge differences in local settings make this impracticable. Many different sets or lists of SDIs consisting of various combinations and selections of indicators can be developed. Therefore, it is important that each local community makes a choice of the most relevant indicators, depending on the values within that community.

Since this book deals with the local level, Elizabeth Kline focuses in Chapter 11 on the development of SDIs for urban areas. The SDI project of the U.S. city of Boston,

Massachusetts, is used as a case study. On the basis of the experience in Boston, Kline discusses eleven helpful hints that offer guidance in developing urban SDIs. The last part of the chapter helps readers understand the nature of urban sustainable development, while capturing relevant community aspects in SDIs. Four community characteristics—to which indicators can be linked—emerge as being most important: ecological integrity, economic security, quality of life, and empowerment with responsibility. For each of these topics sustainability pathways are developed, helping to chart the direction for moving a community toward sustainability.

In addition Mark Roseland presents examples of the use of SDIs at the local level. In particular, the following four well-known North American SDI projects are discussed: the Sustainable Seattle Project, the Willapa Indicators for a Sustainable Community, the Hamilton-Wentworth's Sustainable Community Indicators Project, and the Oregon Benchmarks process.

Sustainability Reporting is a tool that supports and helps other sustainability tools and processes. Reporting is an essential feature of any monitoring system. It brings together relevant information to provide a basis for effective, integrated policy-making, action planning, and awareness raising. Sustainability Reporting is discussed in detail by Dimitri Devuyst in Chapter 12. This chapter also deals with sustainability targets. The development of sustainability targets requires compromises among recommendations by panels of experts, scientific information, political considerations, and the sustainable development vision of the local community. The process of target setting and the experience with the use of targets are discussed. Sustainability targets are not only important for measuring progress, but are also necessary in an impact assessment process in which clear statements have to be made on the acceptability and significance of predicted impacts. Not many local authorities have currently take the step to develop sustainability targets, and where they do exist, very little information is available on their effectiveness. Clearly there is a need for more scientific research on the development of sustainability targets.

In Chapter 13 William E. Rees introduces the Ecological Footprint Analysis, an instrument that can be used both for examining the actual sustainability situation and for monitoring progress. Eco-footprint analysis builds on traditional trophic ecology, assessing the amount in square meters of land and water that would be required for a population to produce all resources consumed and assimilate all waste generated. This exercise shows how divergent lifestyles in different cities of the world result in highly variable footprints: poorer developing countries show footprints of less than one hectare per capita, whereas those of wealthy countries approach nine to ten hectares. These figures should be considered, keeping in mind that there are only 1.5 hectares of ecologically productive land and about 0.5 hectares of truly productive ocean for every person on the Earth (the so-called "fair Earth-share" and "fair sea-share"). Through the eco-footprint analysis, Rees shows that cities are not sustainable on their own and are dependent on the rural hinterland for survival. He discusses what this means for the future of our cities. The last part of the chapter examines the use of the eco-footprint analysis in sustainability assessment and discusses the strengths and methodological limitations of the approach.

Part III provides readers with all the elements needed to introduce a functioning monitoring system. At the local level each community should develop its own set of SDIs that reflect the priorities set in the local community vision on sustainable development. Sustainability targets give us an idea of the desirable direction in which collected data ought to move. Sustainability Reporting is a very important element because it aims to transmit essential information, such as the collected data and the assessment of progress, to policy-makers and decision-makers. Monitoring has an important role to play in sustainability assessment because the predictive capabilities of the instrument are not yet well developed. Since it is still almost impossible to quantify the impact of an initiative on the sustainability of society, monitoring can be helpful in measuring the current situation and tracking changes over time.

Chapter 10

Indicators of Sustainable Development

Thomas van Wijngaarden

Summary

The need for a more sustainable development of our societies also creates the need for means to inform upon and measure our progress towards this common goal. One of the large number of initiatives aiming at providing information on sustainable development is the development and use of indicators of sustainable development. Such indicators provide baseline and monitoring information to decision-makers, but also serve to inform the public on achievements towards sustainable development.

This chapter begins with an introduction of the necessity to monitor sustainable development and then provides an overview of the types of indicators of sustainable development, and how they can be categorized. The text elaborates upon the general problem of valuation of the environment, including environmental, economic, and social goods, while developing and using indicators for sustainable development. The requirements for such indicators and their essential features are discussed, which also brings forward the use of indicator development frameworks. Examples of such frameworks and their use at different levels are provided and discussed. Finally, a sustainable development indicator evaluation method is discussed.

Monitoring Sustainable Development

The call for sustainable development and sustainability is heard nowadays in all domains and at all levels of society. Numerous definitions of what sustainable development entails have been made since the Brundtland report was issued (WCED 1987). Sustainable development was defined in this report as "development that meets the needs of the present generation without compromising the ability of future generations to meet their own needs" and has achieved global recognition. During the United Nations Conference on Environment and Development in Rio de Janeiro in 1992, a political effort to address sustainable development, the Agenda 21 action program was created. Following the agreement of this agenda (United Nations 1992), there has been an increased demand for information on sustainable development for deci-sion-making at all levels. There is a clear need for information at all levels in society and geographic settings. Information is, however, only informative if it is understood by the target groups it is meant for. Statistical charts on emissions of greenhouse gases are not very informative data to the urban inhabitant, although they might be very concerned about the problem. However, such charts are information for a scientist modeling the effects on global warming. The need therefore arises to present information in differ-ent forms to make it understandable for the target group and restrictive to the target area. Indicators, which present information by simplifying, select-ing, and aggregating the original data, are most suitable for this purpose. The term *indicator* traces its roots to the Latin verb *indicare*, meaning "to disclose, point out, make public, announce, estimate, or value." An indica-tor provides insight into a matter of larger significance or makes a trend or phenomenon perceptible that is otherwise not immediately detectable. However, it should be kept in mind that indicators represent an empirical model of reality, not reality itself. In Chapter 40 of Agenda 21, the need to develop indicators of sustainable development was formulated as follows (United Nations 1992):

> Indicators of sustainable development need to be developed
> to provide solid basis for decision-making at all levels, and
> to contribute to a self-regulating sustainability of integrated
> environment and development systems.

Thus, in order to guide the process of sustainable development, informa-tion on sustainable development is important for making decisions that are not only based on economic conditions. The environmental, social, and institutional aspects of sustainability should be taken into account as well, and preferably integrated.

Types of Sustainable Development Indicators

Monitoring the sustainability of development goes beyond the collection and analysis of information on resource use, environmental performance, or consumption patterns. Sustainable development is very much linked to decision-making, which requires sustainability targets and criteria. Sound decision-making is based on appropriate information specific to a level and stage of decision-making, which also involves identification of the users and nature of the subject investigated.

There are many possible ways of classifying the various types of informa-tion, and thus indicators all are somewhat artificial. This is illustrated in the following, using three main criteria: the nature of information, its purpose, and what it measures. Subgroups are also discussed.

Classification Following the Nature of the Information

The degree of analysis and complexity of information is variable, as illustrated by the Information Pyramid in Figure 10-1 (World Resource Institute 1995; UNEP/DPCSD 1995).

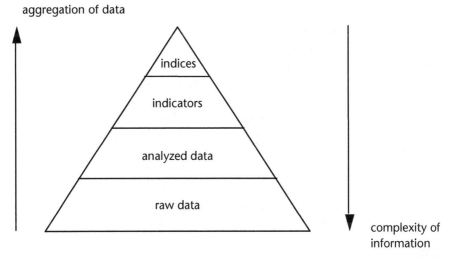

FIGURE 10-1 *The Information Pyramid.*
SOURCE Hammond et al 1995

> ✤ At the bottom of this pyramid are the *primary data*, also referred to as raw data or statistics
>
> ✤ One step higher in the pyramid are *analyzed data*, consisting of analyzed or aggregated primary data, which give a better overview of complex problems and supply decision-makers with more accessible, though still detailed, information
>
> ✤ *Indicators* figure at the next step of the pyramid. They are used to quantify information, show its significance, simplify information on complex phenomena such as sustainable development, and communicate it to potential users, primarily decision-makers. Descriptive indicators, the most common ones, summarize sets of individual measurements of different issues. More aggregated indicators summarize, for example, the findings of cost-benefit and multi-criteria analyses
>
> ✤ At the top of the pyramid are *indices*, which are single factors obtained by aggregating and weighting a number of indicators

The nature of information used depends on the level and stage of the decision-making process. Evaluating policy options often means considering and analyzing a broad range of data and information. Policy implementation involves establishing both broad (policy) goals and specific (technical) targets to be achieved, and/or using and adapting those developed during previous stages of the decision-making cycle. Policy performance needs to be evaluated and the results fed back into the policy-making process. Primary data are used whenever very detailed information is required, typically in the problem identification/awareness raising phase and in the implementation phase. Indicators for policy implementation are either descriptive or aggregated. Aggregated indicators are mainly directed toward specific policy formulation processes. Descriptive indicators provide information on performances achieved in meeting national or international goals, targets and objectives. They are useful for reporting on conditions, trends, and actions that have an impact on sustainable development. Policy evaluation is based on performance indicators and, at high (national) levels of decision-making, on indices.

The higher into the information pyramid, the more room for assumptions, subjectivity, and inaccuracy. A balance must be made between the speed in producing information, its interpretability, accuracy, and reliability, related to the level and stage in the decision-making process (Mortensen 1998).

Classification by Purpose of the Information

The complexity of processes and the overload of information available call for methods to select and simplify information and present the information in another form. Indicators have the function to simplify information about complex phenomena, to quantify information so that its significance is more readily apparent, and to improve the communication of information to the user. In general, terms indicators of sustainable development serve three main objectives:

1. To create agreement about the factors that contribute to the quality of life and the collectivity of human and ecological well-being,

2. To provide a focus and a formal framework for data gathering, data coordination, agreement over methods of collection and interpretation, and mechanisms for continuation, and

3. To generate a policy-integrative approach to administration and political action through which economic, ecological, and social considerations are enriched with shared meanings (O'Riordan 1998).

Indicators basically serve two intertwined purposes: 1) planning: problem identification, allocation of socioeconomic resources and policy assessment and 2) communication: notification (warning), mobilization, and legitimization of policy measures (Kuik and Verbruggen 1991). The development of an environmental index or indicator can follow a scientific method, based mostly on health effects that easily become descriptive or a policy-oriented method, based on target values, preferably the legal standards of environmental quality (De Boer et al. 1991).

Classification by What the Information Measures

Homberg (in Dalal-Clayton 1992) suggests three types of indicators for the measurement of sustainable development: environmental indicators that measure changes in the state of the environment; sustainability indicators that measure the distance between the change in the state of the environment and a sustainable state of the environment; and sustainable development indicators that measure progress toward the broader goal of sustainable development in the national context. The difference between the last two indicators is tenuous, and they are often used interchangeably .

Environmental Indicators

The growing concern for more comparable, comprehensive, and reliable environmental information derives from the growing concern over environmental issues during recent decades. At the national level, many countries and international organizations have initiated "state of the environment" reporting. Initiatives at regional and local level decision-making have followed. The principles and environmental issues of concern considered at these levels are comparable. However, the approach for data collection, priority setting, and dissemination of information can differ substantially, since national level decision-making is mostly organized top-down, whereas local policy concerns often develop in a bottom-up direction.

The goal of environmental indicators is to communicate information about the environment and human activities affecting it, in ways that highlight emerging problems and draw attention to the effectiveness of current policies. Here are some indicators that are currently used for this purpose:

+ Environmental indicators quantitatively describe changes in anthropogenic environmental pressure (pressure indicators) or the state of the environment (environmental effect indicators) (Opschoor and Reijnders 1991)
+ Pressure indicators measure changes in amounts/levels of emissions, discharges, depositions, interventions, and the like. The pressure exerted by society on the environment is commonly categorized as 1) pollution;

2) over-exploitation of resources; and 3) landscape and ecosystems and/or organism modification (biological diversity). Indicators of sustainable development are therefore required for at least each of these three areas

- Environmental effect indicators express the consequences of changes in environmental quality in terms of their effect on specified parameters. For human beings, for instance, effect indicators could include repercussions on the pattern of welfare over time and health indicators

Another way to arrange environmental indicators is by the four interactions between humans and the environment that should be taken into account when monitoring environmental performance:

1. Sources, which include the stocks and flows of natural resources.
2. Sinks, which consider the capacity of ecosystems to absorb pollution and waste.
3. Life support systems, comprising the whole of ecological support systems.
4. Impacts on human welfare and well-being, through pollution pressures on air, water, soil, and so forth.

It is also possible to arrange environmental indicators according to themes or environmental compartments. The compartments could be, for example, air, water, soil, and examples of themes are global warming, acidification, soil contamination, and so on. In this way the indicators can more easily be joined to current policy and focus on current environmental problems.

Sustainable Development and Sustainability Indicators

While environmental indicators provide us with information on the state or development of our environment, sustainable development and sustainability indicators should also inform us on social, economic, and institutional aspects, preferably by linking these aspects. Moreover, they no longer describe a state but inform on progress towards a goal or target, usually defined in a policy, which reflect sustainability criteria for a desired state of society.

Braat (1991) defined sustainability indicators as:

> ...indicators which provide information, directly or indirectly, about the future sustainability of specified levels of social objectives such as material welfare, environmental quality and natural systems amenity.

He defines two types of sustainability indicators: predictive and retrospective. Predictive indicators provide information about the future state and development of relevant socioeconomic and environmental variables. The predictive power results from modeling the human-environmental system, and the information constitutes the basis for anticipatory planning and management. Retrospective indicators provide information on the effectiveness of existing policies or about autonomous developments, and can be sub-divided into policy evaluation indicators and trend indicators. Therefore, there are three options for developing effective sustainability indicators: first, to improve our understanding on the effects of policies and the way they work and system developments, done primarily by extending data collection, so that our responsiveness becomes more precise and more rapid; second, to put effort in modeling in order to create the best possible predictive indicators; and third, a mix of both previous options hoping that results of the one will stimulate the other.

Hammond et al. (1995) phrased three somewhat similar approaches to developing sustainability indicators: 1) through an extension of environmental indicator frameworks into economic and social realms; 2) depart from a few key determinants that reflect the primary policy in environmental, social, and economic domains; and 3) pairing sustainability policy issues with data collection and statistical validity issues.

Whereas environmental indicators are mainly action-oriented, indicators of sustainable development steer us in a different problem-solving approach and help us to reshape the decision-making process as outlined in Chapter 8 of Agenda 21. Sustainable indicators may play an important role in the following action areas:

- Providing a legal, regulatory, and institutional framework
- Making an inventory of the state of environment, development, existing policies and plans
- Adopting flexible and integrative planning approaches that allow the consideration of multiple goals and enable adjustments of changing needs and means
- Monitoring the development process by comparing what has been reviewed to what has been planned
- Cooperating internationally by taking into account both universal principles and differentiated needs and concerns of all countries
- Participating and strengthening the role of nine major groups for moving toward partnership in support of common efforts toward sustainable development
- Reducing the information gap between existing information and availability of data needed to make informed decisions related to environment and development (Gouzee 1998)

Valuation of the Environment

Indicators are selected or developed on the basis of their relevance and feasibility. The higher the aggregation level of indicators, the more crucial these features, as well as the problems of weighting and valuing environmental, social and institutional goods in comparable terms in relation to tradable goods.

Economic Valuation

Many efforts have been made to internalize environmental costs, environmental pollution, or resource depletion into economic accounting. This "Green accounting" has lead to several workable adjustments of macro-economic indicators, such as the Gross National Product (GNP). This GNP measures the value of goods and services produced by a country during a given year, but does not take into account the value of environmental services that were used in the process, nor the current and potential damage resulting from the production process. The Pollution-Adjusted Economy Indicator was developed on the basis of the argument that high carbon dioxide emitting countries use more than their share of the limited capacity of the atmosphere as a sink for their waste, without paying for the use. This indicator is based on the following relationships and variables:

Pollution-Adjusted Economy Indicator = GNP × Average/Actual per capita carbon dioxide emissions. The average per capita carbon dioxide emissions is the world median; the actual per capita emissions is the estimated carbon dioxide emissions per total population.

The first step in covering potential costs or damage of an activity or in finding a balance between environmental loss and economic gain is to express all goods in the same currency; that is, for economic valuation in monetary terms. They can then be compared or aggregated in economic indicators. This is particularly important to compensate for market failures observed where environmental costs are not internalized. Various methods are available: market values, surrogate or estimated prices/shadow pricing, contingent valuation, and macro-economic models (Post et al. 1998). The principle and methodologies of economic valuation of the environment remain, however, controversial from a practical, social, and ethical point of view (Brown 1995a; Winpenny 1997). The main criticisms expressed are:

* ➷ Monetary valuation treats all assets as tradable commodities and thereby undermines other important social values and interests (Brown 1995b)
* ➷ This, in turn, will not favor environmental protection since environmental goods become consumption products

➤ It is also against the principle of equity in sustainable development: the rich can purchase environmental goods from the poor or compensate them for the use and destruction of their environment (e.g. emission trading, estimated pricing)

➤ Valuation of environmental goods is highly subjective: people are willing to pay only for some environmental goods. Moreover, only existing goods are taken into account (contingent valuation method, replacement cost technique)

Ecological Valuation

Ecological valuation of the environment can be done by defining the *carrying capacity* as the human load that a specific system is able to sustain without depletion of its natural resources. The ecological footprint analysis (Wackernagel and Yount 1998) is an area-based indicator quantifying the relationship between actual human load and carrying capacity of a defined region. The currency of this environmental accounting method is the "ecologically productive area," and sustainability is a zero-sum game.

Another method of environmental accounting is the *energy analysis* (Campbell 1998). All environmental, economic, and social resources are produced as a result of energy transformation. Therefore, the energy required for their production can be valued in common terms by converting their energy values into "emergy" (embodied energy, or energy memory). Emergy is used to compare the economy of a region and its supporting environment.

Another example is the *total material consumption*, based on the determination of the ecological rucksack of a product. The sum of the materials used and displaced to produce a good, diminished by the mass of this good, gives insight into the environmental consequences of such production. If an intermediate product is used to produce another good, these data are used as the partial ecological rucksack of the new product. This methodology could be extended to activities as well.

Other valuation methods, based on the importance of *biodiversity*, assess the vulnerability of an ecological system or express the importance of a species or ecological system in terms of its rarity.

Social Valuation

The *social valuation* of environmental and economic goods has so far not received much attention. Examples can be found in sentimental valuation of such goods, and cultural and social norms and values. Lifestyles and consumption patterns can be seen as a measure for the valuation of economic goods and, to a lesser extent, of environmental goods. For some disciplines in environmental impact assessment, such as landscape and cultural heritage, the social or average personal value is sometimes the only value that can be attributed to environmental goods.

Requirements for Indicators of Sustainable Development and Essential Features of Indicators

Indicators are nothing more or less than indicators: they cannot replace a policy, but can only report on its success or failure; they cannot set targets or goals, but only measure our achievements in reaching them. They report only on measurable aspects of the parameters investigated. The availability, the use, or even the favorable evolution of indicators, does not necessarily imply the success of a policy since there is a hidden danger of being blinded by results and vested interests selectively using indicators for sustainable development. Eckman (1993) suggested that measuring "unsustainability" could be less subjective.

The following are some essential features of indicators:

➻ Indicators must be *relevant*, serving their purpose, capturing and measuring the essence of the issue. They must be *understandable* for all members of the target group. Complex indicators developed by academics to measure complex processes or situations lose their relevance unless simplified to be understandable for everyone. Their *conception* must be well founded and easy to interpret. They should be easily *adaptable* to new developments, thus *responsive*. They need to show the *link* between economic, social, environmental, and institutional aspects of society. They should embrace a *long- and wide-range* view and *show trends* over time. Sustainability is indeed defined as long-term community health. Indicators of sustainable development should be *global*: sustainable community indicators should not measure local sustainability at the expenses of a more distant or global community. They must be *reliable* (based on reliable data) although not necessarily precise. *Data* should be available at a reasonable cost-benefit ratio, adequately documented, of known quality, and regularly updated. They must provide information in a timely fashion, so as to be able to prevent or solve problems in due time. The three last features could be summarized in terms of "feasibility": are they appropriately measurable in a time- and cost-effective way? They should also have a *threshold* or *reference value* against which progress could be measured (UN 1996; OECD 1994; Hart 1996, Atkinson *et al.* 1997; Mitchell 1996; Hardi and Zdan 1997)

✦ Indicators should have a clear *use,* specifying *targets to achieve* and *target groups.* Policy indicators must show where society is currently heading, where it would go if, in the course of time, development trends are corrected, and where it should go (reference situation for the future). When referring to sustainable development and sustainability, principles retained need to be specified (Mitchell *et al.* 1995) Braat (1991) pointed out that there is no one sustainable future, but many different ones that combine different levels of the various key variables (population, income, environmental quality). A set of acceptable sustainable futures must be selected from the many possible ones. It is impossible to set a single target, and the use of scenarios seems inevitable, guiding communities and individuals for the choice of indicators. Much of the human behavior is indeed *adapted behavior,* based on previous experience; individuals constantly use scenarios, consciously or not. Our goals in life and our perspectives are constantly re-adjusted targets; the road to attain these goals (our lifestyle) is our policy; and we use reference points and measures, the indicators, to assess our progress. At the local level, community indicators are mainly used for measuring progress, explaining sustainability, educating the community, showing linkages, commitment, and focusing on action. Three main target groups are identified among potential users of sustainable development indicators (SDIs): 1) citizens, that is, the "public, " want and need to have access to understandable information on developments in their environment; 2) policy- and decision-makers need SDIs to weigh options, policy adjustment and development, and for planning and strategy development; and 3) scientists require indicators of a completely different kind and level, either for feeding data into the ultimate index or indicator, or as a framework for investigation and data collection. For the latter group, accuracy will be more important than globality

Although progress in developing indicators for sustainable development has been rapid over the past few years, several constraints for further development remain:

✦ Collected information is still too compartmentalized, with no linkages between data sets (and often incomplete data)
✦ A lack of integration exists between environmental integrity, social well-being, and economic prosperity
✦ The conceptual uncertainty still requires new frameworks

➻ Institutional capacity constraints remain

➻ Barriers exist to information availability, including lack of freedom to access, inadequate or incomplete monitoring and monitoring techniques, data presentation in a form unintelligible to the lay person, and financial resource constraints preventing expansion or clarification

➻ There is a lack of methodologies to use non-quantifiable data and valuation methods for comparison of environmental, social, and economic capital

➻ The global validity of indicators is unclear

Frameworks for Indicator Development

Several approaches have been used to create development frameworks for sustainable development indicators. They can be distinguished in classification by use (developed for specific target groups or a geographic area), by subject, or by causality chains. Another way to classify the indicator development framework is by noting whether they follow a static or a dynamic approach.

The geographic level of application of a framework and target group can be determined at a global, national/regional, and at a local level. Other frameworks arrange their indicators by subject or issues, such as the driving-force-state-response framework of the United Nations (United Nations 1996). The issues maintained here are the four larger issues that have to be addressed in sustainable development: social, economic, environmental, and institutional issues. Another option is to arrange by themes or problem and apply a causal chain throughout the indicators, reporting on the force creating the problem, the state that the problem has created, and the action undertaken to ameliorate the situation. An example of this type of framework is the pressure-state-response framework of the OECD (OECD 1995).

Conceptual Frameworks

The development and use of a conceptual framework that will guide our assessment of progress toward sustainable development is crucial. Indicators can then emerge more naturally, adjustable to the needs of a given system or a set of decision-makers (Hardi and Zdan 1997). A framework provides the analytical context for the formulation, use and presentation of measures of sustainable development, combining or integrating the various aspects of it. An effective framework achieves two goals: it helps 1) prioritize the choice of indicators and 2) identify indicators that may become more important in the future. The distinction between frameworks for indicator formulation and evaluation is somewhat artificial because many frameworks can be used for both. The PICABUE method (Mitchell et al. 1995), originally developed for modeling sustainable development and related indicators in urban areas, is more generally applicable to a wide range of areas and user groups. A more precise description of the methodology is given in the box below.

The PICABUE Methodological Framework for the Construction of Indicators of Sustainable Development

In order to be consistent with widely accepted definitions of sustainable development, considerations relating to the measurement of quality of life and ecological integrity are central to this methodology. The methodological framework has relevance to a variety of spatial scales and to geographically diverse areas (urban and rural, developed or developing countries). Thus, a suite of sustainability indicators can be produced that is tailored to the needs and resources of the indicator user, but which remains rooted in the fundamental principles of sustainable development.

The PICABUE method derives its name from seven principal steps used in the development of sustainability indicators:

1. **P**rinciples and definitions of sustainable development and the objectives of the program in which indicators are developed must be agreed up on by stakeholders;

2. **I**ssues of concern must be identified and selected;

3. **C**onstruct and/or select indicators of issues of concern;

4. **A**ugment indicators for step 3 by the principles of sustainable development identified in step 1;

5. Address **B**oundary issues to modify indicators from step 4;

6. Develop **U**ncertainty indicators from step 4 augmented indicators;

7. **E**valuate and review final sustainability indicators.

To produce strong sustainability indicators, the indicators that are first developed or already in use, and which are often descriptive on the sustainable development concerns, should be augmented with the principles of sustainability identified in the first step. These principles entail social equity (intra-generational equity, identifying issues important to specific social groups), futurity (inter-generational equity, focusing on resource use related to resource limits), and environment (conservation of ecological system integrity, investigating on the limits present in systems which should not be crossed). Thus, progress made to attain goals and objectives or the desired sustainability can be communicated. A fourth principle, public participation, is respected through the process of selecting concerns for indicators (Steps 2 and 3), and in the manner in which indicators are finally used.

Special boundary considerations should be maintained to spatial units to which the indicators relate. Transboundary flows or effects should be integrated and are very different for e.g. an urban indicator compared to an indicator designed for a national territory.

Uncertainty in indicators arises from three causes: limited knowledge on critical system limits; low reliability of data-sets; and unpredictability behavior in systems. Integrating levels of uncertainty into the indicator set respects the pre-cautionary principle.

The sustainability indicators should be evaluated on: their relevance and scientific validity; their sensitivity to be able to change across space, time, and/or groups; data used should be consistent, should be measurable, and undergo appropriate transformation; and target or threshold values should be incorporated as far as possible.

SOURCE Mitchell et al. (1995)

Any SDI program should include the following steps of the method (Mitchell 1996):

1. Clearly define objectives of the indicators' program, specifying the purpose of the indicators and their user group.

2. Define what is understood by sustainable development and the sustainability criteria applied.

3. Define the issues that are important both locally and globally.

4. Evaluate the indicators against desirable indicator features, their objectives, and their target(s).

Five groups of models on which frameworks can be based are: i) models with roots in economics; ii) stress and stress-response models; iii) multiple capital models; iv) various forms of the three themes "social-economic-environment" model; and v) linked "human-ecosystem-well-being" models. The first two models are partial system models, and the last three are full system models, which try to capture all aspects of the system.

Mortensen (1998) distinguished SDI frameworks based on a static or dynamic approach.

Static Frameworks

Static frameworks are most commonly used in developing measures of sustainability. They focus on environmental pressures and actions undertaken to cope with or diminish such pressures at a given time.

The first widely used static framework is the *Pressure-State-Response* (PSR) framework of the Organization for Economic Cooperation and Development (OECD 1995). It consists of a core list of approximately 75 environmental indicators and is based on the concept of causality. Human activities exert *pressure* on the environment and its natural resources, inducing changes in both quality and quantity of the environment (*state*). The society's *response* to these changes consists of environmental, economic, and sector policies. These PSR indicators are structured along environmental issues, such as climate change, waste, and soil degradation.

A further variant of the PSR framework is the *Driving Force-State-Response* (DSR) framework of the Commission for Sustainable Development of the United Nations (United Nations 1996). The DSR framework is not based on causality but focuses more on sustainable development by addressing social, economic, environmental, and institutional indicators. The program to develop this framework is composed of three phases. First, from May 1995 to April 1996, a set of 132 indicators was developed (55 environmental, 39 social, 23 economic, and 15 institutional), all on the basis of a standard methodological sheet. The subdivision by driving-force-state-response indicators results in 42, 53, and 37 indicators respectively. The period from May 1996 to December 1997 was used for improvement of the indicator set and implementation trials. The third phase, from January 1998 to January 2000, will be used to adjust and complete the set in order to compile a final list of sustainable development indicators. The objectives of the program are to use this set of indicators as a common international language in regard to information on sustainable development, and to give assistance in national decision-making (Federal Council for Sustainable Development 1997). Sixteen countries committed to participate in the second phase, the implementation trials.

The Federal Council for Sustainable Development (1997) of Belgium, in charge of screening the list of indicators for their suitability and applicability in the current Belgian context, concluded that until now, only few of these indicators could be utilized directly. The major bottleneck was the availability of suitable data.

In France, data are available for approximately 100 indicators, 58 of which are calculated on a regular basis. However, numerous methodological differences between the national system and the United Nations methodology were observed. Main constraints to the framework are 1) the vague definition of sustainable development, as set out in the Brundtland report; 2) the framework structure, which leaves little room for integrating the three dimensions (economic, social, and environmental) of sustainable development (IFEN 1998); and 3) an orientation toward problem solving, rather than problem prevention, which should be the most important feature of sustainable development.

Eurostat, the statistics department to the European Commission, made a similar effort to apply the United Nations framework at the European Union level, the E.U. being a member of the expert group for implementation trials. Out of the 132 indicators, 21 environmental, 14 social, 9 economic, and 2 institutional indicators were identified to be usable directly. For the use of other indicators, four kinds of obstacles were encountered by Eurostat in selecting SDI's: Some indicators are relevant at the global level, but not at the European Union level; EU member states have different policies implemented on certain issues, which creates problems for the analysis of statistical data; EU statistical data are sometimes not comparable or uniform; and, the accessibility and availability of EU statistical data could be improved for some indicators (European Communities 1997).

The Organization for Economic Cooperation and Development (OECD) uses the driving force-state-response framework for the development of sector specific indicator sets, in agriculture, for example (OECD 1997). To assist in the choice of an operational set of environmental indicators within the framework, each indicator is examined against four general criteria: policy relevance, analytical soundness, measurability, and the level of aggregation. Currently, issues of major relevance to policy makers in OECD countries are soil, water, air, nature, farm financial, and socio-cultural issues. Priority agri-environmental issues for which relevant indicators are being developed include nutrients, pesticide and water use, land-use and conservation, soil and water quality, greenhouse gases, biodiversity, wildlife habitats, landscape, farm management, farm financial resources, and socio-cultural issues. The indicators should be able to quantify the components and issues described in the DSR framework, and agriculture should be a significant component in relation to the issue. With respect to the analytical soundness of the indicator, this concerns the extent to which the indicator can establish and explain links between agricultural activities and environmental conditions and is able to show trends and ranges of values over time, which might be complemented by existing nationally defined targets and thresholds. The "measurability" criterion is mainly determined by data availability. The "level of aggregation" criterion highlights the encapsulation of the spatial and temporal diversity of the environment and the geographical scale of different environmental issues ranging from the farm trough to the global scale.

Static frameworks are very useful for decision-making and facilitate the further development of measures of sustainable development. But because of their weaknesses (applicable only to the past or present, poor ability to analyze linkages), dynamic frameworks have been developed that create the possibility to examine linkages, projections, and scenarios.

Dynamic Frameworks

Dynamic frameworks depart from a systematic modeling of all issues of concern and their linkages, when examining a phenomenon, a problem, or social structures. Integrating sustainability criteria will provide solutions and alternatives toward more sustainable development based on the past and the present. Dynamic frameworks have the advantage that they show

how various measures are interlinked within each issue (cause-effect) and between different issues. They provide insight into the dynamic behavior of measures and enable long-term projections of sustainable development. They also identify critical system variables and provide guidance on how the measures should be selected and aggregated. However, due to the complexity of the subject or issue for which such models are developed, the number of assumptions and simplifications that have to be made indicate that results should be treated with care when used in decision-making (Mortensen 1998). Dynamic models have been developed for numerous applications, but until now, have focused mainly on environmental-economic linkages.

Geographic Application of Frameworks

Apart from dividing frameworks into static and dynamic properties, it is also interesting to distinguish between the geographic application frameworks for which they are designed. The frameworks of the OECD and the United Nations, which were discussed previously, are clearly designed for national and international use.

National/Regional Frameworks

The Department of the Environment of the United Kingdom published a set of indicators of sustainable development for this country (Department of the Environment 1996). This set is set up to monitor progress toward the objectives set out in the Sustainable Development Strategy, published by the government in 1994. The framework specifies, first, broad goals such as "non-renewable resources should be used optimally" or "damage to the carrying capacity of the environment and the risk to human health and biodiversity from the effects of human activity should be minimized." Second, broad environmental issues are identified for each of these aims. For example, the goal of "non-renewable resources should be used optimally," leads to the necessity of identifying "energy, land-use, water resources, forestry and fish resources." Third, the key objectives and issues are specified. For example, for the broad issue of fish resources management, a key objective and issue is "to manage the fishing industry to prevent over-exploitation of fish stocks." Finally, key indicators are provided for fish resource management; that is, maintaining fish stocks, minimum biological acceptable levels, and fish catches. Over 100 indicators are listed, of which sources, references, and details are elaborated in the annex of the report by cross-cutting issues.

In France, the Institute for Environment and the Ministry of the Environment published a list of 31 indicators. These indicators are related to air, water, soil, nature, biodiversity and landscape, waste, noise, risks, and natural resources. The environmental performance monitored by these indicators is related to objectives, norms, and references deriving from political commitments, jurisdiction, or scientific research.

Currently, most national environmental policy plans underline the concept of sustainable development and thus require indicators for it. The last Dutch Environmental Policy Plan was based on scientific research, and illustrated three possible scenarios as to how the Dutch society might develop (RIVM 1997). These three scenarios differ in assumptions concerning global economic development, demographics, socio-cultural and technology development. The scenarios (Divided Europe, European Coordination, and Global Competition) all give a different prospect for the Dutch economy towards the year 2020. National development concerning the environment is foreseen by action, impact, and environmental quality indicators, and results are related to environmental policy targets. These results help to further specify environmental goals and make specific recommendations for policy intensification. Compared to the three other frameworks discussed previously, which were very much related to the static approaches of the OECD and United Nations, this Dutch experience can also be classified as a dynamic approach.

Local Setting

Local initiatives for indicator frameworks are numerous. Indicator sets have been developed in particular for urban environments, mostly for the implementation of Local Agenda 21. An interesting initiative by the Local Government Management Board of the United Kingdom (1996) provided a practical guide for UK local authorities for sustainability reporting. The force of this initiative is that it provides a methodology that might be adopted by all local authorities in the country. In the United Kingdom a framework of Indicators for Sustainable Development was developed in 1996. This consists of a very useful list of indicators, many of which are directly or indirectly linked to mobility issues (Department of the Environment 1996). An example of these indicators is given in Table 10-1.

This contrasts with many other initiatives seen in cities or by local authorities throughout the world. Unfortunately, little methodological harmony is used in measuring sustainable development between initiatives like Sustainable Seattle, Valencia, and others

Evaluating Indicators

In 1996, Hart developed a checklist for evaluating indicators. An indicator is a lot like a compass: it points out the direction to move. But, just like a compass, it is important that an indicator be properly calibrated—that it really points in the right direction. If not, it can lead you somewhere you had no intention of going. The checklist in the box on page 270 is a way to calibrate sustainable community indicators based on a set of characteristics shared by all good sustainability indicators.

TABLE 10-1 *Indicators of Sustainable Development for the United Kingdom, Selected by Relevance to Mobility Issues*

A	The economy	K	Climate change
a4	consumer expenditure	k3	emissions of greenhouse gasses
a8	pollution abatement		
		M	Acid deposition
B	Transport use	k3	road transport emissions of nitrogen oxides
b1	car use and total passenger travel		
b2	short journeys	N	Air
b3	real changes in cost of transport	n1	ozone concentrations
b4	freight traffic	n2	nitrogen dioxide concentration
		n3	particulate matter concentrations
C	Leisure and tourism	n4	volatile organic compound emissions
c1	leisure journeys	n5	carbon monoxide emissions
c2	air travel	n6	black smoke emissions
		n7	lead emissions
E	Energy	n8	expenditure on air pollution abatement
e1	depletion of fossil fuels		
e6	road transport energy use	R	Wildlife and habitats
e8	fuel prices in real terms	r6	habitat fragmentation
F	Land use		
f1	land covered by urban development		
f2	household numbers		
f5	road building		
f6	out of town retail floor space		
f7	regular journeys		
f8	regeneration expenditure		

SOURCE Department of the Environment (1996)

Sustainable Community Indicator Checklist

Does the indicator address the carrying capacity of the community's natural capital?

Natural capital relates to the natural environment on which the community relies. There are three parts to natural capital: the natural resources (both renewable and non-renewable) used by the community, whether local or from distant sources; the services that the ecosystem provides such as wetlands absorbing floodwaters and trees providing windbreaks and shade; and the beauty of nature such as the view from a mountain or seashore, the song of a bird, the peaceful feeling of a park.

Does the indicator address the carrying capacity of the community's social capital?

Social capital relates to people in a society. There are two parts to social capital: (i) skills, education, health and natural abilities of the people in the community; and (ii) connections between people in a community, relationships of families, friends, neighbors and their ability to cooperate and work together.

Does the indicator address the carrying capacity of the community's built capital?

Built capital relates to items created by humans. This includes manufactured goods, equipment, buildings, roads, water supply systems, information resources, and the credit and debt of a community. It is the result of natural and social capital but it also needs to be maintained and enhanced. Capital has to do with the ability of the economic system to provide materials and employment while absorbing any shocks to the system.

Understandable

Is the indicator understandable to the community at large? If it is only understood by experts, it will only be used by experts.

Long-term view

Does the indicator provide a long-term view of the community?

Is there a goal that can be defined for the indicator that fits a sustainable view of the world?

Linkages

Does the indicator link the different areas of the community? Areas to link are: cultural/social, economy, education, energy, environment, governance, health, housing, land use, population, public safety, recreation, resource consumption/use, and transportation (half a point per link, up to seven).

Preserve global sustainability

Does the indicator focus on local sustainability that is achieved at the expense of global sustainability? If the answer to this question is "yes," then the indicator is automatically disqualified.

SOURCE Hart (1996)

Two key components of sustainability are the concept of *community capital* and *carrying capacity*. Community capital is all that a community has that allows its inhabitants to live and interact productively. There are three components to community capital: natural capital, social capital, and built capital. Carrying capacity is the ability of a community's capital to provide for the community's needs on the long-term. "Good" indicators of sustainability will address whether a community is maintaining and enhancing the capital on which it depends.

The Hart checklist is designed to identify indicators that can, in general, be considered "good" indicators of sustainability. However, just because an indicator scores high on the checklist does not mean it is right for every community. The number of salmon is relevant in Seattle, but not in Arizona. The number of subway riders is relevant in urban areas, but useless in rural areas. Each community must decide if a particular indicator is relevant to its own situation and whether there is reliable data for that indicator.

Conclusion

Indicators for sustainable development are very useful in promoting the sustainable development of our societies. However, they should be designed and used properly. Ideally, the design takes place within a methodological framework, adapted to the geographic and social setting in which the indicator set will be used. Coherence in indicator development methodology should be a priority when indicator sets overlap in their settings or application level. For example, a national indicator set and several local sets that are all developed by different methodologies by the respective national and local authorities concerned does not improve communication between these authorities, learning from each others' experiences, and compatibility of results. The development of a universal set of indicators or indices for sustainable development is ideal in national planning and international comparability. However, the huge differences in geographical and economic, as well as socio-cultural, settings are constraints that are difficult to overcome for developing more specific global indicators.

Apart from a uniform methodology for development, indicators for sustainable development should possess many common features. A long-term time frame, integration of social, environmental and economic issues, and integrating sustainability criteria are important features to respect.

Integrating and respecting sustainability criteria provide a link to policies and decision-making. A uniform perception of the concept of sustainable development is not a necessity for developing indicator frameworks, as long as it is defined in the context of the framework. Data is much more accessible in developed countries than in developing countries, and thus is an obstacle to the use of complete sets of indicators. Effective and appropriate mechanisms for the collection and treatment of reliable data should be further developed.

Valuation of the environment and social capital is important to further develop sustainable development indicators and the integration of social, economic, environmental, and institutional aspects. Economic valuation methods, or the expression of environmental or social capital in monetary terms, still receive the most attention, although they are still viable for improvement. Other valuation methods deserve at least as much attention.

References

Atkinson G., R. Dubourg, K. Hamilton, M. Munasighe, D. Pearce and C. Young. 1997. *Measuring Sustainable Development: Macroeconomics and the Environment.* pp. 252. Cheltenham: Edward Elgar .

Braat L. 1991. The predictive meaning of sustainability indicators. In Kuik O. and H. Verbruggen, eds., *In Search of Indicators of Sustainable Development,* pp. 57-70. Dordrecht: Kluwer Academic Publishers.

Brown, D. A. 1995a. Role of ethics in sustainable development and environmental protection decision-making. In J. Lemons and D. A. Brown, eds., *Sustainable Development: Science, Ethics, and Public Policy,* pp. 39-51. Dordrecht: Kluwer Academic Publishers.

Brown, D. A. 1995b. Role of economics in sustainable development and environmental protection. In J. Lemons and D.A. Brown, eds., *Sustainable Development: Science, Ethics, and Public Policy,* pp. 52-63. Dordrecht: Kluwer Academic Publishers.

Campbell, D. E. 1998. Emergy analysis of human carrying capacity and regional sustainability: An example using the state of Maine." *Environmental Monitoring and Assessment* 51: 531-569.

Dalal-Clayton, B. 1992. Modified EIA and indicators of sustainability: First steps towards sustainability analysis. In *Proceedings of the Twelfth Annual Meeting of the International Association for Impact Assessment,* 19-22 August 1992. Washington, D.C.: IAIA.

De Boer J., H. Aiking, E. Lammers., V. Sol and J. Feenstra. 1991. Contours of an integrated environmental index for application in land-use zoning. In O. Kuik and H. Verbruggen, eds., *In Search of Indicators of Sustainable Development,* pp. 107-120. Dordrecht: Kluwer Academic Publishers.

Department of the Environment. 1996. *Indicators of Sustainable Development for the United Kingdom, London:* Department of the Environment, the Government Statistical Service, HMSO.

Eckman, K. 1993. Using indicators of unsustainability in development programs. *Impact Assessment* 11: 275-285.

European Communities. 1997. *Indicators of Sustainable Development. A Pilot Study Following the Methodology of the United Nations Commission on Sustainable Development,* p. 128. Luxembourg: Office for Official Publications of the European Communities.

Federal Council for Sustainable Development. 1997. *Rapport sur la première année de test des indicateurs pour un développement durable en Belgique et propositions pour la coordination et l'efficacité du processus.* Brussels: Belgian Federal Council for Sustainable Development.

Gouzee 1998. *Urban Indicators: Decision-making Tools?* In Proceedings of the Workshop of the Brussels Institute for Management of the Environment (IBGE - BIM), 23 October, Brussels, Belgium.

Hammond A., A. Adriaanse, E. Rodenburg, D. Bryant and R. Woodward. 1995. *Environmental Indicators: A Systematic Approach to Measuring and Reporting on Environmental Policy Performance in the Context of Sustainable Development*. Washington D.C.: World Resources Institute.

Hardi P. and T. J. Zdan, eds. 1997. *Assessing Sustainable Development*. Winnipeg: International Institute for Sustainable Development.

Hart, M. 1996. *Sustainable Community Indicators. A Trainers' Workshop*. Massachusetts: Hart Environmental Data.

IFEN, Institut Français de l'Environnement. 1998. Work undertaken by the French Institute for Environment (IFEN) on sustainable development indicators. In *Proceedings of the Conference Beyond Sustainability,* 19-20 November 1998. Amsterdam: Nederlands Fonds voor Wetenschappelijk Onderzoek (NWO).

Kuik O. and H. Verbruggen, eds. 1991. *In Search of Indicators of Sustainable Development*, pp.128. Dordrecht: Kluwer Academic Publishers.

Local Government Management Board of the United Kingdom. 1996. *A Practical Guide for UK Local Authorities for Sustainability Reporting*. London: The Local Government Management Board of the United Kingdom.

Mitchell, G., A. D. May and A. T. McDonald. 1995. PICABUE: A methodological framework for the development of indicators of sustainable development. *International Journal of Sustainable Development and World Ecology* 2: 104-123.

Mitchell, G. 1996. Problems and fundamentals of sustainable development indicators. *Sustainable Development* 4: 1-11.

Mortensen, L. F. 1998. Measuring sustainable development. In B. Nath, L. Hens, P. Compton and D. Devuyst, eds., *Environmental Management in Practice Vol I Instruments for Environmental Management*. London, New York: Routledge.

OECD, Organization for Economic Cooperation and Development. 1995. *Environmental Indicators*. Paris: OECD.

OECD, Organization for Economic Cooperation and Development. 1997. *Environmental Indicators for Agriculture*. Paris: OECD.

Opschoor H. and L. Reijnders. 1991. Introduction. In O. Kuik and H. Verbruggen, eds, *In Search of Indicators of Sustainable Development*. Dordrecht: Kluwer Academic Publishers.

O'Riordan. 1998. Sustainability indicators as indicators of sustainability. In *Proceedings of the Advanced Study Course on Indicators for Sustainable Urban Development*, 5-12 July1997. Delft: The International Institute for the Urban Environment.

Post R. A. M., A. J. Kolhoff and B. J. A. M. Velthuyse. 1998. Towards integration of assessments, with reference to integrated water management projects in Third World countries. *Impact Assessment and Project Appraisal* 16, 1: 49-53.

RIVM, Rijksinstituut voor Volksgezondheid en Milieuhygiëne, the Netherlands. 1997. *Nationale Milieuverkenning 1997-2020*, p.262. Bilthoven: Rijksinstituut voor Volksgezondheid en Milieuhygiëne.

Rodenburg E., D. Tunstall and F. Van Bolhuis. 1995. *Environmental Indicators for Global Cooperation, Working Paper 11*. Washington D.C.: The World Bank, UNDP, and UNEP.

UNEP/DPCSD, United Nations Environmental Program/Department for Policy Coordination and Sustainable Development. 1995. The role of indicators in decision-making. In *Discussion paper for the workshop on Indicators of Sustainable Development for Decision-Making,* January 1995. Ghent.

United Nations. 1992. *Agenda 21: Program for the Further Implementation of Agenda 21*. New York: United Nations Conference on Environment and Development.

United Nations. 1996. *Indicators of Sustainable Development Framework and Methodologies.* New York: United Nations.

Wackernagel, M. and J. D. Yount. 1998. The ecological footprint: An indicator of progress towards regional sustainability. *Environmental Monitoring and Assessment* 51: 511-529.

WCED, World Commission on Environment and Development. 1987. *Our Common Future.* Oxford: OUP.

Winpenny, J. 1997. Economic valuation of environmental impacts: The temptations of EVE. In C. Kirkpatrick and N. Lee, eds., *Sustainable Development in a Developing World: Integrating Socioeconomic and Environmental Appraisal,* p. 112-124. Cheltenham: Edward Elgar.

World Resource Institute. 1995. *Environmental Indicators—A Systematic Approach to Measuring and Reporting on Environmental Policy Performance in the Context of Sustainable Development.* Washington, D.C.: World Resource Institute.

Chapter 11

Indicators for Sustainable Development in Urban Areas

Elizabeth Kline

Summary

Many people are more concerned about daily survival than meeting their own and their families' long-term needs and dreams. Of necessity, they worry about getting a minimum wage job, escaping violence, drinking unpolluted water, coping with sickness, and hoping that their children will get a good enough education to better their circumstances. How, then, can indicators be relevant to their lives?

The basic test for the appropriateness of specific indicators is that they make sense to the affected persons. Because circumstances vary person-to-person and place-to-place, the development and identification of indicators need to be tailored carefully to resonate with people in language that they understand, in a time frame consistent with their horizons, focused on issues that have significant meaning, and can be applied in ways that help produce tangible improvements in their lives.

This chapter provides an overview of urban sustainable development indicators, describing their multiple values and purposes; identifies key ingredients and a framework for evolving relevant and practical indicators; and concludes by considering the linkage between sustainable developments and environmental assessments.

Throughout this chapter references will be made to the city of Boston, Massachusetts' Indicators Project. The study of one case example allows readers to gain a more in-depth appreciation of choices considered and decisions made.

Background

Boston, the state capital of Massachusetts, is located along the northeastern coast of United States and is considered to be the economic center for New England. The picture emerging from existing data and initial analyses is of a vibrant city, with a thriving downtown and a city as a whole "poised for prosperity in the new global economy," but challenged to find "new ways to push this prosperity into all our neigh-borhoods." Its 575,000 residents live within only 122 square kilometers in distinct neighborhoods (Boston Indicators Project 1998: 5).

Boston's economy has changed dramatically during the last century from being dependent on manufacturing to one based on a variety of economic engines including finance, services, health care, and education. More than 650,000 jobs are located within the city's limits, 275,000 of which are held by residents. While wages have risen steadily for highly skilled workers, incomes for low-skilled workers have fallen. White Bostonians' per capita income (as of 1990) was double that of blacks, twice that of Asians, and more than double of Hispanics. More than 90 percent of the public school children are of color. High housing prices and loss of rent control have reduced the number of moderate income apartments and homes.

The *Indicators of Progress, Change, and Sustainability for an Improved Quality of Life in Boston* project, co-sponsored by the City's Sustainable Boston Initiative and the Boston Foundation's Community Building Network, was initiated in January 1997 as a project of the Boston civic community. It is an outgrowth of the Sustainable Boston Initiative, a multi-year education and community outreach process. The project comprises four steps: (1) identify indicators through a broad civic process; 2) develop neighborhood-level and citywide data to support measurement of these indicators; (3) set neighborhood-level and citywide goals for driving change in several key indicators by the end of the year 2000, 2005, and 2030 (Boston's 400th anni-versary); and (4) work with communities to achieve the first set of goals.

This chapter was written during phase 2 of the project. Most of the refer-ences in this chapter are taken from an in progress draft document entitled *Indicators of Progress, Change, and Sustainability for an Improved Quality of Life in Boston* and dated May 1998 (Boston Indicators Project 1998). Phase 1, identifying indicators, began with public forums to get people acquainted, involved, and talking. A list of ten categories was created from these discus-sions: Jobs and Economy; Civic Health; Environment; Housing; Health; Education; Transportation; Arts, Culture, and Leisure; Public Safety; and Information Technology, Communications and Media. A series of meetings was initiated under the sponsorship of the Boston Foundation and the City of Boston with the goal of developing indicators. At the first meeting in Jan-uary 1997 there were 12 participants. Now, the number ranges from 30 to 50 people for the two-hour lunchtime sessions. Indicators and specific mea-surements for each indicator have been selected, reviewed, and revised in group discussions and through committees established for each category topic. Data experts have joined in to help determine which indicators and measurements are feasible and desirable, and to begin acquiring data bases.

Elaborate plans are underway to link this project and its findings with city and regional organizations' programs. For example, the Metropolitan Planning Area Council is helping to sponsor three seminars focusing on local issues, such as housing, environment, and workforce development, which necessitate regional responses. A Metropolitan Data Center is being established to house the Boston Indicators Project, as well as to provide data, education, and training. The final Boston indicator's report was issued in October 2000 by the Boston Foundation at a major public event. The "Citizens Seminar," a program of the School of Management at Boston College since 1954, will convene gatherings every two years through the year 2030 and then link with the Boston 400 Project's celebration of the city's 400th anniversary.

Value of Sustainable Development Indicators

Even if indicators are understandable and sensible, what makes them valuable? This section describes practical advantages for people discussing, debating, selecting, and applying sustainable development indicators. The next section describes three particular purposes for using such indicators to guide development decisions in urban areas.

As earlier chapters have detailed, indicators in themselves are tools and not end products. Indicators are a vehicle for guiding people's understanding of their community, articulating and weighing options, and helping them make strategic decisions.

The bottom-line desired outcome for most people is to gain more control over their lives. If indicators can help people not only cope with change, but take advantage of changing economic, political, social, and environmental forces to strengthen their own personal lives and further their community's interests, then the indicators will be valued and used. The Boston Indicators Project expresses this fundamental need in its *Guiding Consensus*: "We want to strengthen communities and change the way society functions" (Boston Indicators Project 1998: cover page). Instead of using indicators merely to catalogue impacts, evaluate current conditions, and/or respond to crises, Boston aims to channel development and change.

This proactive strategy is based on choosing sustainable development indicators that reflect community values Indicators that are derived from an extensive, community-based process and are consciously tailored to further commonly held community beliefs, traditions and heritage, and incorporate the special, distinctive qualities of a community serve as more effective screens than others that simply further a particular agenda. The Boston Indicators Project acknowledges the benefit to the city of "measuring what we value as a society and promote change towards a collective vision of where we want to be in the future" (Boston Indicators Project 1998: 1) The group, beginning in early 1997, consciously designed an ever-expanding community outreach and involvement process to create the following ten topics or "categories" for developing indicators:

- Civic Health
- Jobs and Economy
- Environment
- Housing
- Arts, Culture and Leisure
- Education
- Health
- Transportation
- Public Safety
- Information Technology, Communications and Media

Although many of these categories are similar to those of other cities, some are especially important to the values of participants. For example, the last category reflects the dominance of certain industries and businesses and the role Boston plays as a communication center for the region.

With value-laden indicators comes the opportunity to hold people accountable for meeting desired expectations and responsibilities. Unlike theoretically value-neutral indicators or ones proposed by technical experts or a small core group, community-derived indicators give more people a personal stake in their use. The Boston Indicators Project conveys these expectations in brief "Vision" statements accompanying each of the ten categories. For example, the Jobs and Economy category seeks to "develop an economy which promotes economic prosperity that meets the needs of the local population, builds strong sustainable communities, provides a high quality of life, and offers increased opportunity to participate in the competitive global economy" (Boston Indicators Project 1998: 20). The production of jobs is not the desired result; rather, it is essential that local residents be the beneficiaries, and that economic investments translate into a better life for people.

Besides establishing long-term directions and expectations, sustainable development indicators also guide incremental and short-term decisions. They help people working on aspects of complex and interconnected issues to monitor and evaluate progress to ensure that unintended consequences and mistakes are prevented, revealed, and addressed, that hidden agendas can be exposed and dealt with, and that achievements are reinforcing desired goals, values, and outcomes rather than simply producing results. For example, the goal of "ensuring that Boston's neighborhoods and residents participate fully in Boston's economic success" is being measured by the percentage of Boston residents in Boston jobs. As data are collected and analyzed over time, trends can be determined to indicate whether or not the percentage is increasing, at what pace, and for which targeted population group (i.e., teenagers vs. adults or people of color vs. whites).

When viewed collectively, sustainable development indicators can be used to tell a "story" quite different from one based on conventional urban indicators. One of the major objectives of the Boston Indicators project is to "tell the story of Boston's successes at the neighborhood and citywide levels in ways that have been obscured by the conventional social science and mass media approaches to measuring urban change (crime and violence, teen pregnancy, school dropouts, etc.). Problem areas will also emerge, but in the context of our social, political, and environmental capital" (Boston Indicators Project 1998: 2). Moreover, "[t]he data gathered and the goals set for the future will serve to restore hope and faith in the residents of Boston's neighborhoods still recovering from disinvestment" (Boston Indicators Project 1998: 2).

Purposes of Urban Sustainable Development Indicators

The previous section enumerated some ways in which sustainable development indicators can be valuable to people in helping them react to, shape, and evaluate decisions and actions which directly and indirectly affect the quality of their lives and the health of their natural environments. It addressed the "why indicators?" question.

The next section deals more specifically with the *nature* of those efforts. What can indicators accomplish, especially for people living, working, and visiting urban areas?

First of all, urban sustainable development indicators can be used to steer new development decisions by building on a community's assets and furthering community values and interests. Many communities approach development decisions by trying to address problems and/or respond to developers' initiatives. Assets-based development begins with and builds on a community's strengths and opportunities, believing that this approach puts the community more in control. John McKnight and John Kretzmann from Northwestern University's Asset-Based Community Development Institute in the United States have written practical guidebooks, published by ACTA Publications in Chicago, Illinois, on this topic including *Building Communities from the Inside Out* and *A Guide to Mapping Local Business Assets and Mobilizing Local Business Capacities*. This approach is especially appropriate to urban areas because it recognizes that everyone has some-thing to contribute and that all places, including poor neighborhoods, gen-erate and spend money. Given limited and competing outside resources, the authors pronounce that "[t]he hard truth is that development must start from within the community and, in most of our urban neighborhoods, there is no other choice" (Kretzmann and McKnight 1993: 5).

The Boston Indicators Project strives to select indicators that encourage and measure the type of development which meets the community's needs and takes advantage of community assets. For example, under the category of "Civic Health," indicators measure "stability and investment within neighborhoods" by the percentage of people living in their current postal code address over a specified number of years and the extent of stewardship over public spaces. In the "Jobs and Economy" category, one of the indica-tors is "successful business creation and retention," which seeks to measure the number of new business start-ups in Boston and the number of busi-nesses still in operation after five years. These indicators illustrate that quan-tity—total number of people living in Boston; total number of people employed in Boston businesses—does not necessarily tell the story about how well residents are doing nor how well the public, private, and non-pri-vate sector resources are utilizing the time, skills, knowledge, funds of com-munity people and organizations.

A second purpose of urban sustainable development indicators is to restore degraded natural and human environments, with a desire to bring back the quality of life or ecological health of neighborhoods and natural resources which were once thriving. Particularly in urban areas where multi-ple pollution sources exist, land uses have changed on the same site from years of development activity, and significant numbers of people are affected. Instead of viewing contaminated urban land (i.e., "brownfields") as wastelands, government and community organizations are adopting policies and implementing programs that invest in their clean-up and economic reuse. Cities, which already contain substantial infrastructure (i.e., roads, public transportation, sewers, water, housing, schools) can compete success-fully with the suburbs once the tainted image of cities is removed and resources are invested rather than disinvested.

Restoration and reuse of land and water bodies are included in Boston's indicators. The environmental vision seeks to "restore and maintain a high quality of environmental health to support a mosaic of habitats that support a diverse range of human activities, are biologically productive, aesthetically attractive, accessible to all and do not expose any individuals or communi-ties to inequitable health risks" (Boston Indicators Project 1998: 23). Exam-ples of relevant indicators are: "protected and restored functional wetlands in the city" and "clean land in productive use in the neighborhoods" (Boston Indicators Project 1998: 23).

A third purpose is to use urban sustainable development indicators to foster planning and evaluation so that past mistakes are less likely to be repeated and pro-active improvements can be sought. A set of urban sustainable development indicators can not only guide site-specific development decisions, but can also be used as a planning and policy tool to guide development as a whole. Boston, for example, selected as one of its Housing indicators "public sector policies and programs that foster affordability, access, and integrated services" (Boston Indicators Project 1998: 26). As a starting point to measure this indicator, data are being collected and analyzed on the number of public dollars spent on housing by different groups of people (i.e., differentiated by age, income levels, special needs, etc.).

Helpful Hints in Developing Urban Sustainable Development Indicators

Developing authentic sustainable development indicators is a very challenging task. The previous chapter provided some "requirements" for sustainable development indicators and described some indicator frameworks. The task for communities remains how to translate this information into statements that express both *what* they want to measure and *how* they want to chart and evaluate progress. Many, like the Boston Indicators project, are multi-year efforts.

Analysis of lessons learned so far from the Boston project provide 11 helpful hints that other groups might consider in evolving their own urban sustainable development indicators. These key ingredients (listed in no particular order of priority) are couched in generic terms, with examples from the Boston project. Participants in the Boston Indicators project already have concluded that much of the data needed to apply their recommended indicators is not yet available. With that in mind, they selected specific indicators they want to use and then collected and analyzed data based on what can now be measured. The hope is that, over time, data collectors will be persuaded to adjust their studies, surveys, and measurement instruments based on knowing how such information can be used.

Ingredient #1: Focus on Core Concerns Rather Than Symptoms

In early discussions debating what the Boston Indicators project wanted to measure, the concept of targeting core concerns was proposed and immediately embraced. This notion, emanating from the Boston Foundation's Persistent Poverty project, distinguishes between a host of problematic issues and the central cause of those problems. For example, in social and health care fields many researchers, providers, and funding agencies have developed policies and programs to deal separately with teen pregnancy, substance abuse, and domestic violence. Yet the core source of these "symptoms" is poverty. If people better understand the nature of poverty and how to break out of poverty cycles, then we may be more effective in preventing the symptoms that result from poverty. In the environmental arena, the same disconnect often occurs. Policies, laws, regulations, and programs are designed independently to respond to a myriad of pollution sources, such as hazardous waste disposal, solid waste disposal, air toxins, water quality problems, and so on. Yet, only recently have scientists, environmentalists, and regulators begun to focus on ecosystem protection and management—the core concern. Instead of chasing after different sources of human risk to natural systems, the central concerns become what is a healthy wetland or river and how can adverse human impacts be limited to ensure their continued ecological integrity?

A core indicator selected by the Boston Indicators project is: "healthy and accessible water bodies—Boston Harbor, rivers, lakes, and ponds—for swimming, fishing, and other recreation" (Boston Indicators Project 1998: 23). This indicator will be measured through assessments of specific water bodies, such as the Charles, Chelsea, Mystic, and Neponset Rivers to evaluate how they meet various federal and state standards for swimming, boating, and fishing (both for catching and release as well as for eating).

Ingredient #2: Define Issues as "Integrated Systems" and Not Be Limited to Urban Boundaries

One of the key tenets of sustainability is that everything is interconnected. When figuring out how to measure these linkages, the simplest strategy is to start at one point (any point) and trace its links to other points. For example, a water system looks at water sources, their watersheds and aquifers, water distribution systems, consumptive (e.g., drinking water, process industrial water) and non-consumptive (i.e., agricultural, hydropower) uses, and wastewater discharges. In devising appropriate indicators, Boston needs to measure more than its own water supply demands; its sources are managed by a regional entity serving many other communities from reservoirs located miles away, and whose wastewater and storm water drain into Boston's wetlands, ponds, and eventually into Boston Harbor.

Ingredient #3: Measure Community Outcomes in Addition to Process Improvements and Program Accomplishments

Many, if not most, indicators are labeled "bean-counters," meaning that they measure items such as the number of dollars spent, number of open space acres purchased, and numbers of environmental permits issued. Although these figures indicate a commitment to and investment in resources (funds, personnel, time), they do not adequately evaluate how successfully they translate into tangible improvements. Community outcome measures, on the other hand, consider how these process and program improvements affect the well-being of people and the health of natural resources.

Boston wants to measure both types of results, although most of them are oriented toward measuring accomplishments rather than community improvements. For example, the number of pediatricians participating in Reach Out and Read Programs, immunization rates, number of faith-based institutions, and percentage of children enrolled in kindergarten illustrate program and process accomplishments. A few of the indicators try to measure community livability. For example, in promoting "an effective transportation system that meets the needs of people" (Boston Indicators Project 1998: 38), Boston measures the number of people who walk or bicycle to work. In contrast, a process-type measure for this indicator would be the number of parking spaces in the downtown area.

Ingredient #4: Measure Progress at the Neighborhood and Regional Levels as Well as Citywide

The Boston Indicators project looks at the city from three "critical levels"—neighborhood, citywide, and greater metropolitan or region (Boston Indicators Project 1998: 7). Urban areas, especially, need to disaggregate data to understand the disparities among neighborhoods. Citywide information tells the story of Boston as a whole and allows for comparisons with other cities, but only neighborhood data can distinguish where inequities lie among neighborhoods. For example, educational attainment is one of the measures to evaluate progress in the Jobs and Economy category. This measure was chosen by the group because many of the city's current and future jobs require college and advanced degrees. The data from the Boston Redevelopment Authority reveal that Boston has one of the highest levels of educational attainment in the nation, but that some of its poorer neighborhoods are well below that average.

Some of the indicators, such as those dealing with the environment and transportation, place Boston within the context of a metropolitan and eastern U.S. region. Local air quality in downtown Boston and water quality for Boston's industries and residents, for example, are influenced by activities that occur in Massachusetts communities outside of Boston and, indeed, in midwestern and southeastern states.

Ingredient #5: Respond to a Community's Own Sense of Its Priorities

The U.S. city of Seattle, one of the earliest cities to devise sustainable development indicators, chose to highlight the fate of its salmon because it struck a popular chord (Sustainable Seattle 1998). Boston has chosen several indicators that are designed to capture people's immediate attention. After spending millions of federal, state, and local dollars cleaning up Boston Harbor, people now want to have access to use and enjoy its waters and shorelines. Consequently, one indicator measures public transportation access to waterfront recreation (i.e., Boston Harbor islands, beaches, and waterfront parks). Other indicators measure access to and participation in arts, culture, and leisure activities that reflect that Boston is not only the hub for these kinds of events, but also that many people choose to move into or remain in the city because of this richness.

Sparking people's interests through the selection of indicators known to be "hot topics" of special interest to urbanites offers legitimacy, credibility, and hope to people which, in turn, translates into greater interest and participation in community building and improvement.

Ingredient #6: Promote a Clear "Bottom Line" and Encourage Flexible Implementation Choices

Because communities and circumstances change over time, environmental regulators, business managers, and others who establish standards, benchmarks, milestones, target figures, and other markers have learned the importance of creating substantive objectives rather than prescribing specific implementation strategies. For example, Massachusetts acid rain control legislation set a goal of reducing sulfur dioxide emissions by 50 percent from 1990 levels and implemented indicators to track progress related to meeting this target. This approach is preferable to developing an indicator that measures how many fossil fuel facilities are retrofitted with mitigating technologies. The latter is one of many options for meeting the bottom-line target reduction figure.

Sometimes it is easier to measure implementation actions. In these instances, a community may want to adopt several measures to evaluate the effectiveness of meeting the community's desired outcome. For example, Boston's "Arts, Culture, and Leisure" category includes an indicator designed to promote "a full range of opportunities for active engagement in the performing, visual and material arts at the neighborhood and citywide levels." Since there is no simple bottom-line measure for this kind of an indicator, the Boston Indicators project (Boston Indicators Project 1998: 33) proposes to measure the number, location, types, and attendance characteristics of various facilities including studios for training and practicing performing arts (i.e., music recording studios, dance studios, theater workshops); stages and studios available for performing arts (i.e., proscenium stages, concert halls, music clubs); and facilities for the pursuit of visual and material arts (i.e., crafts, painting, sculpture, photography, video).

Ingredient #7: Convert Deficits into Assets

Too many community indicators, even urban sustainable development ones, only focus on what is wrong or what is missing rather than viewing those negative aspects as opportunities for community improvement. For example, instead of measuring the number of welfare recipients, Boston proposes to measure the number of people moving from welfare to work (Boston Indicators Project 1998: 22). The concern is still on people who are publicly subsidized and the desire is to help them become more self-reliant. However, by framing the indicator to measure positive change rather than a deficit, emphasis is shifted toward improvement rather than an accounting of what is not desired.

Sometimes, comparative measures help people understand the relative extent of problems and of constructive responses. For example, Boston's proposed measure to track and evaluate "environmentally clean and healthy neighborhoods" is a comparison between the number of businesses in voluntary compliance versus the number of companies with environmental violations (Boston Indicators Project 1998: 24). The number of firms in compliance could be as or more important to policy-makers, legislators, private sector managers, and the public as is the number of violators. Also, people accustomed to negative images of cities may gain a sense of hope if they learn that aspects of their life are better rather than worse over time.

Ingredient #8: Pay Attention to Maintenance, Replacement, and Reuse

Urban areas tend to be older than suburbs and have more public investments made in infrastructure. Therefore, it is essential that urban sustainable development indicators measure the quality, accessibility, and reliability of existing structures such as water and sewer systems, roads, bridges, harbors, public facilities (i.e., schools, libraries, and government buildings).

One of Boston's Housing indicators is "housing quality and conditions that promote health and safety." Two of its proposed measures are the percentage of home improvement applications approved and the number of permits for major rehabilitation and new construction as compared with the number of permits for demolition (Boston Indicators Project 1998: 26). Other maintenance-oriented indicators focus on the quality and safety of school buildings (Boston Indicators Project 1998: 31) and on the adequacy and repair of transportation facilities (Boston Indicators Project 1998: 38).

Ingredient #9: Address Equity Concerns

All communities have their differences, as in economic class, race, age, and longevity of residents. Urban areas tend to have not only larger populations, but also more people moving in and out of them over shorter time periods than in other community settings. So, understanding who benefits and who is left out become critical concerns to explore through indicators.

The Boston Indicators project seeks to unearth inequities and disparities through a number of its indicators. In some cases, the thrust is to understand differences between and among neighborhoods. For example, one environmental goal is "environmental justice that exposes no groups to inequitable health risk and contributes positively to the long-term quality of life concerns of communities" (Boston Indicators Project 1998: 24). One of the indicators to evaluate progress for this goal is "environmentally clean and healthy neighborhoods" (Boston Indicators Project 1998: 24). Data and analysis on "the decrease in the number of landfills, transfer stations, and hazardous waste sites by neighborhood" (Boston Indicators Project 1998: 24) is proposed as a way to measure how neighborhoods compare with each other in reducing environmental risks and improving public health. Another example under the Jobs and Economy category is "neighborhood unemployment rates" (Boston Indicators Project 1998: 21).

In other instances, the focus is on understanding inequities among people rather than neighborhoods. For example, a proposed measure of "public sector policies and programs that foster affordability, access, and integrated services" is to quantify "the number of dollars in public spending on housing by different groups (income levels, age groups, special needs...)" (Boston Indicators Project 1998: 26). Another example appears within the Arts, Culture, and Leisure category that proposes to measure "the number, location, hours of libraries and range of options for different age groups, ethnic, cultural and language groups" (Boston Indicators Project 1998: 34).

A third kind of equity indicator deals directly with fairness and equal opportunity. For example, three of Boston's proposed measures under the category of Jobs and Economy are: unemployment rates by race, educational attainment by race, and income by race (Boston Indicators Project 1998: 22). These figures are intended to provide details to find out whether race affects people's ability to get jobs, education, and the financial resources to live comfortably in the city.

Ingredient #10: Include Qualitative as well as Quantitative Measures

Although statistical data may be easier to access, often people's perceptions about themselves and their community depends on their feelings, opinions, and beliefs that can be captured only through surveys, discussion groups, interviews, and other such forums. For example, even though crime statistics show a decrease in a city or in a particular neighborhood, negative memories and lingering fears from past experiences may prevent someone from waiting for a bus at night or allowing children to play in a nearby park.

One of the indicators identified in Boston's Public Safety category is "perception of safety versus perception of crime as related to quality of life and social order concerns of citizens" (Boston Indicators Project 1998: 37).

Ingredient #11: Realize that Communities Have Distinctive "Collective Personalities" and Are in Different Stages of Community Development

Indicators need to be tailored not only to fit people's thinking, but also to correspond with the mood of the community as a whole. An assessment of six sustainable community projects, funded by the Ford Foundation in partnership with the New Hampshire Charitable Foundation, the Maine Community Foundation, and the Vermont Community Foundation, found that "like individuals, communities have personalities. A spark of hope can infect an emotional change in a group of people" (Kline 1996: 7). On the other hand, some communities have a negative image of themselves that can translate into less energy and commitment to change, fewer volunteers, smaller amount of outside assistance which, in turn, produces fewer improvements.

Indicators need to be designed to gauge progress based on the psyche of the community at the time. For example, Boston's indicators are explicitly targeted toward helping "government and community leaders find the most effective ways to generate positive change" and "also market Boston and its neighborhoods not only to newcomers but to Bostonians themselves—who, with the help of the media, tend to see our glass as half empty when compared to most American cities, is more than half full" (Boston Indicators Project 1998: 2).

Besides having personalities, communities are also in different phases of community development. This finding was briefly described in the 1996 evaluation and was substantially detailed in the October 1997 assessment, *Northern New England Sustainable Communities Implementation Project:*

Lessons Learned. Communities seem to range from those that are in early stages of community building—have no sense of a community identity; no effective community leaders; very limited institutional resources; dependent-type "personalities"; very limited community financial resources; are adverse to change and hostile to new ideas and non-natives; and are often economically depressed—to those sophisticated communities that have abundant human, natural, and economic capital; are comfortable and effective collaborators connecting within and outside the community; know how to leverage political clout; take bold initiatives and produce tangible community results (Kline 1997: 5). Kline (1997) provides a summary of key characteristics of four basic phases in community building based on the six communities analyzed over a four-year period.

Identifying where a community fits in a community development continuum helps shape the nature of indicators. For example, an urban neighborhood that is chronically economically depressed has many vacant and abandoned lots, has dilapidated and unsafe housing, crowded schools, and many unemployed teenagers and adults. This would be measured by different indicators with different expectations than a community that has a vibrant economy, flourishing cultural and educational facilities, many active non-profit organizations, and attractive landscapes.

Framework for Creating Urban Sustainable Development Indicators

The helpful hints offer guidance of what people who are developing urban sustainable development indicators should take into consideration, and allude to topics and indicator language proposed by the Boston Indicators project. This section more directly helps people understand the nature of urban sustainable development so that all relevant community aspects can be captured in indicators. The conceptual framework was first proposed in *Defining a Sustainable Community* by the author in 1993 and amplified in *Sustainable Community Indicators* in 1995.

A simple definition of a community that strives for sustainable development is one which, over time, becomes "more environmentally sound, economically viable, socially just, and democratic" (Kline 1995: 1). These terms may appear to be straight-forward, but understanding their complexities and making decisions with these results in mind is extremely challenging. Since each community is different, with different contexts, resources, needs, values, and choices, there is no substitute for a community-derived set of sustainable development indicators.

Generic frameworks, such as the one described in this section, can serve as useful references because they enable people to appreciate the full range of community *characteristics* that need to be measured; to have these characteristics defined in ways that reflect sustainability principles and practices rather than conventional theories and models; and to understand *pathways* that define what to measure for each characteristic. Indicators can, then, be created based on knowing where you want to go (i.e., sustainable community characteristics) and how you want to get there (i.e., pathways).

From research and case studies described in *Defining a Sustainable Community*, four community characteristics emerged: ecological integrity, economic security, quality of life, and empowerment with responsibility. Each of these terms will be described and illustrated, along with explanations of the sustainability pathways that help chart directions for moving a community toward sustainability.

Ecological Integrity

Traditionally, environmental policies, programs, laws, and decisions have focused on reducing risks that harm natural resources; that is, environmental protection. Consider, for example, the number of initiatives dealing with waste disposal—hazardous waste cleanup, solid waste, wastewater, medical waste, low level and high level radioactive waste disposal. A sustainable development perspective shifts the emphasis from risks to the health of the natural ecosystems, i.e., *ecological integrity*. The starting point is the functional ability of the natural system rather than the impacts to those wetlands, rivers, and airsheds.

Moreover, environmental impacts are not limited to a project site or to a confined pond or discharge point. Rather, the entire ecological system is of concern, and the challenge is to understand how a particular development not only impacts natural resources on-site, but also how those interconnected land, water, and air resources are affected.

The basic goal of sustainable development in the realm of the environment is to use development to restore, preserve, conserve, and enhance the ability of natural systems to function for the benefit and enjoyment of humans, animals, wildlife, marine life, and other living creatures now and in the future. Development can be an opportunity for community improvement rather than dealt with as a source of environmental problems. How can this lofty goal be pursued?

A community whose development enhances, rather than undermines, ecological integrity is "in harmony with natural systems by reducing and converting waste into non-harmful and beneficial purposes and by utilizing the natural ability of environmental resources for human needs without undermining their ability to function over time" (Kline 1993: 22).

Embedded within this explanation are two key decision-making pathways. The first one aims to produce without polluting and without waste. This approach does not imply that all by-products are eliminated or that emissions are zero. Rather, it reflects the belief that, like natural processes, humans can design and use materials in ways where harmful ingredients are eliminated and non-harmful "wastes" become the building materials for another purpose. Eco-industrial parks are an example of how urban areas can take advantage of the vast volumes of what is now undesired waste such as construction debris, steam, and residuals, as well as the close proximity of people and structures. By identifying these waste streams, revising processes, and finding and creating markets for these wastes or converting them into useable products, people can not only reduce the adverse impacts on the environment and on human health, but they can also generate jobs and income.

Urban sustainable development indicators consider not only how much less waste has been generated, but also the extent of the elimination of harmful materials and the reuse or new beneficial uses of previously discarded materials and products. Meeting government standards may be useful indicators, but they may not adequately cover all relevant sources of pollution and they may not be targeted at thresholds that sufficiently ensure long-term ecological health.

The second pathway seeks to use natural resources within their capacity to continue to function in a healthy state. Environmentally sound utilization of natural resources implies a fundamental change in the way we do business. It means more than pollution prevention, reduction, and reuse. It implies a thorough understanding of how ecosystems function, an ability to set thresholds for their long-term health, and a philosophy of development that seeks to utilize natural resources in ways that keep pace with their productivity.

From an urban sustainable development perspective, this means taking advantage of natural light and air; using solar and wind power; siting structures away from the flood-prone areas; planting trees to reduce heating and cooling costs and to provide scenic amenities, particularly in dense neighborhoods; constructing with wood based on sustainable forestry practices; using water and energy efficient fixtures and recycled products; and a host of other design, construction, and use methods. It also means having a scientific understanding of how different ecosystems operate so that, for example, wetlands and river banks can be effectively used as natural pollution filters and as natural buffers against flood damage.

Urban sustainable development indicators consider the extent of use of environmentally sound products, the benefits derived from using natural resources as compared with non-natural alternatives, and the populations of keystone species such as song birds.

Economic Security

For many communities, especially urban areas, economic development is the highest priority concern. Even people who have jobs worry about future employment or having the skills, knowledge, and ability to advance in their careers or find replacements when their jobs end. The tendency in defining economic development is to focus on jobs and to measure progress by the number of new jobs produced or to focus on revenues or productivity, measured by the number of tax dollars collected or the gross national or domestic product.

However, economic development has a broader scope, measuring individual and community wealth (i.e., income trends, poverty, personal consumption), economic opportunities (i.e., jobs, tourism, environmental businesses), and economic performance (i.e., productivity, cost-effective government infrastructure). From a sustainable community perspective, the term "economic security" may be more appropriate than "economic development" or "growth" because it addresses essential links including education, training, environmental soundness, and occupational safety.

A community whose development promotes *economic security* "includes a variety of businesses, industries, and institutions which are environmentally sound (in all aspects), financially viable, provide training, education, and other forms of assistance to adjust to future needs, provide jobs and spend money within a community, and enable employees to have a voice in decisions which affect them. [It] also is one in which residents' money remains in the community" (Kline 1993: 15-16).

This characteristic, like the other three, are interrelated with each other. For example, financial viability depends, in part, on reduced costs that can be derived from environmental efficiencies and reliance on a local and productive labor pool. Therefore, indicators can be sought that measure several aspects simultaneously. For example, the number of environmental businesses measures both economic development and environmental soundness.

Another helpful hint in devising urban sustainable development indicators to gauge economic security is to look at what to measure in order to know whether or not a community is moving toward economic security. It is important to know that a community is encouraging and producing a diverse economic base, as measured by the variety and number of new locally owned businesses. However, it is even more useful to consider disparities, which measure ranges. For example, the number of employees who work for the largest one to three employers might alert a community to its dependence on a few businesses or industries. This analysis was helpful to the U.S. city of Cambridge, Massachusetts, when economic development officials realized that their economic base was dependent on local government and universities (whose employment were stable, not increasing) and on several industries (whose job base was declining). New policies and incentives were adopted to encourage development of small-scale businesses as the likely growth sector.

Another economic security aspect, besides looking for economic soundness of the economic producers and for disparities, is local wealth. Local wealth comprises both monetary and non-monetary exchanges. It measures the contributions of residents and business persons in their community. All communities, including poor urban neighborhoods, have dollars and human capital to cycle within the community. For example, how many people buy from each other or get their food, stationery, equipment, supplies from outside the community? Are there official exchanges, local currencies, community-assisted agriculture projects? Are there micro-business lending programs and loans available to encourage start-ups and expansions of local businesses?

Another aspect or pathway relating to economic security is mutual assistance, the degree to which people who work together cooperate and share resources that benefit them and the community as a whole. This kind of economic interdependence leads to benefits from economies of scale, job references, psychological support, mentoring, and skills development. Cities offer an abundance of opportunities, including existing and new networks, vacant spaces that can be shared, and access to government services.

Quality of Life

Urban areas are comprised of more than buildings, infrastructure, natural resources, and employment centers. People live in cities and metropolitan regions. Sustainable development strives to improve the quality of life by addressing the universal desire of people to have a decent, safe, enjoyable place to live, work, and visit. Measuring quality of life is more than accounting for material needs, such as housing and health care. It extends into qualitative values such as having a sense of belonging to a community, feeling safe enough to play outdoors, and having a close connection to nature even in the most populated neighborhoods and largest cities.

Some other sustainability frameworks label this characteristic as "society" or "community." The title does not matter; what is important is to develop indicators that help people measure how comfortable, how satisfied, and how special they feel in their community. A community whose development promotes a good quality of life for everyone is one that "recognizes and supports people's evolving sense of well-being which includes a sense of belonging, a sense of place, a sense of self-worth, a sense of safety, and a sense of connection with nature, and provides goods and services which meet people's needs both as they define them and as can be accommodated within the ecological integrity of natural systems" (Kline 1993: 26).

Since these kinds of concerns reflect people's beliefs, opinions, and perceptions, indicators need to be developed that can be measured through surveys, opinion polls, discussion groups, interviews, and other qualitative methods. The Boston Indicators project discovered that many organizations collect quantitative and qualitative data and that it makes sense, strategically, to draw these people early into the process of devising indicators in order to take advantage both of their existing data and to get their support for adjusting their collection instruments.

Clearly, one of the greatest concerns of people in cities all over the world is access to and delivery of basic services. Quality of life, at a minimum, needs to measure the ability of people to get adequate housing, health care, child care, education, and public safety. The standards will vary from place-to-place and country-to-country, but all people deserve basic coverage.

"Basic" in a sustainable development context means more than survival. It means, for example, offering people different housing types to meet their lifestyle needs whether this means apartments for individuals and small families, co-housing for people who want to live more as an intentional community, artist studios, or assisted living for people who have special health care needs. In terms of public safety, it means more than stopping violent crimes. People want to feel safe allowing their children to play in a nearby park or waiting for a bus at an urban street corner at night. Indicators that measure funds invested in public services provide some measure of commitment, but do not reveal the effectiveness of service delivering those services. Other indicators need to be used to evaluate whether or not, for example, children are being inoculated against diseases and not simply that funds have been allotted or spent for such a purpose.

Two other aspects to help measure quality of life are encouraging self-respect and caring. These pathways are connected to individual and community capacity since people who have confidence, respect for each other, and reach out and support others are more willing to take initiatives and produce constructive results. When people feel depressed, anxious, and unhappy, they tend to be critical, resistant to change, and less able to think about the well-being of their neighbors. A pulse of a community can be gauged by reading media descriptions, listening to local commentaries, hearing public testimonies, and analyzing which issues political leaders focus on and how they respond.

Indicators aimed at measuring self-respect and caring are often labeled "civic capacity" or "civic society" indicators because they seek to know how well people know their neighbors; how often they volunteer, mentor, or otherwise assist someone in need; how often they attend multi-cultural events; and how involved they are in community groups and civic associations.

Another related pathway is connectedness. People's quality of life depends on how much they feel that they belong to a group, a neighborhood, a city, a region, a country. The closer the identity, the more likely people are to invest in maintaining and improving their communities. So, indicators such as the length of time people have lived in their neighborhood; their familiarity with their neighbors; knowledge of the watershed in which they live; the number of social/cultural/sports organizations they participate in are valuable measures of connectedness. Ask people in New York City where they live and they will respond with "the Upper West Side," "the Village," "Brooklyn Heights," "near Central Park," or some other neighborhood identity. They also can tell you the places where they shop for fish or specialty foods, especially if they have lived in the area for some time.

Empowerment with Responsibility

The final, and equally important, characteristic deals with the capacity and assertiveness of people to take control over their lives. Empowerment is the "opportunity and capacity for meaningful and effective participation" (Kline 1993: 31). Without empowerment, people and communities are more buffeted and manipulated by outside forces.

Empowerment alone, however, is not sufficient. People's skills, abilities, knowledge, and visions need to be harnessed and converted into actions. They want more say in decisions that affect them, such as how government spends tax dollars, where and what kind of jobs businesses create, what is the nature of health care provided by professionals, and the choice and role of experts who are hired to provide technical assistance in confronting complex community problems.

A community whose development enables people to feel and be empowered and to take charge of their lives is "based on a shared vision, equal opportunity, ability to access expertise and knowledge for their own needs, and a capacity to affect the outcome of decisions which affect them" (Kline 1993: 31).

Four pathways can be considered to measure a community's ability to influence decisions: capacity, equity/fair playing field, "reaching in," and accountability.

In the final assessment of the Northern New England Sustainable Communities Project, the major finding was the "importance of community capacity building as the foundation for achieving constructive results. Quantifiable results are most notable in communities that are the most sophisticated in capacity building—those with seasoned leaders, opportunities for training and educating emerging leaders, effective institutional structures, and widespread public involvement and support" (Kline 1997: 2).

Equity does not mean that everything is equal; rather, it implies equal opportunity and equal access. Equity affects people's sense of empowerment and the degree to which they take responsibility for themselves and for others. Equity indicators measure gaps, but do not presume values. As with many sustainable development indicators, responses may invite more questions and probes to understand why something is happening in a community and what those results mean. For example, the number derived from measuring the percentage of public school teachers of color as compared with percentage of registered school age children of color provides some feedback, but does not interpret whether or not this figure is reasonable, explainable, or acceptable.

"Reaching in" is a term described in *Sustainable Community Indicators* (Kline 1995: 19) to characterize the process of people in communities broadening their base of participation and opening up discussions to attract and listen to diverse constituent perspectives. The common term, "reaching out" implies that someone or some group seeks allies to their position by inviting supporters to join them. "Reaching in," on the other hand, includes strategies such as physically going to where people feel most comfortable (i.e., their home base), asking them to describe their concerns and interests, and building relationships and ideas without having a predetermined solution in mind. Indicators to measure this concept include the number of new participants added to a process over a specified time period; the source of new project ideas and recommendations; and the number of community gardens allotted by the city.

The last pathway, accountability, reflects the belief that promoting sustainable development is everyone's responsibility. No sector including government, no group (elected or appointed), and no individual has the burden of representing everyone's interests. Each person has a special contribution and a responsibility to participate. Indicators can measure the extent to which people and institutions are actively working to make their communities better places and to what extent they are meeting their obligations and are held accountable for their actions.

Conclusion

This chapter focused on helping people appreciate the value and purposes of urban sustainable community indicators and provided guidance to help identify indicators appropriate to their own communities' circumstances. Other chapters deal with how indicators relate to another planning and development tool, environmental impact assessments/strategic environmental assessments. Assessments as currently practiced, at best, evaluate impacts to environmental resources from human activities. They are not designed to analyze community impacts. Standards and thresholds for project reviews may not be appropriate for urban settings since they are based on characteristics such as the number of dwelling units or parking spaces rather than on land-use patterns.

Even the more sophisticated strategic environmental assessments currently used, which evaluate broad policies and program agendas, often fail to capture off-site, cumulative, and long-term impacts and are not oriented toward evaluating non-environmental issues. There is also difficulty in linking policies to project decisions.

So, the question becomes how to adjust the assessment tool so that it can be effectively applied in encouraging more sustainable urban communities? Several recommendations are offered. The substantive scope can be expanded to include economic security and quality of life concerns as well as environmental ones. Thresholds for triggering reviews can be lowered, particularly in special places or circumstances where impacts are most sensitive and of greatest public concern. Site-specific data and information from individual assessments could be pooled into a general data management entity so that information can be shared, data collection and manipulation costs can be reduced, and the ability to forecast impacts can be enhanced. Applicants could, thereby, gain access to a lot more information, data, and analyses which could be included in their own project assessments. Moreover, the data management entity might be able to use computer technology to manipulate data and maps in order to provide alternative impact consequences and various development options.

The usefulness of linking assessments with sustainable development will be greatly improved if the triggers for project review come earlier in the design process. By the time an environmental permit or government funds are requested, developers have invested considerable time and money in a particular site and development scheme. Too often assessments end up being used to oppose or justify projects rather than to guide their design, site location, construction methods and materials, and use. Perhaps some of this problem can be addressed by triggering strategic assessments linked to a community's vision, its land-use plan, or its growth strategy. If these tools are to become more effective in debating options and reaching community and neighborhood agreement on desired uses and siting locations, then the project-level assessments can be focused more narrowly on maximizing cost-effective, environmentally sound practices.

Contacts

For additional information about the Boston Indicators Project, contact Geeta Pradhan (telephone: 617/635-0346 or geeta.pradhan.pfd@ci.boston.ma.us) or Charlotte Kahn (telephone: 617/723-7415 or cbk@tbf.org).

Additional information about the author's sustainable community documents is available via the Internet website: www.tufts.edu/gdae/KLINEPUB.HTML

References

Boston Indicators Project. 1998. *Indicators of Progress, Change and Sustainability for an Improved Quality of Life in Boston*. Boston: City of Boston and The Boston Foundation.

Kline, E. 1993. *Defining a Sustainable Community*. Medford: Tufts University.

Kline, E. 1995. *Sustainable Community Indicators*. Medford: Tufts University.

Kline, E. 1996. *Northern New England Sustainable Communities Implementation Project: An Evaluation*. Concord: New Hampshire Charitable Foundation.

Kline, E. 1997. *Northern New England Sustainable Communities Implementation Project: Lessons Learned*. Concord: New Hampshire Charitable Foundation.

Kretzmann, J. P. and J.L. McKnight. 1993. *Building Communities From the Inside Out: A Path Toward Finding and Mobilizing a Community's Assets*. Chicago: ACTA Publications.

Kretzmann, J.P. and J.L. McKnight. 1996. *A Guide to Mapping Local Business Assets and Mobilizing Local Business Capacities*. Chicago: ACTA Publications.

Sustainable Seattle. 1998. *Indicators of Sustainable Community 1998*. A status report on long-term cultural, economic, and environmental health for Seattle/King County. Seattle: Sustainable Seattle.

Additional Examples of the Use of Sustainability Indicators at the Local Level[1]

Mark Roseland

Moving toward eco-cities is a long-term goal, so it is important that the incremental steps we take in the short-term are leading us in the right direction. There are two kinds of tools available to citizens and their governments for managing community sustainability. Community planning and assessment tools can sometimes be conducted by citizen groups with little training, whereas technical planning and assessment tools more often require the involvement of trained staff or consultants. Technical planning and assessment tools may not lend themselves as readily to public participation, but people can participate more effectively in decision-making if they understand some of the tools available to their communities. One of these tools is developing sustainability indicators.

The steps involved in developing sustainability indicators are to clarify goals (the aim of the evaluation and the type of desired outcome); determine who will lead the process; invite participation (the process of evaluation may be as valuable as the eventual application of the indicators themselves); decide how to choose indicators; collect data by which to measure the indicators; report on the indicators; and update and revise the indicators. (Details on these steps and related issues can be found in Azar et al. 1996, Brugmann 1997, Forss et al. 1994, Kline 1997, Maclaren 1996, McLemore and Neumann 1987, Papineau 1996, Parker 1995, Schön and Rein 1994, Schwandt 1997, and Waddell 1995).

The following initiatives represent a small sample of ongoing and emerging projects to design and use sustainability indicators. They were chosen to represent the spectrum of aims for which sustainability indicators can be used—from the Sustainable Seattle Project, with a focus primarily on community education and empowerment, to the Oregon Benchmarks project, with a greater focus on providing feedback to government agencies. It is still too early to judge the impact of these projects on community sustainability in the long term, but they seem to be helping these communities and regions move in a sustainable direction.

The Sustainable Seattle Project began in 1992 with a meeting of 150 citizens. During this gathering, 99 indicators were proposed. From this initial list, 40 key indicators were selected. The first 20 of these indicators were assessed in a 1993 report. A 1995 report updates those first twenty indicators and also looks into the remaining twenty. The indicators used ranged from total water consumption and per capita waste generation and recycling rates to volunteering in schools and household incomes. The Sustainable Seattle Project plans to update and improve their indicators on an annual basis.

The Sustainable Seattle Project is clear in pointing out the need for action arising from its sustainability evaluation. As they say, "measuring progress is not the same as making it." The project promotes action by encouraging Seattle-area citizens to: employ local media to spread indicator results and analysis; use the political process to promote change in public policy; broaden the information base used for economic decision-making; use indicators in schools for education and as a basis for additional research; form a basis for linking local nonprofit and volunteer groups; and question personal lifestyle choices.

In southwest Washington State, the Willapa Bay indicators project is significant as an attempt to evaluate the environmental, social, and economic sustainability of a rural watershed. A joint effort by the local Willapa Alliance and Ecotrust, a national conservation-based development organization, the *Willapa Indicators for a Sustainable Community (WISC)* report is intended to promote discussion of sustainability issues in the local communities. The WISC indicators explicitly tie the health of the environment to the vitality of the local economy and community. Environmental indicators are divided into three categories: water resource quality, land-use/vegetation patterns, and species populations. Economic indicators are included under the categories of productivity, opportunity, diversity, and equity. Finally, community measures fall under life-long learning, health, citizenship, and stewardship.

The WISC project, while primarily a project for community empowerment and education, is also linking their efforts to other community groups and organizations by publishing *The Directory of Organizations and Services in Pacific County, Including Key Government Officials* as a companion volume to their indicators report. The Willapa Alliance is also involved in several local projects to translate their evaluation exercise into tangible action. Among other projects, the Willapa Alliance has formed the Willapa Science Group. This group of local and regional scientists and educators is encouraging scientific research that is meaningful to local people. Such initiatives are vital to bridge the gap which exists in evaluation for community empowerment between the understandability of indicators and their scientific validity.

In Ontario, *Hamilton-Wentworth's Sustainable Community Indicators Project* arose out of the regional municipality's *Vision 2020* initiative. This vision of a sustainable future was developed by a citizens' task force on sustainable development appointed by the regional council. While the vision process was initiated by the local government, it was able to draw on widespread community participation, with over 400 individuals and 50 community groups taking part (Maclaren 1996). The Sustainable Community Indicators Project is an attempt to measure progress toward the goals outlined in the *Vision 2020* document. As with *Vision 2020*, the municipality recognized and drew on the participation of the community throughout the indicators project. While the final set of indicators are intended to be of use to decision-makers, the prime goal was to develop a set of indicators which were understandable and useful to local citizens. Consideration of the way in which indicators can be used by government agencies begins to move this evaluation process toward an approach that aims at providing feedback to the organizations responsible for strategic planning and implementation.

In 1994, the Santa Monica, California, Task Force on the Environment developed a Sustainable City Program in partnership with the City of Santa Monica Environmental Programs Division. Each of the program's policy areas has clear goals that reflect the city's current and future programs. Specific targets have been established for each goal, and an indicator has been established for each target (ICLEI 1996).

The Oregon Progress Board was established by the Oregon Legislature in 1989. The board, a multi-stakeholder organization, was originally charged with developing strategies and programs to support the state's strategic plan, "Oregon Shines." Out of this came the *Oregon Benchmarks* process. The progress board presented the reporting framework to the state legislature after extensive consultation and the benchmarks process was officially adopted in 1991.

The Oregon framework for reporting consists of 269 indicators. Rather than simply presenting indicators to measure and report trends, however, the Oregon process defines targets, known as *benchmarks*. The benchmarks cover a diverse range of issues surrounding sustainability, including categories such as children and families, education and work force, health and health care, clean natural environment, equal opportunity and social harmony, and economic prosperity. The board publishes a report card every two years to report on progress toward the stated targets.

While the Oregon Benchmarks program has drawn on public consultation and aims to inform the public, its main strength is its ability to promote action and accountability in the state government. Benchmarks are now used as a tool to set program and budget priorities and to encourage interagency cooperation on broad issues. Each state agency has been directed to develop results-oriented performance measures that dovetail with the benchmarks. Through the Oregon Benchmarks program, rational and clear sustainability goals have formed the basis for strategic planning throughout government agencies. The legislature has even passed several bills directing agencies to work toward benchmarks. On a smaller scale, the Oregon Benchmarks are being applied by municipal governments and community organizations and several cities and counties are adopting strategies to complement the state program.[2]

Notes

[1]   The author is grateful to Zane Parker for his contributions to the research on sustainability indicators.

[2]   For more information on the targets linked to the indicators in these programs, see the chapter on Sustainability Reporting and the Development of Sustainability Targets in this volume.

References

Azar, C., J. Holmberg and K. Lindgren. 1996. Socio-ecological indicators for sustainability. *Ecological Economics* 18: 89-112.

Brugmann, J. 1997. Is there a method in our measurement? The use of indicators in local sustainable development planning. *Local Environment* 2,1: 59-72.

Forss, K., B. Cracknell and K. Samset. 1994. Can evaluation help an organization to learn?" *Evaluation Review* 18, 5: 574-591.

ICLEI/IDRC/UNEP, International Council for Local Environmental Initiatives, International Development Research Centre, and United Nations Environment Programme. 1996. *The Local Agenda 21 Planning Guide.* Toronto: ICLEI and Ottawa: IDRC.

Kline, E. 1997. Sustainable community indicators. In M. Roseland, ed., *Eco-City Dimensions: Healthy Communities, Healthy Planet.* Gabriola Island, B.C.: New Society Publishers.

Maclaren, V. W. 1996. *Developing Indicators of Urban Sustainability: A Focus on the Canadian Experience.* Toronto: Intergovernmental Committee on Urban and Regional Research (ICURR) Press.

McLemore, J. R. and J. E. Neumann. 1987. The inherently political nature of program evaluators and evaluation research. *Evaluation and Program Planning* 10: 83-93.

Papineau, D. 1996. Participatory evaluation in a community organization: Fostering stakeholder empowerment and utilization. *Evaluation and Program Planning* 19, 1: 79-93.

Parker, P. 1995. From sustainable development objectives to indicators of progress: Options for New Zealand communities. *New Zealand Geographer* 51, 2: 50-57.

Schön, D.A. and M. Rein. 1994. *Frame Reflection: Toward the Resolution of Intractable Policy Controversies.* New York: Basic Books.

Schwandt, T. A. 1997. Evaluation as practical hermeneutics. *Program Evaluation* 3, 1: 69-83.

Waddell, S. 1995. Lessons from the healthy cities movement for social indicator development. *Social Indicators Research* 34: 213-235.

Chapter 12

Sustainability Reporting and the Development of Sustainability Targets

Dimitri Devuyst

Summary

This chapter gives insight into the sustainability reporting process as developed by the UK Local Government Management Board. Sustainability reporting is a development of State of the Environment Reporting and aims to obtain and interpret information, which can then be used to guide actions and decisions. Periodically repeating the sustainability reporting process makes it possible to get information on progress made over time. Although examples of sustainability reporting at the local level are relatively scarce, the following cases were found: Minnesota Milestones, Oregon Shines, Pierce County Quality of Life Benchmarks, Hamilton-Wentworth's Sustainability Indicators, and the Santa Monica Sustainable City Program. Measuring progress and interpreting data in sustainability reporting and sustainability assessment requires setting sustainability targets. To develop sustainability targets, compromises should be made taking into consideration recommendations by panels of experts, available scientific information, political considerations, and the sustainable development vision of the local community. The process of setting and using targets is detailed in this chapter.

The Assessment of Progress Made Toward Sustainable Development

To assess progress made toward sustainable development, a community must develop a vision for sustainable development. A number of goals may be set on the basis of this vision. The mostly qualitative goals can be made more specific and can lead to quantitative targets. Setting targets is closely linked to the development of indicators. For each specific indicator a target can be set. An assessment of progress toward each goal is possible with the help of these indicators and targets. Moreover, it is informative to consider the indicator measurements of a certain year as the baseline data. It then becomes possible to assess progress by comparing the current situation with both the baseline and the target.

The Charter of European Cities and Towns toward Sustainability, the Aalborg Charter, proposes that local communities establish long-term local action plans toward sustainability, including the establishment of measurable targets (ICLEI 1998). To be able to assess progress, we need to repeat the collection of data regularly (e.g., every year), which makes it possible to discern trends. A yearly report on the progress made toward a more sustainable future can follow a standardized "sustainability reporting" approach.

In this chapter we will present how to use goals, targets, and related concepts in sustainability reporting. We will examine the steps involved in the process, as well as actual examples. The development of goals and targets is discussed, as is the experience with their use.

Sustainability Reporting

The UK Local Government Management Board (LGMB 1996) developed a practical guide for local authorities dealing with sustainability reporting.

Sustainability reporting is a process of:

* Deciding the information on sustainability needed to inform a particular group or to guide a particular activity, plan, policy, or decision
* Obtaining and interpreting the information
* Using the results to guide actions and decisions
* Communicating the results in useful ways to relevant audiences
* Storing the information for future use and to enable access by other users (LGMB 1996)

In other words, sustainability reporting is a way of gathering and assessing information on the sustainability of our societies. The sustainability reporting process consists of seven major steps and takes the form of a loop, as shown in Figure 12-1. It brings together relevant information and builds on previous work to provide a basis for effective, integrated policy-making, action planning and awareness raising. Sustainability reporting is a development of State of the Environment Reporting (SoER). Sustainability reporting and SoER activities follow a trend of the 1990s in which governments emphasize performance measurements. One outcome in the U.S. was the introduction of legislation such as the Government Performance and Results Act. SoER has also grown rapidly in recent years in response to improvements in information technology and to increased demands for information by both the public and politicians. The adoption of Agenda 21 in 1992 has given added impetus to this process (Briggs 1998) and has led to the introduction of a more specific sustainability reporting. Both sustainability reporting and SoER aim to increase knowledge so that the uncertainty in decision-making can be reduced. Information should be given in a way that anyone concerned can use it to make their own judgments. Information about sustainability or environmental conditions should be systematic and logical, and should also be compatible with the planning and decision-making process. Furthermore, information that is simple and understandable will support a high quality of a dialogue and will enhance the ability to reach agreements between different actors (Ryding 1994).

SoER is a process which is mostly practiced at the international or national levels. For example, most European countries now publish information on the state of their environment (Briggs 1998). Sustainability reporting is also very useful at a more local level as a support to the Local Agenda 21 process. Therefore, the UK Local Government Management Board developed a guide for local authorities if they choose to practice sustainability reporting. The following are some important guidelines proposed by the LGMB (1996):

* The single most important rule in sustainability reporting is to be clear from the start as to how the information will be used. Since resources are limited, priorities must be set. It is more useful to do a thorough sustainability reporting on only a few important aspects than to examine many aspects and not be able to go through the whole process

* It should be clear from the start which resources are available and who will manage the sustainability reporting process. It may be appropriate to establish a steering group to oversee the work at a more detailed level. An action plan, timetable, and budget should be agreed upon

* In the selection of topics to be considered in sustainability reporting, the concerns of the local community should be taken into account. Therefore, the local population must be involved in the selection of topics. This process could also be linked to the Local Agenda 21 community consultation processes

* Once a set of topics has been selected, the next step is to decide the data that should be collected to illustrate the current situation and trends. In sustainability reporting, large amounts of data need to be reduced to a smaller number of indicators. One must first examine what simple, measurable data can be used to describe the topics that have been selected; then data must be obtained. Finally, the data must be inter-

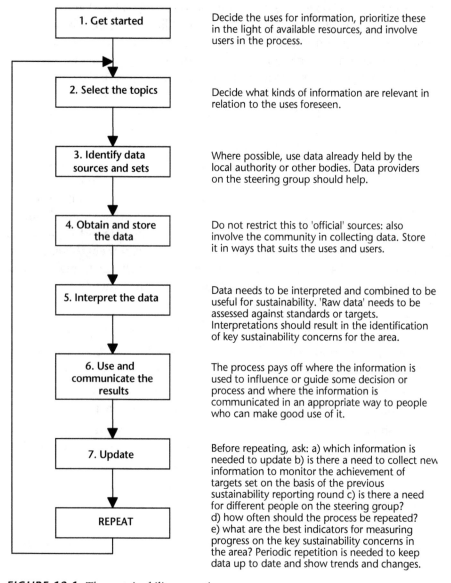

1. Get started	Decide the uses for information, prioritize these in the light of available resources, and involve users in the process.
2. Select the topics	Decide what kinds of information are relevant in relation to the uses foreseen.
3. Identify data sources and sets	Where possible, use data already held by the local authority or other bodies. Data providers on the steering group should help.
4. Obtain and store the data	Do not restrict this to 'official' sources: also involve the community in collecting data. Store it in ways that suits the uses and users.
5. Interpret the data	Data needs to be interpreted and combined to be useful for sustainability. 'Raw data' needs to be assessed against standards or targets. Interpretations should result in the identification of key sustainability concerns for the area.
6. Use and communicate the results	The process pays off where the information is used to influence or guide some decision or process and where the information is communicated in an appropriate way to people who can make good use of it.
7. Update	Before repeating, ask: a) which information is needed to update b) is there a need to collect new information to monitor the achievement of targets set on the basis of the previous sustainability reporting round c) is there a need for different people on the steering group? d) how often should the process be repeated? e) what are the best indicators for measuring progress on the key sustainability concerns in the area? Periodic repetition is needed to keep data up to date and show trends and changes.
REPEAT	

FIGURE 12-1 *The sustainability reporting process.*

preted to derive the key sustainability concerns for the area and the indicators which will be used to measure these

↠ Data can be collected from public sources (from different government bodies and public agencies), from private sources (such as private companies), and from the community. The public can also be involved in the collection of data. Raw data can be stored in different ways, from ordinary paper copies to sophisticated GIS systems

✦ Raw data need to be analyzed and interpreted, and analysis should rely on applying a mixture of expert advice and public values. A vision on where the local community wants to go in the future should lead to sets of standards or targets against which current rates and trends can be measured

✦ Using and communicating the results is a very important step which is often neglected. The following questions should be answered: a) What are the key concerns? b) Why are they occurring? c) What are we doing about it? d) What more should we be doing? and e) How can our proposal contribute?

✦ There should be a regular schedule to the sustainability reporting process. Monitoring and updating should aim to observe the effectiveness of the proposed solutions or responses to the key concerns that were identified. Information should also be provided on the status of the key concerns (problems) and their causes. This approach should lead to an evolution over time in the content of the sustainability reporting process

It is clear that sustainability reporting is a developing field in which much can still be learned. Lessons can be taken from existing SoER initiatives. The availability and quality of essential information is important in the sustainability reporting process. The possibility of creatively involving local communities in collecting data should be examined. Sustainability reporting processes should be evaluated regularly to check if the information generated is used and suitable to guide the decision-making process in a more sustainable way.

Examples of Sustainability Reporting

A number of reporting processes that can be considered examples of sustainability reporting take place at the state level in the U.S., such as in the states of Oregon and Minnesota. Examples at the regional level include: Pierce County in the U.S. State of Washington and the Canadian Hamilton-Wentworth region. Well developed examples of sustainability reporting at the city level are rare. Interesting examples are mostly found in the U.S., such as in Santa Monica, California.

Minnesota Milestones

In 1991 Governor A.H. Carlson established "Minnesota Milestones," a long-range plan for the state and a tool to measure results. Minnesota Milestones set forth a vision with measurable goals for the state's future based on what Minnesotans said were their hopes for the state and their communities. The goals and indicators of Minnesota Milestones were established in 1992. In 1993, the first progress report was published (Minnesota Planning 1994). Further progress reports were released in 1996 and 1998 (Minnesota Planning 1996, Minnesota Planning 1998).

The vision for Minnesota's future is presented in the following box. Directly linked to this vision are 20 goals and linked to the goals are 79 indicators, as indicated in Table 12-1.

Examining Progress in the Santa Monica's Sustainable City
Progress Report for "Transportation"

Adopted indicators/goals

✤ Using 1990 figures as a baseline, increase ridership on Santa Monica Municipal Bus Lines (including shuttles) by 10 percent by the year 2000

✤ Achieve an Average Vehicle Ridership (AVR) of 1.5 for all employers in Santa Monica with over 50 employees by the year 2000

✤ Convert 75 percent of the City vehicle fleet to vehicles using reduced-emission fuels by the year 2000

Progress Towards Goal

✤ Santa Monica Municipal Bus Line (SMMBL) Ridership: the 1990 baseline for this indicator is 19 million riders. The targeted 10 percent increase by the year 2000 is 20.9 million riders. Total ridership in 1995 was 17.875 million riders. This represents a decrease in ridership of 5.9 percent from the baseline

✤ City-wide Average Vehicle Ridership for Employers with more than 50 employees: prior to the implementation of the City's Transportation Management Plan (TMP) Ordinance in 1993, city-wide AVR was 1.29. In 1994, after one year of program implementation the figure had risen to 1.34. By July 1995, city-wide AVR had reached 1.37. Staff from the City's Transportation Management Office expect city-wide AVR to continue rising and believe the indicator goal will be attained, barring any changes in State legislation

✤ Fleet Vehicles: in 1993, 10 percent of the City's fleet operated on reduced-emission fuels (REFs). In 1995, 15 percent of the City's fleet had been converted and by the beginning of 1996 that number had risen to 21.7 percent with 127 out of a total of 585 vehicles operating on REFs. Fleet Management staff indicate that approximately 70 to 80 gasoline or diesel-fueled vehicles will be replaced with REF vehicles in FY 1996-97. With the implementation of the planned Fleet Management Program which began in July 1996 staff fully expect the City will reach the 75 percent indicator goal by the year 2000

Obstacles

❖ Bus Ridership: Ridership on the SMMBL significantly decreased between 1991 and 1994 and is currently 5.9 percent below the 1990 baseline. Transportation Department staff attribute this decrease to effects of the economic recession. In FY1994 and FY1995 ridership began to increase as the economy began to improve. These recent fluctuations indicate that ridership is dependent on several variables, many of which are out of the City's control. These variables may complicate the City's efforts to achieve its indicator goal. In an effort to improve service and increase ridership, the Transportation Department began an outreach program in March 1996. Transportation staff feel that this outreach effort, the development and implementation of service improvements, and increased marketing efforts will help to continue the current upward trend in ridership

❖ Proposed State Legislation Affecting the City's TMP Ordinance: the City's Transportation Management Coordinator has indicated that several bills are pending in the State legislature which would prohibit Air Districts and local governments from imposing mandatory requirements on California businesses for the purpose of reducing air pollution. If any of this pending legislation is signed into law the City may be required to repeal or significantly modify its Transportation Management Plan Ordinance which currently requires employers of more than 10 employees to submit annual emissions reduction plans. If this occurs it would significantly impair the City's efforts to meet the Sustainable City Program goal of city-wide AVR of 1.5 by the year 2000

❖ REF Vehicle Purchase and User Acceptance: since the implementation of the REF Policy for Vehicle Purchases, staff have encountered the following problems related to vehicle purchase and user acceptance: a) staff has found it difficult at times to find REF vehicles that meet the City's specifications and requirements, particularly for specialty vehicles and heavy equipment; b) particularly in the early stages of implementation of the REF vehicle purchasing policy, users reported performance problems such as poor acceleration and limited range from some of the REF vehicles; c) due to performance problems with some of the early vehicles, some users have developed a negative view of REF vehicles and do not wish to use them in the future

Recommendations

❖ Bus Ridership: the Transportation Department's outreach program and Service Improvement Plan should be given high priority and support, as its purpose is to improve service and increase bus ridership. Because the City is behind schedule on this goal due to past ridership decreases, this indicator should be carefully monitored and future decreases should be analyzed to pinpoint the cause and identify changes necessary to reverse the trend

❖ Transportation Management: City staff should investigate various avenues to oppose the pending State legislation that threatens to significantly impair the City's efforts to reach its city-wide AVR target. This might include passage of a motion by Council expressing the City's opposition to the legislation and support of lobbying efforts in Sacramento

❖ Fleet Management Program: as this program is designed to ensure that the City meets its 75 percent REF vehicle goal, it should be given full support and implemented as a high priority. The program should be annually monitored to ensure that it meets its annual targets and remains on course

➼ Bicycle Master Plan: in light of the fact that this plan represents a scaled back version of the original plan, every effort should be made to fully implement the plan's recommendations. In addition, the City should conduct an assessment to determine the city-wide change in bicycle ridership due to implementation of the Bicycle Master Plan

SOURCE City of Santa Monica (1996)

TABLE 12-1 *Minnesota Milestones 1996 Goals and Indicators*

Goals	Indicators
1. Our children will not live in poverty	➼ Percentage of children living in a household below the poverty line ➼ Percentage of parents who receive full payment of awarded child support
2. Families will provide a stable environment for their children	➼ Teen pregnancy rate ➼ Runaways ➼ Percentage of 12th-graders who have ever attempted suicide ➼ Apprehensions of children ➼ Percentage of children who use alcohol or illegal drugs at least monthly ➼ Rate of divorces involving children ➼ Percentage of students who move more than once a year
3. All children will come to school ready to learn	➼ Percentage of sixth-graders watching television or videos more than 40 hours per week ➼ Percentage of parents satisfied with their child-care arrangements ➼ Percentage of children who have healthy diets ➼ Abused or neglected children
4. Minnesotans will excel in basic academic skills	➼ Achievement test scores ➼ Number of school districts with a 12th-grade drop-out rate over 10 percent
5. Minnesotans will be healthy	➼ Infant mortality rate ➼ Percentage of low birthweight babies ➼ Percentage of children who are adequately immunized ➼ Percentage of Minnesota adults who do not smoke ➼ Life expectancy

TABLE 12-1 *Minnesota Milestones 1996 Goals and Indicators (Continued)*

Goals	Indicators
6. Our communities will be safe, friendly and caring	✦ Percentage of people who feel they can rely on another person in their community for help ✦ Violent crimes reported ✦ Percentage of people who feel safe in their communities ✦ Percentage of people who have been crime victims ✦ The rate of violent and injury-related deaths ✦ Percentage of Minnesotans who volunteer for community activities ✦ Percentage of youths who volunteer at least an hour a week
7. People who need help providing for themselves will receive the help they need	✦ Number of people using homeless shelters ✦ Percentage of recipients of Aid to Families with Dependent Children on assistance more than 24 consecutive months ✦ Percentage of unemployed remaining unemployed more than 26 weeks ✦ Quality of life for people with long-term limitations
8. People with disabilities will participate in society	✦ Percentage of public facilities that are accessible
9. We will welcome, respect and value people of all cultures, races and ethnic backgrounds	✦ Number of discrimination complaints filed in Minnesota ✦ Percentage of people who say they have been discriminated against in the past year ✦ Percentage of state legislators and constitutional officers who are members of an underrepresented racial or ethnic group ✦ Percentage of state legislators and constitutional officers who are female
10. Minnesota will have sustained, above-average, strong economic growth that is consistent with environmental protection	✦ Minnesota's per capita gross state product as a percentage of US per capita gross national product

TABLE 12-1 *Minnesota Milestones 1996 Goals and Indicators* (*Continued*)

Goals	Indicators
11. Minnesotans will have the advanced education and training to make the state a leader in the global economy	➻ College graduation rates of various systems ➻ Cost of college tuition ➻ Percentage of high school graduates who are pursuing advanced training, apprenticeships or higher education one year after high school ➻ Percentage of recent technical college graduates employed in a job related to their training ➻ Percentage of Minnesotans who use public libraries
12. All Minnesotans will have the economic means to maintain a reasonable standard of living	➻ Minnesota median family income as a percentage of US median family income ➻ Percentage of population living in households with incomes at least 200 percent of the poverty line ➻ Percentage of Minnesotans with health-care insurance
13. All Minnesotans will have decent, safe and affordable housing	➻ Percentage of low-income housing units with severe physical problems ➻ Percentage of low-income renters paying more than 30 percent of their income for housing ➻ Home ownership rate
14. Rural areas, small cities and urban neighborhoods throughout the state will be economically viable places for people to live and work	➻ Percentage of Twin Cities population living in census tracts with poverty rates 1.5 times the state average ➻ Percentage of population living in countries with per capita income less than 70 percent of US nonmetropolitan per capita income ➻ Minnesota nonmetropolitan per capita income as a percent of US nonmetropolitan per capita income ➻ Primary-care physicians per 10,000 people in nonmetropolitan Minnesota ➻ Minnesota's rank in telecommunications technology ➻ Percentage of nonmetropolitan population in communities served by two or more options for shipping freight
15. Minnesotans will act to protect and enhance their environment	➻ Average annual energy use per person ➻ Highway litter ➻ Total water use ➻ Solid waste produced and recycled ➻ Percentage of students passing an environmental education test

TABLE 12-1   *Minnesota Milestones 1996 Goals and Indicators (Continued)*

Goals	Indicators
16. We will improve the quality of the air, water and earth	→ Air pollutants emitted from stationary sources → Number of days per year that air-quality standards are not met → Percentage of river miles and lake acres that meet fishable and swimmable standards → Percentage of monitored wells showing ground water contamination → Soil erosion per acre of cropland → Toxic chemicals released or transferred → Quantity of hazardous waste generated → Number of Superfund sites identified and cleaned up
17. Minnesota's environment will support a rich diversity of plant and animal life	→ Diversity of songbirds → Number of endangered, threatened or special-concern native wildlife and plant species → Acres of natural and restored wetlands → Acres of forest land → Land area in parks and wildlife refuges
18. Minnesotans will have opportunities to enjoy the natural resources	→ Miles of recreational trails → Number of public access sites on lakes and rivers
19. People will participate in government and politics	→ Percentage of eligible voters who vote in gubernatorial elections → Percentage of dollars contributed to campaigns coming from small contributions
20. Government in Minnesota will be cost-efficient and services will be designed to meet the needs of the people	→ Percentage of the state budget for which goals and outcome measures have been established → Percentage of the local government budgets for which goals and outcome measures have been established → Percentage of Minnesotans who say they get their money's worth from their local and state taxes

SOURCE Minnesota Planning (1996)

The 1998 progress report (Minnesota Planning 1998) shows clear progress for the following goals:

→ Academic achievement

→ Health

❖ Support for independent living

❖ Sustainable economic growth

❖ Skilled work force

❖ Standard of living

❖ Outdoor recreation

and a decline in the following two goals:

❖ Stable, supportive families

❖ Participation in democracy

Goals with stable or mixed results include:

❖ Safe, caring communities

❖ Decent, affordable housing

❖ Viable rural and urban economies

❖ Conservation of natural resources

❖ Healthy ecosystems

Goals which could not be measured because of inadequate data include:

❖ Child poverty

❖ School readiness

❖ Inclusive communities

❖ Responsive, efficient government

❖ Quality of the environment

Table 12-2 shows detailed results for the goal "Stable, supportive families." This goal is measured by five indicators: "Satisfaction with child care"; "School transfers"; "Child abuse and neglect"; "Teen pregnancy"; and "Runaways." The figures show problems of increasing school transfers, child abuse, and neglect and children running away from home (Minnesota Planning 1998).

Oregon Shines

In the late 1980s, Governor N. Goldschmidt challenged the citizens of Oregon to take control of their economic destiny. In response, more than 150 businesses, government, and community leaders came together to develop a vision for the future that included creating good jobs and a strong economy while enhancing quality of life. Called "Oregon Shines," that 1989 vision has guided the state toward a more diversified economy and created a workforce

TABLE 12-2 *Detailed Figures for the Indicators Included in the Goal: "Stable, Supportive Families"*

Goal: Families will provide a stable, supportive environment for their children

Indicator	Results
Satisfaction with child care	Percentage of parents satisfied with the quality of child care (formal and informal) their children receive 1995 1997 84% 97%
School transfers	Public school transfers during the school year, as a percentage of total enrollment 1993 1994 1995 1996 1997 14% 15% 16% 17% 17%
Child abuse and neglect	Abused or neglected children (per 1.000 children under age 18) 1990 1991 1992 1993 1994 1995 1996 7.8 8.6 9.3 9.0 8.4 8.3 8.2
Teen pregnancy	Teen pregnancy rate (per 1.000 girls age 15 to 17) 1990 1991 1992 1993 1994 1995 1996 33.7 32.3 30.3 29.3 28.3 27.7 27.5
Runaways	Runaways (per 1.000 children under age 18) reported by sheriffs and police 1990 1991 1992 1993 1994 1995 1996 8.7 8.9 9.0 10.1 11.4 13.3 14.7

SOURCE Minnesota Planning (1998)

with the skills necessary for success today and a continuing high quality of life. In 1996, Governor J. Kitzhaber convened a 46-member citizen task force to assess the state's progress toward the goals set for Oregon Shines in the beginning phase and to update the original vision. This resulted in the Oregon Shines II report (Oregon Progress Board 1997). As shown in Figure 12 2, Oregon Shines has a structure in which the vision is translated into goals and these are linked to benchmark topic areas.

For each of the three interconnected goals—quality jobs for all Oregonians; safe, caring and engaged communities; and, healthy, sustainable surrounding—the Oregon Shines II report asks the following important questions:

➤ Where are we?

➤ Where do we want to be?

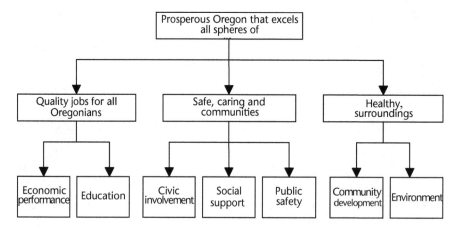

FIGURE 12-2 *"Oregon Shines" vision, goals, and benchmark.*
SOURCE Oregon Progress Board 1997

➤ How will we get there?

➤ How will we monitor progress?

Within each benchmark topic area, indicators were introduced and benchmark targets set. While the Oregon Shines II report (Oregon Progress Board 1997) does not give a detailed description of trends for the different benchmarks, it does elaborate on how benchmarks were set and evaluated. This is further discussed in following sections of this chapter.

Pierce County Quality of Life Benchmarks

The purpose of the Pierce County Quality of Life Benchmark Project is to gather and publish information about the well-being of the community as a whole. The Pierce County benchmarking effort uses information gathered from a variety of sources to track changes in different aspects of residents' lives. The intention of the benchmark process is to make very generalized statements about whether life in the county is getting better, worse, or staying the same. The idea is that by making objective measurements of past trends and current conditions, it is possible to shape future public policy in an informed and holistic way. The Benchmark Project combines over 200 discrete pieces of data into 80 indicators, organized into 29 different subgroups. The 80 indicators are divided into 9 goal categories, and those are ultimately combined into one composite "quality of life" measurement. Data for the benchmarks have been gathered for 1989–1996, with 1990 serving as the "benchmark year" against which all other years are measured. This historical perspective allows for the identification of trends in the data. Those trends can then be analyzed in terms of the ability to influence positive changes in successive years (Pierce County 1998). In annual reports the important information is presented in tables, trends are illustrated in graphs, and issues are discussed.

Overall, of the 80 indicators tracked since 1990, 46 have improved, while 26 have declined, and 8 show no real change. As a result, the Quality of Life index shows a 6 percent improvement over the last 7 years. If the volatile land-use indicators are removed, the index shows the same general pattern, but results in an improvement of 9 percent. The housing and land-use indicators taken together show some improvement since 1990. The economy has been strong, showing steady improvement over the past seven years. After peaking in 1992, the environmental indicators have dropped below their 1990 level. The decline has been driven by increases in toxic chemical releases relative to manufacturing employment. Pierce County's public transportation system has contributed to an overall improvement in the transportation indicators. Infrastructure indicators have shown a mix of positive and negative trends—waste-water treatment indicators are positive, whereas solid-waste indicators have moved in the opposite direction. Cultural and recreational opportunities for area citizens have improved, and problems have been encountered with the collection of data for indicators of secondary education (Pierce County 1998).

Hamilton-Wentworth's Sustainability Indicators

In 1993, the Hamilton-Wentworth Regional Council adopted its Vision 2020. The vision established hundreds of goals and strategic actions developed by citizen working groups. To help the region in judging progress, a set of signposts called *indicators* were identified. The indicators represent each of the eleven theme areas of Vision 2020. The aim is to observe indicators over time and report trends annually, making it possible for citizens, political leaders, business and community groups to understand the impacts of some of the decisions and actions of the community. The indicators are things that can be measured directly and which provide information about changing conditions and trends. Over a hundred individuals representing different interests in the community worked along with a project team to choose the indicators. The indicators were selected from dozens of possibilities because they were meaningful to the people involved in choosing them and were based on data that was already being collected (Regional Municipality of Hamilton-Wentworth 1995).

The Hamilton-Wentworth approach is characterized by its clarity. Information presented can be easily understood by the public. The report consists of 28 report cards, one for each signpost. As is shown in Figure 12-3 for the category "water resources," each report card contains a detailed description of an indicator and the topic it represents. A graph illustrates the trend or condition being measured. The text also describes how the community can make an improvement and where the data was obtained. The face in the upper right corner tells readers at a glance whether the indicator is improving or not since 1993. A heavy arrow on the graph points in the direction of positive change (Regional Municipality of Hamilton-Wentworth 1995).

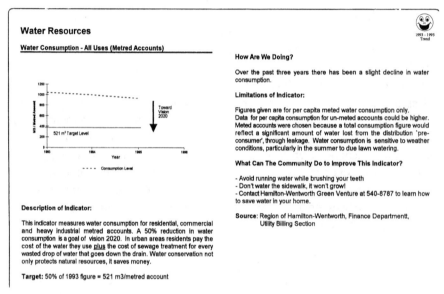

FIGURE 12-3 *An example of a report card of the Hamilton-Wentworth's sustainability indicators.*
SOURCE Regional Municipality of Hamilton-Wentworth 1995

The 1995 Background Report (Regional Municipality of Hamilton-Wentworth 1995) indicates that of the 28 signposts, 10 show a positive trend for the period of 1993-1995, five show a negative trend and 13 are neutral (not evolving in a positive or negative way). Negative trends are shown for the following indicators (and related theme areas): the number of complaints about air quality per year (theme area: air quality); the amount of road salt used on regional roads (theme area: water resources); office vacancy rates (theme area: urban form); annual transit ridership per capita (theme area: transportation); number of farms participating in the environmental farm plan program (theme: agriculture).

Santa Monica's Sustainable City Program

In September 1994, the Santa Monica City Council adopted the Santa Monica Sustainable City Program. This program was developed by the city's task force on the environment and city staff as a way to create the basis for a more sustainable way of life. The Sustainable City Progress Report (City of Santa Monica 1996) assesses the city's progress in implementing the Sustainable City Program and meeting the program's goals. For each policy area, an indicator is introduced to which a target is linked. The progress teport gives the adopted indicator and goal for each policy area, discusses progress toward the goal, examines obstacles, and develops recommendations. The following box gives an example for the "transportation" policy area.

Examining Progress in the Santa Monica's Sustainable City Progress Report for "Transportation"

Adopted indicators/goals

↠ Using 1990 figures as a baseline, increase ridership on Santa Monica Municipal Bus Lines (including shuttles) by 10 percent by the year 2000

↠ Achieve an Average Vehicle Ridership (AVR) of 1.5 for all employers in Santa Monica with over 50 employees by the year 2000

↠ Convert 75 percent of the City vehicle fleet to vehicles using reduced-emission fuels by the year 2000

Progress Towards Goal

↠ Santa Monica Municipal Bus Line (SMMBL) Ridership: the 1990 baseline for this indicator is 19 million riders. The targeted 10 percent increase by the year 2000 is 20.9 million riders. Total ridership in 1995 was 17.875 million riders. This represents a decrease in ridership of 5.9 percent from the baseline

↠ City-wide Average Vehicle Ridership for Employers with more than 50 employees: prior to the implementation of the City's Transportation Management Plan (TMP) Ordinance in 1993, city-wide AVR was 1.29. In 1994, after one year of program implementation the figure had risen to 1.34. By July 1995, city-wide AVR had reached 1.37. Staff from the City's Transportation Management Office expect city-wide AVR to continue rising and believe the indicator goal will be attained, barring any changes in State legislation

↠ Fleet Vehicles: in 1993, 10 percent of the City's fleet operated on reduced-emission fuels (REFs). In 1995, 15 percent of the City's fleet had been converted and by the beginning of 1996 that number had risen to 21,7 percent with 127 out of a total of 585 vehicles operating on REFs. Fleet Management staff indicate that approximately 70 to 80 gasoline or diesel-fueled vehicles will be replaced with REF vehicles in FY 1996-97. With the implementation of the planned Fleet Management Program which began in July 1996 staff fully expect the City will reach the 75 percent indicator goal by the year 2000

Obstacles

↠ Bus Ridership: Ridership on the SMMBL significantly decreased between 1991 and 1994 and is currently 5,9 percent below the 1990 baseline. Transportation Department staff attribute this decrease to effects of the economic recession. In FY1994 and FY1995 ridership began to increase as the economy began to improve. These recent fluctuations indicate that ridership is dependent on several variables, many of which are out of the City's control. These variables may complicate the City's efforts to achieve its indicator goal. In an effort to improve service and increase ridership, the Transportation Department began an outreach program in March 1996. Transportation staff feel that this outreach effort, the development and implementation of service improvements, and increased marketing efforts will help to continue the current upward trend in ridership

❧ Proposed State Legislation Affecting the City's TMP Ordinance: the City's Transportation Management Coordinator has indicated that several bills are pending in the State legislature which would prohibit Air Districts and local governments from imposing mandatory requirements on California businesses for the purpose of reducing air pollution. If any of this pending legislation is signed into law the City may be required to repeal or significantly modify its Transportation Management Plan Ordinance which currently requires employers of more than 10 employees to submit annual emissions reduction plans. If this occurs it would significantly impair the City's efforts to meet the Sustainable City Program goal of city-wide AVR of 1.5 by the year 2000

❧ REF Vehicle Purchase and User Acceptance: since the implementation of the REF Policy for Vehicle Purchases, staff have encountered the following problems related to vehicle purchase and user acceptance: a) staff has found it difficult at times to find REF vehicles that meet the City's specifications and requirements, particularly for specialty vehicles and heavy equipment; b) particularly in the early stages of implementation of the REF vehicle purchasing policy, users reported performance problems such as poor acceleration and limited range from some of the REF vehicles; c) due to performance problems with some of the early vehicles, some users have developed a negative view of REF vehicles and do not wish to use them in the future

Recommendations

❧ Bus Ridership: the Transportation Department's outreach program and Service Improvement Plan should be given high priority and support, as its purpose is to improve service and increase bus ridership. Because the City is behind schedule on this goal due to past ridership decreases, this indicator should be carefully monitored and future decreases should be analyzed to pinpoint the cause and identify changes necessary to reverse the trend

❧ Transportation Management: City staff should investigate various avenues to oppose the pending State legislation that threatens to significantly impair the City's efforts to reach its city-wide AVR target. This might include passage of a motion by Council expressing the City's opposition to the legislation and support of lobbying efforts in Sacramento

❧ Fleet Management Program: as this program is designed to ensure that the City meets its 75 percent REF vehicle goal, it should be given full support and implemented as a high priority. The program should be annually monitored to ensure that it meets its annual targets and remains on course

❧ Bicycle Master Plan: in light of the fact that this plan represents a scaled back version of the original plan, every effort should be made

> to fully implement the plan's recommendations. In addition, the City
> should conduct an assessment to determine the city-wide change in
> bicycle ridership due to implementation of the Bicycle Master Plan
>
> SOURCE City of Santa Monica (1996)

The general findings of the progress report (City of Santa Monica 1996) can be summarized as follows:

✤ Santa Monica is currently implementing numerous effective sustainable programs and policies in most areas of city operations

✤ Significant progress has been made toward meeting the indicator targets for Water Usage, Landfilled Solid Waste, City Fleet Vehicles Using Reduced-Emission Fuels, Wastewater Flows, Average Vehicle Ridership of Employers with over 50 Employees, Dry Weather Stormdrain Discharges to the Ocean, Deed-Restricted Affordable Housing Units, and Public Open Space

✤ Little or no progress has been made toward meeting the indicator targets for Energy Usage, Ridership on Santa Monica Municipal Bus Lines, Community Gardens, and the Implementing of a Sustainable Schools Program

✤ Progress cannot be adequately measured or outside variables affect the city's ability to meet the indicator targets for Post-consumer Recycled Content of City Paper Purchases, Trees in Public Spaces, Use of Hazardous Materials city-wide, and Known Underground Storage Tanks Sites Requiring Cleanup

✤ Despite the progress made toward meeting the various indicator targets, sustainable policies and programs are still being undertaken on a piecemeal basis within the city. Coordinated implementation of the Sustainable City Program within the city has not yet been achieved. Little or no effort has been made to merge the goals and objectives of the Sustainable City Program with the goals and objectives presented in the elements of the city's other plans

✤ To date, little or no effort has been made to involve the business community, schools and colleges, NGOs, and residents in the program (City of Santa Monica 1996)

Goals and Targets for Sustainable Development

One of the important steps in any process of impact assessment is the stage of attaching a value to predicted impacts. Once the magnitude of the impacts have been predicted, this must be put into a context that makes it possible to state whether or not the impacts will be significant, or to what degree they can be considered positive or negative. Similarly, in measuring progress, data must be interpreted that requires introducing value judgments. Concepts that can have a function in the interpretation of raw data which are often encountered in progress reports include, among others, benchmarking, goals, targets, and baseline data. The following box gives an overview of some important definitions.

Terms Related to Interpreting Data

❧ An action goal: is a specific aim that the community wishes to strive towards to achieve its vision for the future. Goals are used to guide organizations, experts, or professional staff to develop specific programs, and in this way they serve as an intermediate step between a Community Vision statement and specific measurable targets for improvement of conditions related to sustainability (ICLEI 1996)

❧ A target: a measurable commitment to be achieved in a specific time frame (ICLEI 1996)

❧ A benchmark: a point of reference, a criterion (Brown 1993)

❧ A trigger: is a unique form of a target. Triggers are agreed-upon future conditions that trigger further action by stakeholders when addressing a problem (ICLEI 1996)

❧ An action strategy and commitment: it is essential that an Action Plan specifies the action strategies and commitments of different stakeholders in order for them to work as partners in achieving the different objectives of an Action Plan. Action strategies and commitments should be very precise and contain specific projects, time schedules for implementation, and commitments to allocate money, time, and human resources (ICLEI 1996)

❧ A threshold: a lower limit of some state, condition, or effect; the limit below which a stimulus is not perceptible or does not evoke a response; the magnitude or intensity that must be exceeded for a certain reaction, phenomenon, result, or condition to occur or be manifested (Brown 1993)

❧ A standard: a required or specific level of excellence, attainment, wealth, etc. (Brown 1993)

❧ A norm: a standard (Brown 1993)

SOURCES Brown (1993) and ICLEI (1996)

Figure 12-4 shows how goals, targets, target-based indicators, and triggers play an important role in sustainable development planning.

In a benchmarking process, specific points of reference are set. Benchmarks can take many forms. Targets can be used as benchmarks against which the long-term performance of an activity can be evaluated. For instance, the data from a certain year can be used as the "benchmark year."

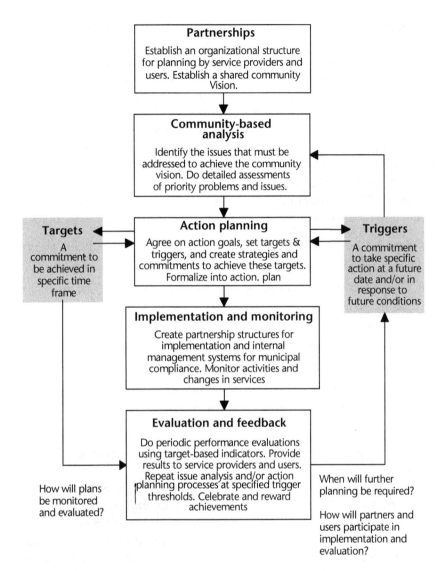

FIGURE 12-4 *The elements of a sustainable development planning.*
SOURCE ICLEI (1996)

In the Pierce County Quality of Life Benchmarks (Pierce County 1998), for example, 1990 serves as the "benchmark year" against which all other years are evaluated.

Targets can be used to check progress made toward a certain goal. This is being done in practice, for example, in the Santa Monica Progress Report (City of Santa Monica 1996). For each indicator a target has been set and the city reports on meeting the target. If the target has not been met, obstacles are discussed and recommendations are made. Targets for sustainable development should support long-term planning initiatives.

Introducing targets and mechanisms for performance review leads to two important features: direction and dynamism. The setting of a target, with an appropriate and realistic date by which it should be met, can have an important catalyzing effect (Pearce 1993). Targets, as they are used today in the sustainability discussion, can be quite general and descriptive or very specific and numeric. The more detailed and quantitative the target, the easier it becomes to use it in measuring progress.

Targets are linked to goals and objectives. For example, the Sustainability Plan for the City of San Francisco (Department of the Environment 1996) has a general structure in which the plan sets out the following for each topic:

* Broad, long-term social goals—these are very general and deal with the basic human and ecosystem needs that are to be addressed
* Long-term objectives to reach sustainability—these describe the state of the city when it reaches sustainability
* Objectives for the year 2001—the aim is to have quantified objectives that can be reached within a five-year time frame. They include objectives for businesses and individual residents as well as for city programs
* Specific actions to be taken to achieve the objectives—these are actions that can be taken by the government, business, the non-profit community and individuals

Table 12-3 gives an example of this San Francisco approach for the transportation sector.

In Environmental Impact Assessment (EIA) or Health Impact Assessment (HIA) we often make use of environmental and health standards to assess the importance of impacts. Analogously, if we want to develop a system for sustainability assessment, it is important that we think about standards for sustainability. The concept of sustainable development may, however, be too "soft" for the use of standards (Goeteyn 1996). The result is that sustainability standards cannot be found in the literature today. If developing standards for sustainability is difficult or even impossible, it may prove easier to develop thresholds for unsustainability. It may be easier to state clearly what is certainly not sustainable and which practices cannot be tolerated any longer.

While a *target* is very much indicative of where we want to go, a *threshold* signals a problem situation. A *trigger* can be considered a threshold which is very much action related. While several cities have developed goals and targets for sustainable development, no examples could be found of sustain-

TABLE 12-3 *Example of Goals, Long-Term Objectives, Objectives for the Year 2001, and Actions Included in the Sustainability Plan for the City of San Francisco*

Long-term objectives to reach sustainability	Objectives for the year 2001 (5-year plan)	Actions
Goal: To reduce dependence on automobiles		
A. 90 percent of all San Francisco trips, including at least 90 percent of commuter trips downtown, are made by means other than the private automobile	1. 85 percent of commuter trips to the downtown and 45 percent elsewhere are made by means other than a drive-alone private automobile	1-a. Create shuttle services from existing garages to key outlying destinations
	Non-commute-generated automobile trips have been reduced by five percent	1-b. Create a weekend and holiday Golden Gate Park and Museum shuttle from nearby garages and transit
B. 100 percent of all San Francisco-generated automobile trips are made in renewable-energy-powered vehicles		1-c. Provide incentives for businesses that provide commuter vans for employees
		1-d. Study the effects of eliminating private through traffic on Market Street and the methods of implementing such a plan
C. "Traffic calming" projects on a majority of City streets have been implemented		1-e. Study the effects of eliminating automobile traffic from small lengths of other streets in addition to Market Street
D. Market Street is closed to private automobiles east of Van Ness Avenue		1-f. Provide package storage lockers in transit stations and transit centers
E. A number of sections of streets have been closed to private automobiles		1-g. Provide access through doors and elevators in space on vehicles for baby carriages and large packages
		1-h. Use teleconferencing to reduce long-distance travel, and home businesses and telecommuting to reduce local travel (for business and government)
		1-i. Develop additional delivery services (for private sector)
		1-j. Advertise and promote transit use as a means of avoiding and not contributing to traffic congestion

ability thresholds or sustainability triggers. Triggers could be very forceful instruments, which can lead to immediate action if a certain state of sustainability or unsustainability is reached. City officials do not yet seem ready to take immediate action upon negative sustainability progress results; in this case thresholds and triggers would be very appropriate

In the same way that it is impossible to develop a standard set of sustainability indicators which is applicable in any city, it is not feasible to set universal standard targets. Indicators and targets for sustainable development at the local level have to be developed within a local context and have to be linked to the sustainability vision developed by the local community. For example, the human health chapter of the Sustainability Plan for the City of San Francisco includes the target: "reduction by 20 percent of the incidence of weapon-carrying by adolescents aged 14-17." While this is a very important sustainable development objective in a U.S. context, it may be irrelevant and far removed from the local context in Europe and Asia.

How to Develop Targets - the Process of Setting Targets

At the community and state levels, indicators and goals have usually resulted from the participation of major interest groups in a process to envision a desired future. The resulting indicators and goals thus represent a diversity of interests and reflect a community's values and priority concerns (Corson 1995).

Setting a target for sustainable development is to some degree comparable to setting an environmental or health standard. In both cases, we not only need to take into account the guidelines established by international panels of experts, but also to look at technical, economic, social or political considerations as deemed necessary (Hens and Vojtisek 1998). Setting targets to reach absolute sustainability seems an impossible task because it is the result of a process of making compromises. Again, this process is comparable to setting health standards in which a compromise is made between known health effects and what are practical and operational levels. In this way health standards will not often be a guarantee for a completely safe environment, free of any health threats (Janssens en Hens 1997). Thus, achieving sustainability targets does not necessarily indicate that we have reached a completely sustainable state. It is more likely to mean that we have reached a certain acceptable goal and that we should get ready for a new step toward a more sustainable society.

Targets should be challenging, yet realistic. This approach is taken by, among others, the cities of Santa Monica, California, and Vancouver, Canada (for its Southeast False Creek Sustainable Neighborhood project). Santa Monica states that targets are to be aggressive yet realistic and achievable

(City of Santa Monica 1996), while Vancouver says that its targets have been intentionally set high, so as to encourage innovative and creative solutions (City of Vancouver 1998).

The Oregon Shines II report gives some indications on how targets can be developed (Oregon Progress Board 1997): benchmark targets are set empirically and reflect the most complete knowledge of the factors and processes that will influence the achievement of the desired condition. The Oregon Progress Board (1997) has adopted a methodology for setting targets based on historical trends as a starting point for the process. In cases where benchmark trends have moved in a positive direction, these targets were carried forward. When the Board decided that Oregon needed to be more aggressive about achieving a future target, the trend was doubled to establish a higher target. In cases where Oregon has been moving backward, away from the established target since 1990, the aim was set to return to the best previous level by 2000. When available, national or regional norms were consulted for comparison. Oregon's position relative to the norm helped determine the amount of progress that is possible (more progress is possible when one is well below a norm and less when one is already near the top) and helped moderate some of the targets. In most cases, policy experts were then consulted to assure that the targets are aggressive, but realistic (Oregon Progress Board 1997).

When trend analysis is used in the Oregon Shines process, one of the four following approaches is followed (Oregon Progress Board 1997):

1. Standard-Positive method—the method consists of establishing the percentage change using the longest data time series available (starting with either 1980 or 1990). Apply the percentage change to the most recent data in order to set the 2010 target. Assume a straight line between 2010 and most recent data to set interim targets. If the 2010 target is better than the current value for the best state in the nation, the value for the best state is substituted. This method is applied to benchmarks that have shown satisfactory progress (this method is used for benchmarks showing a positive trend).

2. Standard-Negative method—the aim is to return to best level in time series by the year 2000 and improve by 0 percent between 2000 and 2010 (this method is used for benchmarks showing a negative trend).

3. Aggressive-Positive method—the method consist of establishing the percentage change using the longest data time series available (starting with either 1980 or 1990). To set the 2010 target, apply this percentage change to the most recent data, then double that value and add it to the

most recent data. Assume a straight line between 2010 and the most recent data to set an interim target. This method is applied to benchmarks that have shown unsatisfactory progress or benchmarks that have been targeted for special attention by the state government (this method is used for benchmarks showing a positive trend).

4. Aggressive-Negative method—the aim is to return to best level in time series (either 1980 or 1990) by 2000 and improve by 20 percent between 2000 and 2010 (this method is used for benchmarks showing a negative trend).

For each target developed in Santa Monica, a short explanation is given on how the target was set (City of Santa Monica 1996). The following are reasons given for setting a target:

❧ Target is mandated by state law

❧ Target represents a 20 percent reduction from the 1990 baseline. This is felt to be aggressive yet achievable based on results of existing programs and anticipated impacts of planned programs

❧ A new target is currently being developed

❧ Target reflects what was thought by the Task Force on the Environment and city staff to be aggressive yet achievable. It will likely be revised based on findings and recommendations

❧ Target reflects council-adopted ordinance requirements

❧ Target based on preliminary analysis

❧ Recommendations of Environmental Programs Division staff

❧ Target represents a 15 percent reduction in city-wide wastewater flows relative to the 1990 baseline

❧ Target based on council-adopted affordable housing targets

❧ Original target has been revised upward to reflect Recreation and Parks Commission standards

❧ Target is being reviewed as part of the Open Space Element/Parks and Recreation Master Plan process and may be revised upon review and final approval of those documents

❧ Target is mandated by federal law

❧ Revised target is based on recommendations by the Task Force on the Environment

❧ Revised target uses the Recreation and Parks Commission standard

It becomes clear that setting a target can be based on:

* Existing, adopted or mandated goals (e.g., based on a state law)
* More informal goals of the local authority, such as a sustainable development vision
* A choice made by the local authority. (This can be based on the recommendations of, for example, a Sustainable Development Task Force or experts)
* An analysis, such as a trend analysis

Because of this wide range of possibilities for setting targets, it is very important that authorities always clearly indicate why and how a certain target was chosen and how this may evolve over time.

We also have to watch out for possible conflicting targets: for example, in Oregon a conflict was indicated between the target for maintaining open space in the Portland metropolitan area and the target for maintaining urban growth boundaries (Oregon Progress Board 1997).

In general, the Oregon Progress Board (1999) proposes the following procedure in selecting targets (what they call benchmarks or goals):

1. Review the goal and make sure it is realistic (or sufficiently ambitious).

2. If possible, identify the payoffs from achieving this goal in terms of:
   * Reduced costs for future budgets
   * Improved lives for Oregonians
   * Improved productivity for the economy
   * Other measures

3. Examine recent efforts to address this problem, such as:
   * Programs and budgets, both by the state and other entities.
   * Who have been the key players?
   * What were the successes or setbacks?
   * Have strategies already been developed to achieve these goals?

4. Examine the best practice from other states, and especially, from around the world.

5. Propose a strategy to accomplish this goal:
   * Programs
   * Organizational change
   * Incentives
   * Budgets

6. Summarize what it will take to achieve the goal and what different levels of effort can be expected to achieve.

Examples of Existing Targets

Targets for sustainable development can be set at many different levels: that is, at international, national/regional, local or household levels.

At an international level, many goals have been formulated through the series of recent United Nations conferences, on education (Jomtien 1990), children (New York 1990), the environment (Rio de Janeiro 1992), human rights (Vienna 1993), population (Cairo 1994), social development (Copenhagen 1995), and women (Beijing 1995). A number of targets have been identified as a result of these conferences to measure the progress of development in particular fields. They reflect broad agreement in the international community and are achieved with the active participation of the developing countries. The following box gives an overview of the specific numerical targets set in Agenda 21. By examining these targets it becomes clear that they are mostly applicable to sustainable development in developing countries, not to urban areas in more developed countries. The number of quantitative targets in Agenda 21 is quite limited. The document contains many more descriptive and general goals, such as "the overall human settlement objective is to improve the social, economic and environmental quality of human settlements and the living and working environments of all people, in particular the urban and rural poor," but lacks specific targets for its goals (UNCED 1992).

Targets Mentioned in Agenda 21

For municipal and local governments overwhelmed by urban health problems, the global objective is to achieve, by the year 2000, a 10 to 40 percent improvement in health indicators for infant mortality, maternal mortality, percentage of low birth-weight in newborns and specific indicators (e.g. tuberculosis as an indicator of crowded housing).

Agenda 21 calls for a supply of 40 liters of safe water per person per day and the environmentally sound collection, recycling, or disposal of 75 percent of the solid waste from urban areas.

By 2000, more than 50 percent of each country's youth should have access to secondary education or equivalent vocational training.

Developed countries should reaffirm their commitment to reach as soon as possible the United Nations target of 0.7 percent of Gross National Product (GNP) annually for official development assistance; some have agreed to reach the target by the year 2000.

By the year 1995, in industrial countries, and by the year 2005, in developing countries, ensure that at least 50 percent of all sewage, waste waters and solid wastes are treated or disposed of in conformity with national or international environmental and health quality guidelines.

By the year 2025, provide all urban populations with adequate waste services.

By the year 2025, ensure that full urban waste service coverage is maintained and sanitation coverage achieved in all rural areas.

Ensure universal access to basic education, achieve primary education for at least 80 percent of girls and boys of primary school age through formal schooling or non-formal education and to reduce the adult illiteracy rate to at least half of its 1990 level.

Goals for communicable diseases:

❖ By the year 2000, to eliminate guinea worm disease.

❖ By the year 2000, eradicate polio.

❖ By the year 2000, to reduce measles deaths by 95 percent and reduce measles cases by 90 percent compared with pre-immunization levels.

❖ By the year 2000, reduce the number of deaths from childhood diarrhea in developing countries by 50 to 70 percent.

❖ By the year 2000, reduce the incidence of childhood diarrhea in developing countries by at least 25 to 50 percent.

❖ By the year 2000, reduce mortality from acute respiratory infections in children under five years by at least one third.

❖ By the year 2000, to provide 95 percent of the world's child population with access to appropriate care for acute respiratory infections within the community.

❖ By the year 2000, achieve an overall reduction in the prevalence of schistosomiasis and of other trematode infections by 40 percent and 25 percent, respectively, from a 1984 baseline.

SOURCE UNCED (1992)

At a local level it becomes easier to set specific sustainability targets because we can focus on the specific sustainability problems of a limited area and set a target on the basis of a compromise between available information and the different values of the local population. The following box gives an overview of targets set by the UK Bedfordshire County Council (Barton et al. 1995).

Sustainability Targets Set by Bedfordshire County Council

❖ Reduce the level of households/population not within easy walking distance of local services by 20 percent.

❖ Go for 50 percent of all journeys to work within urban areas by public transport, walking and cycling.

❖ Reduce the need to travel and overall distances traveled by passenger transport.

❖ Reduce CO_2 emissions from buildings, industry and transport to their 1991 level by the year 2001 and then by a further 15 percent.

❖ Increase the level of energy produced from non fossil fuel sources by 100 percent.

- ❖ Double the area of native broadleaf woodland by the year 2015.

- ❖ Provision and maintenance of public squares and green spaces at 25-35 percent of total land within the main urban areas.

- ❖ Increase the number of miles of rivers and waterways that are class 1 by 25 percent.

- ❖ Reduce the level of air pollution; suspended particles and gases; in residential/mixed use urban areas and town centers by 25 percent and to meet all EC guide values for local air quality.

- ❖ Make 80 percent of all new built development to be situated within urban areas and sites within the growth corridors identified.

- ❖ Reduce loss of greenfield land to development by 50 percent compared with 1986-1991 period.

- ❖ Reduce amount of derelict land and vacant buildings/housing stock by 50 percent.

- ❖ A doubling of recycled aggregates as a proportion of total aggregate products.

- ❖ Reduce the amount of waste disposal by landfill by 25 percent.

- ❖ Increase the extent of land covered by designations, managed by the County Council or covered by management agreements by 25 percent.

- ❖ Reduce noise pollution by 50 percent.

- ❖ Reduce the overall level of deprivation in Bedfordshire and reduce the disparity between the most deprived and least deprived areas.

- ❖ Reduce long term unemployment to 50 percent of 1991 level.

- ❖ Reduce the percentage of homeless and households in need by 50 percent.

- ❖ Reduce road accident casualties by one-third by the year 2000.

    Source: Barton et al. (1995)

The following box gives an example of targets set at an even more local level, in this case in a new urban development in the U.S. city of Tucson, Arizona.

Performance Targets in the Civano-Tucson Solar Village

The Civano-Tucson Solar Village is a major real estate development in the US desert city of Tucson, Arizona that is being built using the principles of sustainable development and traditional neighborhood design. Specific objectives are to: a) build a model sustainable community for an arid climate; b) use sun energy, conserve water, use land efficiency, reduce waste products and conserve time; c) promote a sense of community and place; and d) demonstrate that conservation can be both economically viable and socially relevant.

Key to the Civano Village concept are the ambitious performance targets developed for the village. The original targets, as expressed in the 1992 Master Plan, are to:

- ☀ Reduce energy consumption by 75 percent
- ☀ Reduce the consumption of water by 65 percent
- ☀ Reduce the production of solid waste by 90 percent
- ☀ Reduce air pollution by 40 percent
- ☀ Provide one job for every two residential units built
- ☀ Limit auto access by developing an international transportation circulation pattern that encourages pedestrians, bicycle, and (with permit) electric golf cart
- ☀ Provide affordable housing within the village

These targets will be updated every two years. An "IMPACT"-system approach (referring to Integrated Method of Performance and Contribution Tracking) was developed as a means of organizing resource efficiency goals and stakeholder cooperation for sustainable community development, and for measuring progress toward those goals over time. The "Performance Target Compliance Workbook", based upon the "Measurement and Implementation Plan", is intended to help land developers and builders achieve the goals and targets of Civano. Performance targets are also translated into performance standards. Builders and designers who meet these standards can be certified by the city for work on Civano.

SOURCE Concern et al. (1998)

Evaluating Sustainability Reporting and Sustainability Target Initiatives

It is important to realize that reporting on sustainable development is a subjective business, insofar that choices have to be made on the set of goals, targets, and indicators which will be reported on. When we deal with the question of "sustainability," we have to address "whose sustainability." We should never uncritically accept the results of a Sustainability Reporting study, and we need to realize that its results will be different depending on who has an influence on it: the ideas of politicians with widely differing ideologies, lobby groups, organized and unorganized community groups may be more or less represented.

Therefore, we should encourage input in Sustainability Reporting or even separate Sustainability Reporting initiatives by people or institutions representing different interests. Depending on the choice of goals and indicators, a very different picture of progress toward sustainable development may emerge. Consequently, it is important to know which basis goals, targets, and indicators have been selected and who has been participating in this selection. It would be positive if Sustainability Reporting were not only performed by local authorities, but also by NGOs (sustainable development for environmental groups, women's groups, minorities, consumer groups, etc.), academics, and corporations.

As reported by Ranganathan (1999), astute businesses are starting to recognize that there are many benefits in measuring and reporting on sustainability, such as heightened internal awareness and enhanced reputation and relations with outsiders. Although Sustainability Reporting is taken up by corporations, major limitations remain: the current approach does not help companies describe and disclose their accountability for the social, environmental, and economic impacts of their operations around the globe, but is instead very fragmented (Ranganathan 1999). Moreover, many corporations discuss environmental, social, and economic aspects of their operations separately, without integration of information in a sustainability framework.

Experience with sustainable development planning at local level shows a real need for targets. For example, in the UK the environmental group Friends of the Earth (FoE) issued a press release in January 1998 titled, "Local Agenda 21 Must Have Hard Targets and Timetables." FoE expects to see the government setting real targets and timetables for implementation of LA21 local plans, the provision of some financial support, and a process for evaluating the implementation of plans. FoE is convinced that achieving sustainable development at the local and community levels will only succeed if real targets and timetables are set. Targets proposed by FoE are: 30 percent cuts in CO_2 emissions over the next 10 years; traffic reduction targets of 30 percent by 2010; zero purchase of tropical timber products; and a 30 percent reduction in energy use over 10 years (FoE 1998).

Targets used as stepping-stones to long-term goals of securing sustainable development should lead to a discussion on sustainability issues and action within society and result in a more sustainable society. We should examine if this happens in reality.

Very little information is available to date on the effectiveness of sustainability targets. One reason is that measuring progress toward a sustainable society is a new activity and most cities have not yet started or are still struggling with the introduction of indicators and targets. Those who do have a system for measuring progress have yet to formulate an evaluation process. Also, we have to take into account that targets are set for the future. Evaluations of the target-based approach and of the targets themselves will become inevitable once the target dates have been reached.

A strategic plan for a sustainable future that includes an evaluation of benchmarks and targets is the Oregon Shines II (Oregon Progress Board 1997). After a review of the benchmarking system introduced in the original Oregon Shines plan of 1989, the Oregon Shines Task Force made eight recommendations to the Oregon Progress Board regarding changes to the benchmarking system. Seven all-day hearings were held to hear from interested parties regarding the effectiveness of the benchmarks. A special data analysts working group provided technical recommendations on improving the integrity of the data used in benchmarking.

The major conclusion of this review exercise reads as follows (Oregon Progress Board 1997):

> "The benchmarks are an excellent tool for encouraging collaboration among different interests, engendering long-term thinking and developing results-oriented management systems. These changes in the benchmarking system reflect an understanding that the current system is maturing, but is far from perfect."

The following recommendations were made to improve the system in Oregon (Oregon Progress Board 1997):

* Reduce the number of benchmarks from the current 259 to approximately 100
* Develop a system that shows how benchmarks are related
* Adopt only benchmarks that can and will be measured
* Adopt benchmark targets that are statements of realistic outcomes that can be achieved by the government, social institutions, businesses, and citizens acting over a given period of time
* Develop accurate, understandable, and timely local data for all benchmarks by 2002 in consultation with local data users
* Encourage state agencies to play a leadership role in achieving the benchmarks
* Facilitate the development of strategies that impact the benchmarks
* Overhaul the reporting format, using a graphical format in place of the existing tabular format

In addition, the UK national practice with targets and reporting mechanisms can also teach us something in regard to the future use of targets at local level. ENDS (1993) cited in Pearce (1993) reveals the following on the UK practice:

> "The UK has a number of environmental targets enshrined within legislation, but there is a need for many more. Since the 1990 White Paper (ENDS 1993), the government has produced two updates which report the progress made against the original 350 commitments. The two updates also contain further pledges, and report the progress in achieving these. This rolling program of target setting and

reporting now looks set to become an annual process. However, many of the targets contained in these documents are
vague and unambitious, or they simply require actions that
should have been done anyway—implementing EC Directives for example. The reports themselves are largely self-
serving and lack self-criticism. The parallel annual depart-
mental reports are also difficult to assess, for they also contain few explicit objectives such as that for improving
energy efficiency. Targets need to be set consistently across
all departments for a number of key parameters such as
waste recycling, waste emissions and air pollution. Without
these, it is very difficult to differentiate short-term, incremental measures from purposeful and deliberate long-term
progress towards sustainability."

In other words, working with targets requires us to be very clear, ambitious, and critical about what we are doing, while keeping the explicit objectives in mind.

It is very important that in the future, more research is carried out in relation to the effectiveness of targets. Moreover, it is advisable that local authorities are supported and guided in setting targets by specialized agencies.

Conclusion

Sustainability reporting is a very important activity because it provides us
with a standard way of measuring progress toward a sustainable society.
Measuring results of government programs is increasingly being accepted as
the correct way to examine progress in society. Local authorities that have
introduced sustainability reporting are still scarce, and those that have are
currently in a learning process. Examples of sustainability reporting show
that each one uses its own terms and methodology. Terms such as goals,
objectives, benchmarks, milestones, and signposts are used interchangeably
and this shows the need for a detailed definition of reporting practices.

Setting sustainability targets is, to a large degree, dependent on the sustainability vision of a local community. Setting standard targets which would be
applicable in many different cities is impossible. Targets should reflect the
varying importance each community attaches to the different issues. The targets also represent varying degrees of commitment to their achievement.

Sustainability targets are not only important for measuring progress, but
are also necessary in an impact assessment process in which clear statements
have to be made on the acceptability and significance of predicted impacts.
Therefore, sustainability assessment certainly benefits from clear sustainability targets.

Because of the wide range of possible methods in setting targets, it is very important that local authorities indicate why and how a certain target was chosen and how it may evolve over time. To date, very little information is available on the effectiveness of sustainability targets. Scientific research in sustainable development often focuses on indicators, but literature available on studies relating to setting targets is rare. Clearly there is a need for additional attention to sustainability targets in sustainable development research programs.

References

Barton, H., G. Davis and R. Guise. 1995. *Sustainable Settlements: A Guide for Planners, Designers and Developers*. Bristol: University of the West of England, Faculty of the Built Environment and the Local Government Management Board.

Briggs, D. J. 1998. State of environment reporting. In B. Nath, L. Hens, P. Compton and D. Devuyst, eds., *Environmental Management in Practice. Volume 1. Instruments for Environmental Management*, pp. 90-107. London: Routledge.

Brown, L. 1993. *The New Shorter Oxford English Dictionary*. Oxford: Clarendon Press.

City of Santa Monica. 1996. *Sustainable City Progress Report. Initial Progress Report on Santa Monica's Sustainable City Program*. Santa Monica: Task Force on the Environment.

City of Vancouver. 1998. Southeast False Creek sustainable neighborhood. *City of Vancouver* Website http://www.city.vancouver.bc.ca/commsvcs/planning/ sefc.htm (8 Jan. 1999).

Concern, the Community Sustainable Resource Institute and the Jobs & Environment Campaign. 1998. *Sustainability in Action. Profiles of Community Initiatives Across the United States*. Revised and updated 1998 edition. Washington DC: US EPA and Concern Inc.

Corson, W. H. 1995. Linking sustainability levels to performance goals at national and subnational levels. In T. C. Trzyna, ed., *A Sustainable World: Defining and Measuring Sustainable Development*. IUCN, ICEP, Development Alternatives. London: Earthscan.

Department of the Environment. 1996. *The Sustainability Plan for the City of San Francisco*. San Francisco: City and County of San Francisco.

ENDS. 1993. Latest Whitehall reports lack environmental targets - UK environmental policy. Environmental Data Services Report 205, 3. London: Environmental Data Services.

FoE, Friends of the Earth. 1998. Local Agenda 21 must have hard targets and timetables. *Friends of the Earth* Website http://www.foe.co.uk/pubsinfo/ infoteam/pressrel/1998/ 19980114145619.html (25 Nov. 1998).

Goeteyn, L. 1996. Measuring sustainable development at the national and international level. In B. Nath, L. Hens and D. Devuyst, eds., *Sustainable Development*, pp. 161-180. Brussels: VUBPress.

Hens, L. and M. Vojtisek. 1998. The establishment of health and environmental standards. In B. Nath, L. Hens, P. Compton and D. Devuyst, eds., *Environmental Management in Practice. Volume 1. Instruments for Environmental Management*. London: Routledge.

ICLEI, International Council for Local Environmental Initiatives. 1996. *The Local Agenda 21 Planning Guide. An Introduction to Sustainable Development Planning*. International Council for Local Environmental Initiatives, the International Development Research Centre and the United Nations Environment Programme.

ICLEI, International Council for Local Environmental Initiatives. 1998. Charter of European cities and towns towards sustainability (the Aalborg charter). *ICLEI* Website http://www.inclei.org/europe/ac-eng.htm (25 Febr. 1999).

Janssens, P. and L. Hens. 1997. *Mens en Milieu. Onze Gezondheid Bedreigd?* Antwerpen: Monografieën Stichting Leefmilieu, Stichting Leefmilieu, Uitgeverij Pelckmans.

LGMB, Local Government Management Board. 1996. *Sustainability Reporting. A Practical Guide for UK Local Authorities.* Bedfordshire: The Local Government Management Board.

Minnesota Planning. 1994. *Minnesota Milestones: 1993 Progress Report.* St. Paul: Minnesota Planning.

Minnesota Planning. 1996. *Minnesota Milestones: 1996 Progress Report.* St. Paul: Minnesota Planning.

Minnesota Planning. 1998. *Minnesota Milestones: 1998 Progress Report.* St. Paul: Minnesota Planning.

Oregon Progress Board. 1997. *Oregon Shines II: Updating Oregon's Strategic Plan.* Salem: Oregon Progress Board.

Oregon Progress Board. 1999. Selecting benchmarks. *Oregon Progress Board* Website http://www.econ.state.or.us/opb/strategy.htm (18 Jan. 1999).

Pearce, D. 1993. *Blueprint 3. Measuring Sustainable Development.* Centre for Social and Economic Research on the Global Environment. London: Earthscan.

Pierce County. 1998. *Pierce County Quality of Life Benchmarks. Annual Report. 3rd Edition.* Pierce County: Department of Community Services.

Ranganathan, J. 1999. Signs of sustainability. Measuring corporate environmental and social performance. In M. Bennett and P. James, eds., Sustainable Measures. Evaluation and Reporting of Environmental and Social Performance. Sheffield: Greenleaf Publishing.

Regional Municipality of Hamilton-Wentworth. 1995. *Hamilton-Wentworth's Sustainability Indicators. 1995 Background Report.* Hamilton: Environment Department.

Ryding, S. O. 1994. *Environmental Management Handbook. The Holistic Approach - From Problems to Strategies.* Amsterdam and Boca Raton: IOS Press and Lewis Publishers.

UNCED, United Nations Conference on Environment and Development. 1992. *Agenda 21: Program for Action for Sustainable Development.* New York: United Nations Conference on Environment and Development.

Chapter 13

Global Change, Ecological Footprints, and Urban Sustainability

William E. Rees

Summary

This chapter uses ecological footprint analysis to assess the prospects for urban sustainability and the conditions necessary to achieve it. Up to 80 percent of the human populations of high-income countries "live" in cities, and half of humanity will be urbanized early in the new millennium. Some people see this trend as evidence that technological "man" is leaving nature behind. This reveals a culturally distorted perception of biophysical reality—urbanites actually weigh even more heavily on the natural world. In strictly ecological terms, cities are incomplete ecosystems, the human equivalent of cattle feedlots. They are highly disturbed nodes of consumption in the total vast and widely dispersed human habitat. Indeed, eco-footprint analysis shows that urban populations are absolutely dependent on the biophysical output of an extra-urban ecosystem area up to 1,000 times the size of cities themselves. Meanwhile, excessive resource demand and waste discharge threatens the productivity of those same ecosystems. Today we live in an era of accelerating global change and are dependent on insecure external sources of supply. This situation demands sound and risk-averse strategies to ensure urban sustainability. In this light, we must rethink the concept of "city," reduce material consumption, and expand the scope of urban land-use planning with a view toward increasing regional self-reliance. Such ideas fly in the face of prevailing emphasis on growth, expanding trade, and global economic integration.

Elements of Human Ecology

A central premise of this chapter is that human-induced global environmental change is a symptom of fundamental or systemic ecological dysfunction. It follows that if we are to develop "new popular ideas" for urban sustainability (Girardet 1992), we must do so from an explicitly human ecological perspective.

This perspective requires that we examine humans as integral components of the ecosystems that support them.[1] Grappling with this relationship is initially difficult for technological "man." How many of us recognize that, in ecological terms, humans are the most ecologically significant marine carnivore (not to mention terrestrial herbivore)? Most people are simply unaccustomed to thinking of themselves as ecological entities in quite this way. The "Cartesian dualism" that permeates modern industrial society has created a mental barrier between people and the rest of nature that prevents us from fully knowing ourselves—and this perceptual distortion is magnified by spatial and psychological effects of urbanization itself. The time has come, therefore, to acknowledge a simple ecological truth: despite our technological, economic, and cultural achievements, the relationship of humankind to the rest of the ecosphere is fundamentally no different from that of millions of other species with which we share the planet (Rees 1990). Like all other organisms, humans still depend on other species, on intact ecosystems, and on the so-called "life support functions" of the ecosphere for survival.

The Ecological Worldview

If we hope to understand the ecological niche of humans and the ecological properties of cities, we must study people much the same way as we would any other plant or animal species. The important question is: What are the critical material relationships between humans and the other components of their supportive ecosystems?

To begin, ecologists would classify people as macro-consumers, with reference to their place within the global food web. In general, macro-consumers are large organisms, mainly animals, that depend on other organisms, either green plants or other animals, which they consume to satisfy their metabolic needs. There is of course a major material difference between humans and other macro-consumers. In addition to our biological metabolism, the human enterprise is characterized by an industrial metabolism (Ayres and Simonis 1994). Like our bodies, all the artifacts of industrial culture—buildings, equipment, infrastructure, tools, and toys (the human-made "capital" of economists)—require continuous flows of energy and material to and from "the environment" for their production, operation, and maintenance. Both our bio-metabolic and industrio-metabolic appropriations from global supplies have to be accounted for in any human ecological assessment.

Humans are also producer organisms. However, there is an important difference between production in nature and production in the economy. In nature, green plants are the "factories." They are called "primary producers" because they use the simplest of low-grade inorganic chemicals (mainly water, carbon dioxide, and a few mineral nutrients) and an extraterrestrial source of relatively dispersed energy—light from the sun—to assemble the high-grade fats, carbohydrates, proteins, and nucleic acids upon which most other life forms and the ecosphere at large are dependent. By contrast, human beings are strictly secondary producers. Production by our factories requires enormous inputs of high-grade energy and material resources extracted from the rest of the ecosphere. That is, all economic output requires the consumption of a larger quantity of available energy and material first produced by nature (Rees 1999). Seen through our ecological lens, the economy appears as an open, growing, wholly dependent subsystem of a materially closed, non-growing, finite, ecosphere (Daly 1992) (Figure 13-1).

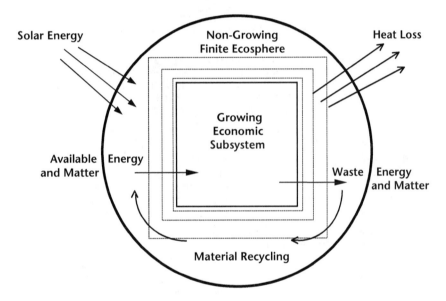

The human economy, in all its diversity and complexity,
is a wholly dependent sub-system of the ecosphere

FIGURE 13-1 *The ecological perspective.*

Because economic production depends on production in nature, the hierarchical relationship between the economy and the ecosphere is critically relevant to urban sustainability in light of modern interpretations of the second law of thermodynamics (see following box for full details). The second law states that all complex, self-organizing systems are subject to forces of spontaneous disintegration. That is, any isolated system becomes increasingly unstructured and dissipated in a inexorable slide toward thermodynamic equilibrium or maximum entropy. (This is a state in which "nothing happens or can happen" [Ayres 1994].) However, open systems, like cities, can maintain themselves and grow by importing high-grade energy and material from their host environments and by exporting entropy (degraded energy and material) back into those environments. Thus, our cities can produce "the wealth of nations" only by consuming the products and services of the ecosphere. This suggests two basic criteria for ecological sustainability: 1) consumption of renewable and replenishable energy and resources by the economy must not exceed their production in nature, and; 2) production of degraded energy and matter (entropic waste or pollution) by the economy must not exceed the assimilative capacity of local ecosystems or the ecosphere.

The Second Law, Cities, and the Ecosphere

➤ The second law of thermodynamics states that the "entropy" of any isolated system spontaneously increases. That is, concentrations of material are dispersed, available energy is dissipated, gradients disappear, and structural order and integrity break down. Eventually, no point in the system can be distinguished from any other

➤ Open systems are subject to the same forces of entropic decay as isolated systems. However,...

➤ Complex self-organizing, self-producing systems can maintain or increase their internal order by importing available energy/matter (essergy) [2] from their host environments and exporting degraded energy matter back into them. That is...

➤ Complex systems develop and grow 'at the expense of increasing the disorder [entropy] at higher levels in the systems hierarchy' (Schneider and Kay 1994)

➤ Systems that maintain themselves in dynamic non-equilibrium through the continuous dissipation of essergy extracted from their host systems are called "dissipative structures"

➤ Cities are prime examples of highly-ordered, far-from-equilibrium, dissipative structures. As major components of the human economy, they are also sub-systems of the materially closed ecosphere. Thus, in thermodynamic terms, cities exist in a potentially parasitic relationship to the rest of nature. (A parasite is any organism that gains its own vitality is purchased at the expense of the vitality of its host.)

➤ It follows that with continuous population and material growth of urban economies, a point will be reached when the disordering of the ecosphere (e.g., biodiversity loss, ecosystems collapse, climate change, toxic contamination, ozone depletion, etc.) becomes unsustainable, perhaps irreversible

N.B. The second law and hierarchy theory seem to provide a sufficient physical explanation for the onset of global ecological change

SOURCE Revised from Rees (1997)

Measuring Human Load: Ecological Footprint Analysis

As noted, human economic activity is a transformational process in which utility (goods and services) are refined, at a great entropic cost, from resources produced by nature. It follows that in spatial terms, the city is a node of intense material consumption and waste discharge embedded in a vastly larger human-dominated production system. In short, the bio-productive processes supporting an urban population and its economy actually occur outside the boundary of the city itself.

This raises two questions at the heart of sustainability: First, how large is the productive area needed to support the material demands of a typical city? Second, are the productive capacities of the ecosphere sufficient to support the anticipated demands of the human economy into the next century while simultaneously maintaining general life-support functions?

A Note on Carrying Capacity

Carrying capacity is usually defined as the largest population of a given species that can live in a defined area without permanently damaging the ecosystem(s) that sustain it (i.e., without destroying its habitat). However, economists generally argue that the limits to growth implied by this concept do not apply to people. Since local resource shortages can be relieved by trade and by technology (e.g., through substitution), regional population or economic growth need not be constrained by local resource scarcities.

In the global context, this is an ironic error. It ignores the fact that regardless of trade or level of technological sophistication, production/consumption by humans imposes a measurable *load* somewhere on the ecosphere.

Total human load is a function not only of population but also of *per capita* consumption and the latter is increasing even more rapidly than the former due (ironically) to expanding trade and technology. As Catton (1986) observes: "The world is being required to accommodate not just more people, but effectively 'larger' people..." For example, in 1790 the estimated average daily energy consumption by Americans was 11,000 kcal. By 1980, this had increased almost twenty-fold to 210,000 kcal/day (Catton 1986). In 1995, the "extra-somatic" energy being used *per capita* in rich countries like Canada and the United States, was equivalent to the work that might be done by 200 '"ghost slaves" (Price 1995). Energy use is, of course, strongly correlated with the consumption of many other resources. As a result, *load* pressure relative to carrying capacity is rising much faster than is implied by mere population increases. Shrinking carrying capacity may therefore soon become the single most important issue confronting humanity (Rees 1996, 1997).

Eco-footprint analysis estimates human load and overturns the economists' objections to carrying capacity, by inverting the standard carrying capacity ratio. Rather than asking what population a particular region can support sustainably, the relevant question becomes: How large an area of productive land and water is needed to sustain a specified population indefinitely wherever that land may be located?

SOURCE extracted and revised from Rees (1996)

These questions are fundamentally about human carrying capacity (see Box 2) and are potentially answerable using "ecological footprint analysis" (Rees 1992; Rees and Wackernagel 1994; Rees 1996, 1997; Wackernagel and Rees 1996; Walker and Rees 1997, Wackernagel et al. 1999). Eco-footprinting recognizes that human systems are component subsystems of the ecosphere, that most of the goods and services "produced" by the economy require energy and material extracted from ecosystems, and that these same ecosystems serve as sinks for wastes.

Eco-footprint analysis builds on traditional trophic ecology. We construct what is, in effect, an elaborate "food-web" for a specified study population, quantifying the material and energy flows supporting both the human population and its industrial metabolism. Eco-footprinting is further based on the fact that many of these material and energy flows can be converted into land- and water-area (ecosystem area) equivalents. Thus, the ecological footprint of a specified population is the area of land and water required to produce the resources consumed, and to assimilate the wastes generated by that population on a continuous basis, wherever on Earth that land is located. In theory, the ecological footprint includes both the area appropriated through commodity trade and the area needed to produce the referent population's share of certain "free" land- and water-based services of nature (e.g., the waste sink function).

Basic Eco-Footprint Calculations

In theory, to calculate an individual's ecological footprint, we would begin by estimating the land/water area ("a") required to produce the quantity of each major resource or consumption item ("i") consumed by that person in a year. We do this by dividing the individual's annual consumption ["c"] of each item by its average annual productivity or yield from the land ["p" in $kg \times ha^{-1}$]. Thus:

$$a_i = c_i/p_i = (kg_i/capita)/(kg_i/ha)$$

This yields the productive area required for item "i" in hectares for that person.

We would then obtain the individual's total ecological footprint ("f") by summing all the ecosystem areas appropriated by the "n" individual items in his/her annual shopping basket of consumption goods and services. That is:

$$f = \sum_{i=1}^{n} a_i$$

In practice, however, monitoring an individual's actual material consumption and waste output for a year is extremely laborious and time-consuming. We therefore usually use aggregate annual consumption data to estimate the land area required to produce each commodity (A_i) for the entire referent population. The total ecological footprint of the population, (F_p), is then calculated by summing the population footprints for the "n" individual items:

$$F_p = \sum_{i=1}^{n} A_i$$

The average per capita ecological footprint, "f_c," is then obtained by dividing the total population footprint by population size, "N":

$$f_c = F_p/N$$

We use average world productivity for the terrestrial component of most general eco-footprint assessments, since it is difficult to determine the origins—and corresponding land productivities—of trade goods or to apportion the latter when there are several sources.

A rough estimate of the marine footprint attributable to ocean fish consumption can be made by dividing seafood consumption by the average productivity of that 8.2 percent of the oceans (about 29.7 million square kilometers) which produces about 96 percent of the world's fish catch (see Wackernagel and Rees 1996 for details).

The land/water requirements for certain waste flows (e.g., carbon dioxide, nutrients in domestic sewage) can similarly be estimated. For example, we account for direct fossil energy consumption and the energy content of consumption items by estimating the area of carbon-sink forest that would be required to sequester the carbon dioxide emissions associated with burning fossil fuels ([carbon emissions/capita]divided by [carbon assimilation rate/hectare]). (An alternative is to estimate the area of land required to produce the biomass energy equivalent [ethanol]of fossil energy consumption. This produces a larger energy footprint than the carbon assimilation method.)

Depending on the purpose of the analysis, various corrections can be applied at each stage of the assessment to increase accuracy (see Wackernagel et al. 1999 for examples). Beyond that, we make every effort to avoid double-counting in the case of multiple land-uses and where there are data problems or significant uncertainty regarding consumption or productivity, we err on the side of caution. Also, most published ecological footprint estimates to date account for only carbon dioxide emissions on the waste side. In any event, many forms of toxic waste and hormone mimics, for example, cannot readily be converted to a "land" area equivalent. Accounting fully for waste sink demand would add considerably to the ecosystem area appropriated by economic activity (see Folke et al. 1997 for an example). In short, calculated eco-footprints are usually significant underestimates of total human load on the ecosphere.

The Ecological Footprints of Cities

> Great cities are planned and grow without any regard for the fact that they are parasites on the countryside which must somehow supply food, water, air, and degrade huge quantities of wastes. (Eugene P. Odum 1971)

Our early estimates, accounting for just food, fiber, and fossil energy consumption, showed that the eco-footprints of average residents of high-income countries range as high as five and six hectares *per capita* (Rees and Wackernagel 1996; Wackernagel and Rees 1996). More recent analyses bump wealthy country footprints up to 10 hectares *per capita* while people in the poorer developing countries have footprints of less than one hectare (Wackernagel et al. 1997, 1999).

Since it is difficult to find city-specific data on consumption, national statistics adjusted for known local or state/provincial variations can be used to estimate the ecological footprints of individual cities. Whether using direct calculation or extrapolation, assessments show that the ecological footprints of high-income cities are typically two to three orders of magnitude larger than their geographic footprints. For example:

➤ British researchers have estimated London's ecological footprint for food, forest products, and carbon assimilation to be 120 times larger than the city's geographic area (IIED 1995) or about nine tenths the area of the entire country. This means that in the absence of trade and natural material cycles, and assuming the inter-convertibility of farm and forest-land, the entire ecologically productive land base of the U.K. would be required to sustain the population of London alone

➤ The author's home city of Vancouver, Canada, has a political-geographical area of 11,400 ha and about 472,000 residents. Using the Canadian average ecological footprint for food, fiber, and carbon sinks of about 7.7 hectares (ha) per capita (Wackernagel et al. 1999), we can estimate that the aggregate eco-footprint of Vancouver city is 3,634,000 ha, or 319 times its nominal area

➤ Using data for the Toronto, Canada, region, Onisto et al. (1998) estimate a per capita ecological footprint of 7.6 ha. The 2,385,000 residents of the city of Toronto proper therefore have an aggregate eco-footprint of 18,126,000 ha, an area 288 times larger than the city's political area (63,000 ha)

➤ In the more comprehensive study using region-specific data, Folke et al. (1997) estimate that the 29 largest cities of Baltic Europe appropriate for their resource consumption and waste assimilation, an area of forest, agricultural, marine, and wetland ecosystems 565 to 1,130 times larger than the area of the cities themselves

Such data dramatically confirm the shift in the spatial relationships between modern wealthy urbanites and the land that sustains them. Even in the simplest case, well over 99 percent of the biophysical resources and processes upon which these typical modern cities depend lies outside their own political boundaries. Thus, while cities are among the "brightest stars in the constellation of human achievement," ecological footprint analysis shows that they also act as "entropic black holes, sweeping up the output of whole regions of the ecosphere vastly larger than themselves" (Rees and Wackernagel 1996).

Most significantly in the present context, these cases emphasize what should be obvious—but which is often forgotten in our technological age—that no city (as presently defined) is self-sustaining. High-income cities and urban regions have their ecological feet firmly planted in the global rural hinterland. Urban populations get their food, water, fresh air, natural fiber, and most other material and energy from the extra-urban environment; they also use the latter as a carbon sink and general waste dump.

Rural populations obviously also benefit from their relationships with cities. They sell their products in urban markets, enjoy the technologies developed in cities, and use many of the goods and services produced by urban industries. However, while the countryside could, in a pinch, survive without the city, there could be no cities without the countryside. As noted, in eco-thermodynamic terms, cities are increasingly concentrated nodes of consumption within the increasingly human-dominated global landscape. The ecologically productive and certainly co-equal component of the system is the rural landscape. It seems that if we want to maintain the ecological security of urban populations, we must maintain the viability and productivity of field and forest. Indeed, in an increasingly urbanized world it might be wise to consider the rural hinterland as a fully integrated functional component of the total human/urban system.

Implications of Ecological Footprint Analysis for Cities

In 1995, the well-known technological optimist Julian Simon wrote: "Technology exists now to produce in virtually inexhaustible quantities just about all the products made by nature..." and "We have in our hands now...the technology to feed, clothe, and supply energy to an ever-growing population for the next seven billion years..." (Simon 1995, cited in Bartlett 1997).[2] The forgoing analysis shows that despite such ebullient confidence in human ingenuity, modern society is far from being able to duplicate all the goods and services of nature. In fact, ecological footprint analysis suggests that the role of technology has been mainly to extend humanity's reach and to accelerate the rate at which we dissipate the vast stocks of natural capital (resources) that have accumulated in nature. This creates the illusion of freedom from ecological constraints even as burgeoning populations and economic activity threaten to consume the ecosphere from within (Rees 1999).

It is mainly urban populations and their consumption, particularly in wealthy cities, that drive human-induced global change. At the same time these populations are increasingly at risk as global change threatens not only ecological stability but also geopolitical security. Recent studies suggest that "in many parts of the world, environmental degradation seems to have passed a threshold of irreversibility" and "that renewable resource scarcities of the next 50 years will probably occur with a speed, complexity, and magnitude unprecedented in history" (Homer-Dixon et al. 1993: 38). If this proves true, the increasing disordering of regional ecosystems and the ecosphere will likely be accompanied by increasing social entropy–the breakdown of civil order within countries and increasing turbulence in international relations.

This implies that we should be thinking of new forms and policies for urban governance. For example, self-reliance, once a noble virtue, has become anathema to the free-trading world of today. However, in an era of real or incipient ecological change, it is time to reconsider the assumptions of the prevailing development model. The data presented in this chapter suggest that circumstances may already warrant consideration of the potential benefits for cities—indeed, for the world—of greater intra-regional self-reliance and ecological independence. How economically and socially secure can a city of ten million be if distant sources of food, water, energy, or other critical resources are threatened by accelerating ecospheric change, increasing competition, dwindling supplies, or civil/international strife? Does any development pattern that increases inter-regional dependence on vital but vulnerable resource flows make ecological or geopolitical sense? (Tight interdependence that might be security-enhancing in an ecologically stable world would likely become destabilizing in a rapidly changing one.)

Achieving significant progress in this area will be difficult for some large urban regions with limited natural capital assets. (As noted above, supplying London alone would require virtually the entire biophysical output of the

U.K. [see IIED 1995]). An alternative (or supplementary) response is for vulnerable urban regions to negotiate more formal long-term relationships (including international treaties through the facilities of national governments) with politically stable producer territories to help ensure reliable supplies of biophysical goods and services. In effect, such agreements acknowledge and formalize portions of the distant ecological footprints of the consumer regions. This is not as far-fetched as it seems. One example: partially in anticipation of the effect of global warming on its domestic production, Japan currently has nearly 1.5 percent of the farmland of Ontario, Canada, under long-term contract to supply a variety of bean crops to the Japanese market.

To reduce their dependence on external flows, more politically powerful urban regions may chose to implement policies to rehabilitate their own natural capital stocks and to promote the use of local fisheries, forests, agricultural land, and so forth. This makes sense in several ways. First, it would result in a more reasonable and manageable match in scale between the management unit and the ecosystems being managed. Local people, over time, acquire intimate knowledge of local systems and have a better chance of maintaining control over a discrete number of key management variables. Second, people who depend directly on a resource system are more likely to manage it for the long term. Off-shore owners of local resources who are attempting merely to maximize returns to capital have a strong incentive to liquidate natural capital stocks if alternative investments produce a higher return. Everyone is eventually impoverished by this process. Third, a philosophy of locally based resource management may enable the establishment or reestablishment of effective common-property management regimes at the community level for mobile resources such as fisheries or for community forests.

There is now a rich literature describing how small-scale, self-organizing, and self-managed common property regimes have flourished around the world for decades providing a model of sustainable management (Vogler 1995, Berkes 1989). In addition to sustainable management of resource stocks, such regimes build community and a commitment to place. This is clearly lacking in prevailing circumstances in which transnational corporations own and control local resource stocks, and harvests are oriented more toward maximizing production and profit than toward sustaining people and community. Finally, if each significant urban region were to manage its own territorial resources in a sustainable manner and enter into only ecologically balanced and socially fair exchanges with other regions, then the aggregate effect would be global sustainability.

Policy in support of increased local self-reliance would obviously require revision of existing and planned international agreements designed to liberalize commodity trade and capital flows. The point, of course, is that as presently conceived, these agreements are the antithesis of sustainability. This is not to say that trade is inherently bad. Indeed, ecological footprint and similar analyses show that many nations could not survive in anything like their current forms without trade. However, not all trade is sustainable trade (Rees 1994). Sustainable trade would restrict trade to necessary or desirable flows that do not deplete critical natural capital stocks. (This is commerce in true biophysical surpluses.) I have elsewhere advocated negotiation of a General Agreement on the Integrity of (Ecological) Assets—a GAIA agreement—to regulate economic activities affecting global life support systems, including patterns of commercial trade in ecologically significant commodities (Rees 1994). Such an agreement would set the ecological "bottom line" for such existing instruments as the GATT and the World Trade Organization.

In the final analysis ecological footprint analysis provides a strong argument for a restoration of balance between the forces of local cohesion and globalization. The increase in welfare from enhanced food security, improved environmental quality, increased local control, and stronger communities will offset any loss in gross economic product and is therefore good economics. In any event we want to reduce the present scale of energy and material throughput—it's part of the problem. Accordingly, the United Nations Environment Program's recent "Global Environment Outlook 2000" report argues the need for a tenfold reduction in resource consumption by high-income countries. Even the Business Council for Sustainable Development seems to have accepted that "Industrialized world reductions in material throughput, energy use, and environmental degradation of over 90 percent will be required by 2040 to meet the needs of a growing world population fairly within the planet's ecological means" (BCSD 1993: 10). In this context, increasing the self-reliance of urban regions would result in significant net savings of energy and material currently used in the global transportation of goods. This would be a valuable contribution to sustainability worth making even in the absence of any threat to urban security per se.

Should we not also be reconsidering how we define city systems, both conceptually and in spatial terms? "Sustainable city"—as we currently define cities—is an oxymoron (Rees 1997). Perhaps it is time to think of cities as whole systems—as such, they comprise not just the node of concentrated activity as presently conceived, but also the entire supportive hinterland. Many city regions could be reorganized formally to incorporate much of their supportive hinterland into their political realms. Short of so great a conceptual leap, there is much that can be done incrementally to increase the sustainability of our cities. For example, in the domain of land-use planning, planners and politicians should find ways to (from Rees 1997):

✦ Integrate planning for city size/form, urban density, and settlement (nodal) patterns in ways that minimize the energy, material, and land-use requirements of cities and their inhabitants

✦ Capitalize on the multifunctionality of green areas (e.g., aesthetic, carbon sink, climate modification, food production, functions) both within and outside the city

✦ Integrate open space planning with other policies to increase local self-reliance respecting food production, forest products, water supply, carbon sinks, and so on. For example, domestic waste systems should be designed to enable the recycling of compost back onto regional agricultural and forest lands

✦ Generally protect the integrity and productivity of local ecosystems both to sustain local populations and reduce the ecological load imposed on exporting regions and the global common pool

✦ Strive for zero-impact development. The destruction of ecosystems and related biophysical services due to urban growth in one area should be compensated for by equivalent ecosystem rehabilitation in another

If these ideas seem naïve or impossible, let us recall that sustainable development is often accompanied by the term "paradigm shift" even in the speeches of politicians. A paradigm shift implies a sea-change in world-view, in the fundamental beliefs, values, and assumptions underpinning industrial society. Ecological footprint analysis suggests that nothing less than this is required if global society is to achieve sustainability. Ironically, however, the "radical" changes required for sustainability would merely bring humans back to living in the real world.

Conclusion: Using Eco-Footprint Analysis in Sustainability Assessment

Ecological footprint is a conceptually simple but powerful tool that translates human impacts (human load on the ecosphere) into a simple index—appropriated land area—that ordinary people can understand. Everyone, rich or poor, uses biophysical goods and services produced by real-world ecosystems. Thus, the relevant question—how much ecosystems area is required to produce the resources used and to assimilate the wastes produced by a given population—suggests itself to most people without too much prodding.

It also leads to a series of secondary questions relevant to sustainability assessment. Once we have some sense of how much ecosystem area is needed, it is natural to ask just where it is. How much is local or at least under the political control of domestic governments? How dependent are we on (potentially insecure) extraterritorial resources? What are the trends?

It is particularly useful in this context to recognize that there are only 1.5 ha of ecologically productive land and about .5 ha of truly productive ocean for every person on Earth. (These might be called the "fair Earth-share" and "fair sea-share," respectively.) In effect, in an equitable world, each person would be entitled to the biophysical "goods and services" of about 2 ha of productive mixed ecosystems. On the other hand, we have shown that the residents of high-income cities and countries generally appropriate two to four times this amount. As a result, it is not just cities that overshoot the productive capacity of their local ecosystems. Because of their large per capita footprints, most highly urbanized high-income countries have an ecological footprint several times larger than their national territories (Wackernagel and Rees 1996, Wackernagel et al. 1997,1999). In effect, these countries are running massive "ecological deficits" with the rest of the world (Rees 1996).

Now, it should be self-evident that not all countries can run ecological deficits indefinitely—other countries or regions must have surplus accounts if the total human enterprise is to operate sustainably. Unfortunately, ecological footprinting reveals that in important dimensions, consumption by the present human population already exceeds the long-term productivity of the ecosphere. In other words, we are not living entirely on surplus "natural income," but rather we are depleting our natural capital. According to Folke et al. (1997), the carbon dioxide emissions of just 1.1 billion people (19 percent of humanity) living in 744 large cities exceed the entire sink capacity of the world's forests by 10 percent. Ozone depletion, falling water tables, and the widespread collapse of fish stocks provide other well-known examples. Wackernagel and Rees (1996) and Wackernagel et al. (1997) estimate that, overall, the present world population exceeds long-term global carrying capacity by up to one third using prevailing technologies and average consumption levels.

A Family of Area-Based Sustainability Indicators

❖ **Ecological Footprint** - The productive area of land/water required on a continuous basis to support a defined population (i.e., to produce its resource needs and assimilate its wastes) at a specified material standard of living, wherever on Earth that land is located

❖ **Appropriated Carrying Capacity (ACC)** - The biophysical resource flows and waste assimilation capacity appropriated from global totals by a specified population (individual, city, country, etc.). ACC is sometimes used synonymously with "ecological footprint"

> ⇝ **Fair Earthshare** - The amount of ecologically productive land "available" *per capita* on Earth, currently (1998) 1.5 - 1.6 hectares. Similarly, a fair seashare (area of ecologically productive ocean—mostly coastal shelves, upwellings, and estuaries—divided by world population) is about 0.5 hectares
>
> ⇝ **Ecological Deficit** - The level of resource consumption and waste discharge by a population in excess of locally/regionally sustainable natural production and assimilative capacity (In spatial terms, the eco-deficit is the difference between that population's effective ecological footprint and the geographic area it actually occupies)
>
> ⇝ **Sustainability Gap** - Related to the ecological deficit. A measure of the decrease in consumption (or the increase in material and economic efficiency) required to eliminate the ecological deficit (Can be used at any spatial scale, from local to global)
>
> SOURCE Revised from Rees (1996)

The most critical conclusion from this is that the affluent material lifestyles people lead in developed countries are simply not extendible to the entire world population along the present development path. Indeed, to raise the present world population to Canadian material standards using prevailing technologies would require nearly four Earth-like planets. (Six billion people living at Canadian standards would require about 46 billion ha of productive land and marine ecosystems. There are only about 12 billion ha of such land and water on Earth.) To put today's ecological inequity in perspective, consider that the 2.4 million residents of Toronto, at over 7.6 ha per capita, have a total ecological footprint comparable to that of the 18 million residents of Bombay, India (India has an average per capita footprint of approximately one ha). It seems clear that if the world's poor are to achieve even a materially acceptable standard of living, energy and material use by the rich will have to decrease markedly through changes in either technology or lifestyles.

Eco-footprint analysis thus shows that the current global development model is unsustainable, but it also suggests various ways to address the conundrum. For example, we have argued that the spatial organization of global urban society should not be based on assumptions of "business as usual" or a mere extension of historic trends. Human-induced global change may threaten the long-term security of megacities in a globalizing world. Thus, urban sustainability and the economic security of the populations— both urban and rural—that depend on cities may well be enhanced through deliberate policies to enhance local self-reliance. There is a need to restore balance between the forces of localism and globalism. This is not an argument against either urbanization or globalization. Rather it suggests the means to reconcile both these trends with ecological reality and to ensure the survival of a human socioeconomic system whose vitality is increasingly tied to the vitality of cities.

Eco-footprinting suggests several other ways to achieve and monitor progress toward sustainability at all spatial scales (see *A Family of Area-Based Sustainability Indicators* for definitions):

- ✦ The model reveals which products and activities have the greatest ecological impact and therefore offer the greatest leverage for total footprint reduction

- ✦ Individuals committed to sustainability can determine the effect of alternative consumption patterns and lifestyles on their personal eco-footprints or can contrast their eco-footprints with their "fair Earth-shares" in setting personal targets for footprint reduction

- ✦ At the most general level, cities (or countries) can use periodic eco-footprint analyses simply to monitor whether policies are working to reduce their ecological footprints. More specifically, they can assess the efficiency and effectiveness of alternative policies designed to reduce consumption

- ✦ Specific reduction targets or goals can be set for many jurisdictions. For example, it would be reasonable for some countries to attempt to eliminate their "ecological deficits"

- ✦ Individual cities might compare their actual ecological footprints to one based on the "fair Earth-share" and "fair sea-share" (city population multiplied by two ha), and devise policies to reduce the difference

- ✦ Eco-footprint analysis can be used for technology assessment, that is, to compare the relative effectiveness of alternative material technologies, land-use patterns, transportation modes, and so forth for footprint reduction. For example, Walker and Rees (1997) found that shifting to higher density three-story walkup apartments or other multiple unit dwellings from single-family houses reduced that portion of the per capita eco-footprint associated with housing and urban transportation requirements by up to 40 percent

- ✦ Cities can enter into friendly competitions to be first in achieving a predetermined eco-footprint reduction

Whatever the specific target or strategy, the overall goal should be to reduce the "sustainability gap" both nationally and globally.

Strengths of Eco-Footprint Analysis

A major strength of ecological footprint analysis is that it addresses many of the concerns about humankind-environment relationships raised by other analysts. For example, it incorporates George Borgstrom's 1960s notion of "ghost acreage" (referring to the extra-territorial foodlands required to support densely populated regions and countries); it is conceptually related to the embodied energy (emergy) analyses of Howard Odum (see Hall 1995);

and it reflects the crucial role of the second law of thermodynamics in human affairs as stressed by Nicholas Georgescu-Roegen (1971) and Herman Daly (Daly 1991).

Because it accounts for both population size and resource consumption, ecological footprint analysis is fully compatible with Catton's concept of human load (the greater the load, the larger the footprint). The ecological footprint also corresponds closely to an earlier conceptual definition of human impact on the environment formulated by Ehrlich and Holdren (1971): I = PAT, where "I" is impact, "P" is population, "A" is affluence, and "T" is technology. The ecological footprint (F) corresponds to impact (I) in the latter formulation and is itself a function of population size and consumption. However, because consumption is, in turn, a function of income (affluence, "A") and the state of technology ("T"), F becomes an area-based analogue of PAT.

Eco-footprinting does, however, have several practical advantages over other indicators of gross human impact. As noted, a major strength is conceptual simplicity. Eco-footprinting provides an intuitive and visually graphic tool for communicating one of the most important dimensions of the sustainability dilemma—energy and material consumption. It takes real data on a wide variety of energy and material flows and represents them in terms of a single concrete, variable land area. Land itself is a particularly powerful indicator because it (like consumption) is readily understood by ordinary people. As noted above, this facilitates useful comparisons of ecosystems supply and demand in public discussions of sustainability policy options. Most important, in the sustainability context, the aggregate human footprint seems already to be larger than the entire planet.

Since the supply of land is finite at any scale, such comparisons suggests several secondary indices of (un)sustainability that can be used as measurable policy targets. For example, if we are running a significant ecological deficit, the question becomes Is this a major problem? and, if so, What must be done to reduce it? (Note that humanity's ecological deficit may ultimately be more important than any fiscal deficit, yet the former is totally ignored in the current frenzy to reduce the latter in many countries.)

Finally, ecological footprint analysis enables us to dispel one of the prevailing myths of our age, namely that knowledge-based economies are somehow more ecologically benign that resource-based or manufacturing economies. This appealing myth is rooted in the belief that the appropriate economic niche for the most advanced of the developed countries in a globally restructured economy is as purveyors of high-priced technology, of so-called "intellectual capital." Our wealth is now to be generated through nonpolluting products of the mind. However, those countries that do manage to make the shift so that most of their citizens earn their living in computer software, engineering services, or similar high-tech industries will remain at the upper end of global income distribution. Indeed, the average incomes of their citizens will increase. The problem is, so will their per capita consumption.

Basic ecological footprint calculations for individuals or countries are based on final demand. The land-related impacts of production are credited to the consumers, not the producers, of goods and services. Thus, it is not how one earns a living, but rather what and how much one consumes that determines one's personal ecological footprint. The eco-footprint of a logger earning $50,000 per year is typically a fraction of that of a $200,000 account executive. If wealthy North Americans continue to build large houses with three-car garages and four bathrooms, well-stocked with all the big-ticket consumer appliances and electronic gadgetry that their high incomes command, then their average contribution to ecological destruction may well increase as their economies become increasingly knowledge-based.

Methodological Limitations

For all its conceptual and operational strength, eco-footprint analysis does not solve all our ecological problems, nor is it free of scientific uncertainty. Much of the latter weakness, however, can be ascribed to the availability and reliability of the input data rather than to problems with the theory and methods. Many jurisdictions have poor statistics on resource and material flows, and it is often difficult to make adjustments for trade in commodities and manufactures. Similarly, there is dispute over such matters as the carbon dioxide sink capacity of various ecosystems. Other methodological limitations are as follows:

Ecological footprints do not capture all consumption and pollution impacts. Some critics suggest that the term "ecological footprint" is misleading because the method does not, in fact, capture the full range of ecologically significant impacts on the ecosphere. This is true–for example, eco-footprinting does not account for the cause or effects of ozone depletion or endocrine (hormone) mimicry. In fact, calculations to date do not even tell the whole consumption/pollution story on land and water. Only major categories of consumption have been included and we are only beginning to examine the spatial implications of waste discharges other than carbon dioxide and essential nutrients.

Actually, it is unlikely the method will ever cope satisfactorily with toxic waste discharges. The biological effects of chronic low-level chemical or pseudo-hormone contamination, for example, cannot readily be translated into a simple appropriated land area. However, this may not be as problematic as it seems. Environmental economists already agree that society should have zero tolerance for highly toxic or biologically active chemical wastes and radioactive substances for which the ecosphere has no measurable assimilative capacity. Such substances should simply be banned or phased out (as in the case of chlorinated pesticides and ozone-depleting CFCs).

In any event, the limited scope of eco-footprinting in no way invalidates the model for what it does do. Rather it suggests that eco-footprint calculations are almost certainly underestimates of actual ecosystem appropriations and that extensions and methodological improvements will result in considerably larger footprints. Most significantly, ecological footprinting already produces unambiguous conclusions and clear policy directions. Improvements that increase the scope of the analyses will add to our sense of urgency, but will not likely shift the direction of needed policy change.

Eco-footprinting is static and does not reflect technological change. While acknowledging its power to communicate a fundamental message, some commentators have suggested that the eco-footprint concept is too simplistic. For example, the model is static, whereas both nature and the economy are dynamic systems. Ecological footprinting therefore does not seem to take into account such things as technological change or the adaptability of social systems and does not allow predictions.

It is true that footprint analysis is not dynamic modeling and has no predictive capability. However, prediction was never the intent. Ecological footprinting acts, in effect, as an ecological camera—each analysis provides a snapshot of our current demands on nature, a portrait of how things stand right now under prevailing technology and social values. This in itself is an important contribution. Eco-footprinting suggests that humanity has significantly exceeded carrying capacity and that some people contribute significantly more to this ecological "overshoot" than do others. Ecological footprinting can also estimate how much we have to reduce our consumption, improve our technology, or change our behavior to achieve sustainability.

Moreover, if used in a time-series study (repeated analytic "snap-shots" over years or decades), ecological footprinting can help monitor progress toward closing the sustainability gap as new technologies are introduced and consumer behavior changes. (After all, even a motion picture is a series of snap-shots.) Eco-footprint analysis can also be used in static simulation studies to test, for example, the effect of alternative technologies or settlement patterns on the size of a population's ecological footprint. To reiterate, ecological footprint analysis is not a window on the future, but rather a way to help assess both current reality and alternative "what if" scenarios on the road to sustainability.

Eco-footprinting ignores the implications of complex systems theory: Some critics have correctly noted the impossibility of mapping the behavior of a complex process or system with a single indicator and that eco-footprint analysis simply ignores the implications of complexity theory.

This criticism seems to miss two points. First, we do not claim that eco-footprinting captures the entire systems behavior. It is advanced simply as one indicator of humanity's "engagement" with the rest of nature. Eco-footprint analysis should be used in conjunction with economic, social, or any other indicators that bear on the issues at hand.

Second, there is no denying that science has gained important insights from complex systems theory. As we have come to better understand the behavior of dynamical, self-organizing, adaptive systems, we are being forced to abandon simplistic approaches to ecosystems management. On the other hand, there are still many phenomena that have simple explanations and, in these circumstances, Occam's razor applies: *there is no need to contrive a complex explanation if a simple one suffices.*

For example, the 747/flight-crew/air-traffic control system is complex and prone to unpredictable if not chaotic behavior in certain circumstances. However, the whole system will certainly crash (the 747 literally so), if the plane runs out of fuel in full flight. (This is absolutely predictable.) While there may be a complexity-based explanation for the "systems failure" that led to insufficient fuel being loaded and no one noticing, the proximate cause of the plane crash is clear and uncomplicated.

Eco-footprints have something of this quality. The survival of any population depends ultimately on the availability of adequate ecosystems support (i.e., on the availability of "fuel"), regardless of political regime or prevailing socio-cultural values. Once this point is accepted, the eco-footprint debate comes down to scope and details of method and the (geo)political implications of apparent overshoot.

Ecological footprints do not track biodiversity loss. The loss of biodiversity—species, critical habitats, unique ecosystems—is among the most pressing of sustainability-related problems, and it is certainly true that ecological footprinting does not measure biodiversity directly. However, one of the ways human populations have spread over the earth is by displacing competing species from their habitats and ecological niches. In effect, we divert energy and material flows through ecosystems from other species to ourselves. The increasing appropriation of the products of photosynthesis by humans is therefore fundamentally in conflict with the imperative of biodiversity conservation—in other words, what people consume, other species cannot. (This is a variation of the ecologists' "competitive exclusion principle.")

Since eco-footprinting does measure urban land, domestic crop and grazing lands, and land in commercial forest (lands taken out of production, or producing mainly for humans), it may well be possible in future to index footprint calculations to track impacts on biodiversity. Hypothesis: Biodiversity is invariably negatively correlated with the size of the human ecological footprint.

The concept of an "energy footprint" (e.g., carbon sink forests) is too abstract or hypothetical—it is not a "real" land use. For many high-income industrialized countries, the fossil energy component may make up half of the total eco-footprint as presently calculated. Because science cannot account fully for the global carbon budget, and there is dispute over the relative roles of terrestrial and marine ecosystems in carbon sequestration, some critics see energy footprint calculations as unreliable or hypothetical. Others don't regard carbon sequestration as a real land use.

This latter criticism reflects our cultural bias that land "use" requires physical occupancy or the consumption of some tangible product of the land. We are unaccustomed to thinking of the provision of other (particularly non-market) biophysical services of nature, no matter how essential to sustainability, as effective uses of landscape. In terms of thermodynamic law and mass balance, it is this cultural bias that needs adjustment, not eco-footprint theory. All energy and material use is "throughput," implying the generation of wastes, and to the extent that the assimilation and recycling of those wastes (including carbon dioxide) requires an exclusive dedication of land or water, it is legitimate component of eco-footprint calculations.

Returning to the first point, while the details of the carbon budget may be in dispute, there is no question that carbon dioxide levels are increasing in the atmosphere and that these increases represent about half of current carbon emissions from fossil fuel and biomass combustion. This implies that available land and water carbon sinks are insufficient to sequester all anthropogenic carbon dioxide at current rates of emission. In eco-footprint terms, our current global energy footprint is excessive. We are running a global carbon sink deficit in exactly the same sense that many countries run a food-land deficit.

Energy footprint calculations simply estimate the area of dedicated carbon-sink forests that would be necessary to assimilate a study population's carbon emissions. The estimates are based on the average estimated carbon sequestration rates of the world's growing forests. We use forest land rather than ocean surface in eco-footprint estimates because land sequestration rates are higher (producing smaller footprints) and because we can readily manipulate land to provide dedicated terrestrial sinks. Dedicated carbon sink forests are exclusive uses insofar as they cannot be used for purposes (such as pulp/paper production) that would quickly release carbon back into the atmosphere.

In short, the notion of carbon sink land is not hypothetical even if the real thing is in short supply. Indeed, various electric utilities are now planting dedicated carbon sink forests to offset the carbon emissions of their fossil fuel-burning generating plants. Similarly, entire countries are contemplating developing carbon sink forests as an alternative to reducing their carbon emissions and to be part of a future system of tradable carbon emission rights. (For example, planting a carbon sink forest would release an equivalent quantity of carbon emission rights for sale on the market.)

Finally, we should recognize that any alternative to fossil fuels will also generate an energy eco-footprint. For example, the ethanol equivalent of fossil fuel, generated from biomass, would require a fuel-crop growing area considerably larger than the energy footprint based on carbon sinks. Even solar-based alternatives require that a large land area be set aside for collector surfaces (though this land need not be biologically productive). The point is that all energy sources and related sinks have a corresponding land equivalent and the present method of calculating an energy footprint may actually be fairly conservative.

Eco-footprinting seems to promote self-sufficiency (anathema to free-market advocates and free-traders). In fact, ecological footprinting in itself does not encourage anything in particular. It simply provides an indicator of the current population-environment state of affairs. It can reveal, of course, that certain populations and countries are presently highly dependent on extra-territorial productivity and life-support services; that is, on both commercial and natural "trade." In a world experiencing rapid population growth, rising material expectations, resource depletion, and the threat of accelerating global change, such dependent relationships may well be interpreted as a potential threat to ecological or geopolitical security. In this case, increased self-reliance (if not self-sufficiency) may well be a prudent risk-averse strategy.

The counter-argument seems to rest on the notion that self-reliance (self-sufficiency) is less economically efficient than specialization and trade. Any move toward self-reliance would therefore reduce economic output, income growth, and trade flows. Underlying this concern, in turn, seems to be the assumption that Gross Domestic Product (GDP) growth per capita can always be equated to increased welfare. This is false and misleading. Economists recognize that many factors in addition to growth contribute to human well-being. Thus, to the extent that enhanced self-reliance can contribute more to overall welfare (by, for example, increasing security, diversifying the economy, and reducing natural capital depletion) than is lost by any consequent reduction in growth rates, it would be a good thing.

To reiterate, ecological footprint analysis per se does not advocate self-reliance or self-sufficiency. However, by providing a new input into the multivariate decision-making required for sustainability, it may well significantly affect the way we look at things. If, in this light, self-reliance acquires an attractive new patina, then so be it.

Eco-footprinting appears to ignore the human factors at the heart of sustainability. This is perhaps the most frequent "criticism" of ecological footprinting. The analysis does not formally consider the distribution of political and economic power or the responsiveness of the political process. Nor does it account for the myriad direct and indirect social effects of errant production/consumption. On the other hand, is it realistic to expect ecological footprinting to tell the entire sustainability story? Any overly comprehensive single index can be seriously misleading (consider the problems with GDP!). The ecological footprint is designed to be just that—a land-based index of a limited range of ecologically significant biophysical flows. It is simply not structured to reveal the sociopolitical dimensions of the global change crisis.

But this does not mean that eco-footprinting is entirely useless to social impact analysts. The method may not reveal the disruption of traditional livelihoods and damage to public health which are often the most interesting local impacts of expanding economic activity in developing nations. However, the ten- or twenty-to-one ratio of high to low income eco-footprints does speak volumes about the growing distributional inequity that sullies the mainstream approach to global development. Indeed, as the concept spreads, the term "footprint" is increasingly being used to encompass the overall impacts of high-income economies on the developing world. (See, for example, "Citizen Action to Lighten Britain's Ecological Footprint," a report prepared by the International Institute for Environment and Development for the U.K. Department of the Environment [IIED 1995]).

In short, the original objective of eco-footprint analysis was to bolster the ecologists' critique of the prevailing development paradigm and to force the international development debate beyond its focus on GDP growth to include biophysical reality. To the extent this has been achieved the model has served its purpose. Certainly eco-footprinting has contributed substantially to our understanding of urban ecological sustainability. It is a tribute to the power of the basic concept that others are now extending it to include sociopolitical factors as we search for more holistic approaches to sustainability.

Notes

[1]   Human ecology therefore differs from most "environmental studies" and "environmental assessments" which analyze the impacts of human activity "in here" on an external environment "out there."

[2]   Professor Simon later corrected this statement to read "for seven million years," but as physicist Albert Bartlett (1997) has shown, even 1 percent annual growth over this much reduced time-frame would result in a human population 30,000 orders of magnitude larger than the estimated number of atoms in the known universe. Some optimists haven't done the math!

References

Ayres, R. U. 1994. *Information, Entropy and Progress: A New Evolutionary Paradigm.* Woodbury, NY: AIP Press.

Ayres, R. U. and U. Simonis. 1994. *Industrial Metabolism: Restructuring for Sustainable Development.* New York: United Nations University Press.

Bartlett, A. 1996. The exponential function XI: The new flat earth society. *The Physics Teacher* 34: 342-343.

BCSD, Business Council for Sustainable Development. 1993. Getting eco-efficient. *Report of the BCSD First Antwerp Eco-Efficiency Workshop*, November 1993. Geneva: Business Council for Sustainable Development.

Berkes, F., ed. 1989. *Common Property Resources: Ecology and Community-based Sustainable Development.* London: Belhaven Press.

Catton, W. 1986. Carrying capacity and the limits to freedom. *Paper prepared for Social Ecology Session 1, XI World Congress of Sociology, August 1996.* New Delhi, India.

Daly, H. E. 1991. *Steady-State Economics.* Washington, DC: Island Press.

Daly, H. E. 1992 Steady-state economics: Concepts, questions, policies. *Gaia* 6, 333-338.

Ehrlich, P. and J. Holdren. 1971. Impact of population growth. *Science* 171:1212-1217.

Folke, C., A. Jansson, J. Larsson, and R. Costanza. 1997. Ecosystem appropriation by cities. *Ambio* 26: 167-172.

Girardet, H. 1992. *The Gaia Atlas of Cities: New Directions for Sustainable Urban Living.* London: Gaia Books (also: Anchor/Doubleday).

Georgescu-Roegen, N. 1971. *The Entropy Law and the Economic Process.* Cambridge, MS: Harvard University Press.

Hall, C., ed. 1995. *Maximum Power: The Ideas and Applications of H.T. Odum.* Niwot, Colorado: University Press of Colorado.

Homer-Dixon, T. F., J. H. Boutwell, and G. W. Rathjens. 1993. "Environmental Change and Violent Conflict." *Scientific American* February, 1993.

IIED, International Institute for Environment and Development. 1995. *Citizen Action to Lighten Britain's Ecological Footprints.* London: International Institute for Environment and Development.

Odum, E. P. 1971. *Fundamentals of Ecology* (3rd ed). Saunders, Philadephia.

Onisto, L., E. Krause, and M. Wackernagel. 1998. *How Big Is Toronto's Ecological Footprint?* Toronto: Centre for Sustainable Studies and City of Toronto.

Price, D. 1995. Energy and human evolution. *Population and Environment* 16: 301-319.

Rees, W. E. 1990. The ecology of sustainable development. *The Ecologist* 20:1:18-23

Rees, W. E. 1992. Ecological footprints and appropriated carrying capacity: What urban economics leaves out. *Environment and Urbanization* 4:2: 121-130.

Rees, W. E. 1994. Pressing global limits: Trade as the appropriation of carrying capacity. In T. Schrecker and J. Dalgleish, eds, *Growth, Trade and Environmental Values*. London, ON: Westminster Institute for Ethics and Environmental Values.

Rees, W. E. 1996. Revisiting carrying capacity: Area-based indicators of sustainability. *Population and Environment* 17: 195-215.

Rees, W. E. 1997. Is 'sustainable city' an oxymoron? *Local Environment* 2: 303-310.

Rees, W. E. 1999. Consuming the Earth: The biophysics of sustainability. *Ecological Economics* 29:23-27.

Rees, W. E. and M. Wackernagel. 1994. Ecological footprints and appropriated carrying capacity: Measuring the natural capital requirements of the human economy. In A-M. Jansson, M. Hammer, C.Folke, and R. Costanza, eds., *Investing in Natural Capital: The Ecological Economics Approach to Sustainability*. Washington, DC: Island Press.

Rees, W. E. and M. Wackernagel. 1996. Urban ecological footprints: Why cities cannot be sustainable and why they are a key to sustainability. *EIA Review* 16: 223-248.

Schneider, E. and J. Kay. 1994. Life as a manifestation of the second law of thermodynamics. Preprint from *Advances in Mathematics and Computers in Medicine*. Working Paper Series, Faculty of Environmental Studies Waterloo, Ontario: University of Waterloo Faculty of Environmental Studies.

Simon, J. 1995. The state of humanity: Steadily improving. *Cato Policy Report* 17:5. Washington, DC: The Cato Institute.

Vogler, J. 1995. *The Global Commons: A Regime Analysis*. Chichester: John Wiley.

Wackernagel, M. and W. E. Rees. 1996. *Our Ecological Footprint: Reducing Human Impact on the Earth*. Gabriola Island, BC and New Haven, CT: New Society Publishers.

Wackernagel, M., L. Onisto, A. C. Linares, I. S. L. Falfán, J. M. Garcia, A. I. S. Guerrero, and M. G. S. Guerrero. 1997. *Ecological Footprints of Nations*. Report to the Earth Council, Costa Rica.

Wackernagel, M., L. Onisto, P. Bello, A. C. Linares, I. S. L. Falfán, J. M Garcia, A. I. S. Guerrero, and M. G. S. Guerrero. 1999. National natural capital accounting with the ecological footprint concept. *Ecological Economics* 29:375-390.

Walker, L. and W. E. Rees. 1997. Urban density and ecological footprints: An analysis of Canadian households. In M. Roseland, ed., *Eco-City Dimensions*. Gabriola Island, BC and New Haven, CT: New Society Publishers.

WCED. 1987. *Our Common Future*. Report of the [UN] World Commission on Environment and Development. New York and Oxford: Oxford University Press.

Part IV

Tools for Sustainability Assessment at the Household Level

Dimitri Devuyst

The last part of this book examines sustainability assessment at the most local level, which is the individual or household level. Reaching a more sustainable state will require effort from all levels of government and from various sectors. Individuals will also be confronted with requests and demands to lead more sustainable lifestyles. The goal of this part is to examine how sustainability aspects of lifestyles can be measured and how the sustainable nature of personal or household decisions can be assessed.

In Chapter 14, Paul Harland and Henk Staats examine the EcoTeam Program, a program that has been developed by Global Action Plan to offer households a method to design a more sustainable way of living. Harland and Staats describe the results of their research, which assesses whether the EcoTeam Program reached its goals. To get an idea of the effects of the EcoTeams on the individual behavior of the participants, people were asked to complete questionnaires at three distinct points: directly before participation, directly after completion of the EcoTeam Program, and two years after participation. The results show that the EcoTeam Program succeeds in its aim: participants show improvements in a considerable part of their environmentally relevant household behavior. In addition, participation leads to a decrease in the use of natural resources. This chapter not only shows how similar programs can be developed to steer household behavior toward a more sustainable direction, but it also presents a research project design that allows the actual behavior changes to be measured.

In her contribution "Global Action Plan Research in Switzerland," Susanne Bruppacher describes the impact of the EcoTeam program in Switzerland. Bruppacher concludes that moderate behavior change is taking place, especially in the areas of waste and consumption, and that there is still room for improvement when it comes to transport and energy aspects. Bruppacher stresses the need for an extra effort to attract people who have not yet adopted a more sustainable lifestyle in comparison to the average population. These are the people who would make the greatest behavioral improvements by participating in the program, but are currently not reached by Global Action Plan.

Every day individuals and households make important decisions and take part in routine activities that have consequences for the sustainable development of our societies. Therefore, developing tools to assess the sustainable nature of our daily life decisions is crucial. In Chapter 15, Dimitri Devuyst and Sofie Van Volsem introduce the new term "Sustainable Lifestyle Assessment" and develop its methodological framework. The authors first define the term "sustainable lifestyle" and look into the current movements that try to encourage people to take positive action in relation to sustainable consumption, such as lifestyle simplification and the introduction of more sustainable economic systems. Further, a distinction is made in Sustainable Lifestyle Assessment (SLA) between a "self assessment" approach and a "research" approach. The "self assessment" approach aims to encourage individuals or households to evaluate the sustainability of their own behavior. The "research" approach is a scientific examination resulting in information that can be used by individuals and governments to make choices. A case study on personal decision-making in relation to travel behavior reveals the complexity and various factors that influence the choices made by individuals. Devuyst and Van Volsem also propose a seven-step approach to a personal or household SLA tool. The main goal of the approach is to provide meaningful information that can be used by individuals or households to make decisions, while conscious of their impact on the sustainability of their lifestyle. Finally, a proposal is made for a new concept called "Household Agenda 21," following the successful example of Local Agenda 21. Moreover, the integration of SLA in scientific disciplines such as Family Resource Management is discussed.

Part IV shows how individuals, households, and authorities can be approached to stimulate creative thinking about leading more sustainable lifestyles. SLA is an important component of sustainability assessment in general, but is still largely undeveloped. Scientists should get more involved in the development of methodologies that are meaningful in personal decision-making.

Chapter 14

Striving for Sustainability via Behavioral Modification at the Household Level: Psychological Aspects and Effectiveness

Paul Harland and Henk Staats

Summary

The EcoTeam Program has been developed to offer households a method to design a more sustainable way of living. In this program, people organize themselves in groups of six to ten neighbors or friends to evaluate and if possible to improve their environmentally relevant household behavior. What are the psychological mechanisms in this program that would help participants to succeed in their striving for sustainability? To what extent do participants actually improve environmentally relevant household behavior, and do these improvements lead them to decrease their use of natural resources? This chapter focuses on these questions by offering a long-term view on the effectiveness of the EcoTeam Program in the Netherlands.

The results show that the EcoTeam Program succeeds in its aim: participants in the EcoTeam Program show improvements in a considerable part of their environmentally relevant household behavior. In addition, participation leads to a decrease in the use of natural resources. These improvements are not apparent in the observed behavior of non-participants. Finally, it appears that these effects are sustained for up to two years after the end of the EcoTeam Program. In order to make use to a greater extent of the effectiveness of the EcoTeam Program, future development of the EcoTeam Program should be aimed at acquiring larger numbers of participants. This chapter concludes by focusing on further research that might lead to a higher participation rate in the EcoTeam Program.

Introduction

Large proportions of our everyday behavior affects the environment, either directly or in the long run: from regularly turning off the tap while brushing our teeth to using the car to go to work, and from using energy-saving light bulbs to making the choice between single- and double-paned windows. These are the behavioral choices that affect the impact our households have on the environment. Attempts to modify these environmentally relevant household behaviors in order to lessen their environmental impact can be seen as one of the most local ways of striving toward sustainability. In the environmental domain several ways to stimulate environmentally relevant household behavior have been studied. However, interventions that make use of groups to promote pro-environmental behavior seem potentially valuable, but are largely ignored (Dwyer et al. 1993). Since 1990, the international organization Global Action Plan (GAP) for the Earth disseminates a behavior change program aimed at environmentally relevant household behavior: the EcoTeam Program. The EcoTeam Program is relatively new and integrates several behavioral-modification interventions. It therefore raises a number of questions, the most important of which deals with the effectiveness of the program in leading households to a greater extent of sustainability. This question is addressed in this chapter.

This chapter is organized chronologically. We will first highlight the psychological aspects and expected effectiveness of participation in the EcoTeam Program. The main part of this chapter will deal with the results of the research project carried out by the Center for Energy and Environmental Research (Department of Social and Organizational Psychology) of Leiden University (Staats and Harland 1995, Harland and Staats 1997). This multiphased or longitudinal project lasted three years. It evaluated the effectiveness of the EcoTeam Program (for psychological backgrounds of environmentally relevant behavior, which is also included in this project, see Harland et al. in press). After a brief description of the research design, we will address the short-term and the long-term effectiveness, considering attempts for behavioral change. In addition, a short-term and long-term view on participants' attempts to save environmental resources is given. Finally, we will look ahead and see what might be needed to fully utilize the merits of the EcoTeam Program. But let us first introduce the reader to the EcoTeam Program itself.

Global Action Plan and the EcoTeam Program

Global Action Plan for the Earth is an environmental-action organization that disseminates the EcoTeam Program. The Global Action Plan organization is aimed at the preservation of the environment on earth. It was founded in 1990 by an international group of behavioral and environmental scientists. In 1991 Global Action Plan became active in the Netherlands, and by 1999, it was operating in fifteen countries. Global Action Plan's main assumption is that many people want to help create a better environment, but they often do not know where to start. Furthermore, many people are of the opinion that, on the whole, their individual effort will be negligible. These are the people whom the Global Action Plan organization wants to target for participation in the EcoTeam Program.

The EcoTeam Program is aimed at behavioral change. It consists of three parts that are described next.

The EcoTeam

The Global Action Plan organization sets up EcoTeams: small groups of six to ten neighbors, friends, club members, church members, and the like. These groups of people, representatives of their households, meet once a month. During these meetings their experiences, ideas, and achievements related to the EcoTeam Program are discussed. Following the EcoTeam workbook, the EcoTeams subsequently concentrate on each of the following six themes for four consecutive weeks: garbage, gas, electricity, water, transport, and consumption. The program lasts approximately eight months. Each team is supported by a coach or by the report center.

Information

EcoTeam members are provided with a personal workbook and a logbook in which team results are recorded. The EcoTeam workbook includes a short introduction to each of the six themes previously mentioned and an explanation of the goals GAP pursues. This introduction is followed by a list of a large number of pro-environmental actions that can be undertaken in the household. The workbook contains a questionnaire that enables the individual team members to check their progress in terms of environmental actions. In addition, team savings of gas, energy, waste reduction, and the like are recorded in the EcoTeam logbook. Team members may thus gain insight into their own behavior with regard to the six mentioned themes and track their progress, on an individual and team level.

Feedback

In each EcoTeam the group results are recorded and sent to a central database at the national GAP office in The Hague. The results of all active EcoTeams in the Netherlands are compiled, and individual teams can receive feedback about the savings volume. The Dutch EcoTeams also receive feedback about the accumulated results of all EcoTeams in the Netherlands and in other countries via the "EcoTeam-Newsletter," which is distributed every three months. The newsletters provide feedback for the teams and give them insight into the growth of the EcoTeam Program and its numerical results.

The goal of the combination of these three elements of the EcoTeam Program is to empower people to take pro-environmental action. *Empowerment* is an important underlying idea of the EcoTeam Program. Empowerment is intended to help people concentrate on what they can achieve instead of barriers to pro-environmental behavior.

From a social-psychological point of view, the EcoTeam Program is interesting because it (1) clarifies the relationships between household behavior and its effects on the environment; (2) gives practical and psychological support; and (3) increases commitment to behavioral changes by committing oneself to goals and to team members. The next section provides a more detailed view on these social-psychological aspects of the EcoTeam Program.

Psychological Characteristics of the EcoTeam Program and Expectations about Its Effectiveness Based on Previous Research

What do we know about the three elements of the EcoTeam Program— teamwork, information and feedback—from previous research in environmental psychology? The answer to this question is based on an earlier discussion of the psychological characteristics of the EcoTeam Program (Staats and Weenig 1995). Two of the three elements of the EcoTeam Program, *information* and *feedback*, have been studied extensively during the last decades. Working in a group that has organized itself to discuss and ameliorate its own environmentally relevant behavior has never before been documented, so no specific knowledge on this topic is available, although some of the psychological mechanisms inherent in this approach have been studied before.

The Divulgence of Information

Information is one of the most widely used means to bring about change. Information can be given on a scale ranging from mass media campaigns to close personal friends. Mass media information campaigns, commonly employed by the government, usually do not have a direct influence on behavior but may be helpful in putting an item on the public agenda (see McGuire 1985 for an extensive overview or Staats et al. 1996, for a specific application on effects of mass media information about the greenhouse effect). Information will only change behavior directly when the behavior is relatively easy to perform, when it is specified very precisely on a behavioral level, when convenient pro-ecological alternatives are available, when the message is delivered in close proximity to the target behavior, and when the message does not constitute a threat to an individual's perceived freedom

(Geller 1989). Given these conditions, more than information alone is usually needed to change in behavior. Nevertheless, information is an important prerequisite for other techniques to build upon. Information is helpful in increasing problem awareness, in developing a sense of personal responsibility, and in giving specific behavioral information. The latter is necessary because attitude changes that lead to behavior change are dependent on the perceived likelihood and evaluation of outcomes of behavior (Fishbein and Ajzen 1975). Unknown outcomes of environmentally unfriendly behavior have to be pointed out, and alternatives have to be indicated. A perceived lack of control to carry out pro-environmental behavior can be reduced or taken away completely by practical information (Ajzen and Madden 1986). In the EcoTeam Program both kinds of information are provided in the workbook. It contains information about the backgrounds to environmental problems and points out how individual behavior carried out collectively, leads to the problems that humankind is now facing. It also contains practical advice about how to change behavior with low behavioral costs. Given the broad scope of the EcoTeam Program, the workbook is a valuable source of practical information for pro-environmental behavior in the household.

Feedback

Apart from information, the EcoTeam Program aims to provide feedback. Feedback is the provision of information about the effects of previously performed behavior in order to adjust future behavior. Feedback has to be provided close in time to the performed behavior, in meaningful terms, and in comparison to a relevant criterion. Based on the results of an earlier study (Weenig et al. 1994), the Dutch EcoTeams are provided with feedback that stresses the collective accomplishments of all active EcoTeams. This type of feedback probably increases the sense of individual and collective efficacy (Bandura 1977). Feedback is provided each month which is reasonably fast and is based on self-registration. Other research has shown that self-registration imposes a barrier for participation but is quite effective for people who are willing to make this effort (De Boer and Ester 1985). In general, feedback is helpful in changing behavior. One drawback is that these effects usually disappear when feedback is no longer provided (Dwyer et al. 1993). The EcoTeam Newsletter, which is also distributed to former EcoTeam members and contains information about the accomplishments of all present and past EcoTeams, may therefore be important as an additional source of motivation to persist in the changes that have been realized in the period when individuals were active team members.

The EcoTeam

As previously mentioned, no studies have been made of groups that organize themselves on a voluntary basis to improve environmentally relevant
behavior. Similar organizations only exist in other domains, as in Weight
Watchers and Alcoholics Anonymous. However, the comparison is somewhat skewed as these groups are based on self-interest and are not organized
out of concern for a societal problem. Yet despite the lack of previous experience with the team phenomenon, some of the psychological processes that
are likely to happen within EcoTeams have been studied extensively and
have been identified as powerful mechanisms of change and action. The first
process is commitment. EcoTeam members commit themselves to keep
appointments for meetings with their team. These appointments refer to
attendance of EcoTeam meetings, collection and distribution of the environmental household data, discussion of experiences and declaration of intentions to change behavior. *Commitment* is one of only a few mechanisms that
are known to produce behavior changes that last, after the period in which
the intervention takes place (De Young 1993). A second important characteristic is the *social support* that is likely to be received by EcoTeam members.
Changing behavior, especially behavioral routines, is difficult and demands,
apart from good reasons and specific behavioral information, a social environment that welcomes these changes and is willing to help when practical
problems emerge. A third and related effect of working as a team may be the
reduction of uncertainty: doubts about efforts that other people are willing to
take are reduced (Wit and Wilke 1995).

Nonetheless, there are possible drawbacks to working in groups. There is
the risk of undue social pressure on team members to increase their efforts
beyond what they are willing to do. An atmosphere of competition to obtain
the most spectacular results, within or among teams, may emerge. Members
may also feel their privacy threatened if they feel obliged to disclose information about their household which they consider personal. A pilot study
(Harland et al. 1993) has indicated that the benefits of working in teams are
likely to be dominant when EcoTeams have organized themselves, while the
perceived drawbacks prevail in people who are not yet engaged in the
EcoTeam Program. The latter finding has been confirmed in a quantitative
study on attitudes and intentions toward participation in the EcoTeam Program (Staats and Herenius 1995).

Quantitative data concerning effects of participation in the EcoTeam Program are detailed in the remaining of this chapter. The main research question is this: Is the EcoTeam Program an effective way for households to
obtain a higher level of sustainability? This question is addressed by answering the following sub-questions: (1) Do people improve their environmen-

tally relevant behavior?; (2) Do they lessen their use of natural resources as a consequence of participation in the EcoTeam Program?; and (3) To what extent do these improvements persist?

Design of the Research Project

In order to study short- and long-term behavioral changes, a longitudinal research design is needed. Within this project, measurements were taken at three moments in time: directly before participation, directly after completion of the EcoTeam Program, and two years after participation.

All people who were ready to begin the EcoTeam Program in the period from January to March 1994 received a request to participate in the research. A total of sixty EcoTeams were involved, consisting of 445 participants, 289 (65 percent) of whom cooperated prior to participation in the EcoTeam Program by completing the first set of questionnaires (pre-test). In October 1994, these people were approached for a second time (first post-test). Of the original sample of 289 respondents, 205 (71 percent) completed the first post-test questionnaires. In the winter of 1996/1997 this group was approached again. The non-response rate was small: 150 ex-participants cooperated with this second post-test. Non-response was neither related to socio-demographic characteristics nor to general environmental concern (Staats and Harland 1995; Harland and Staats 1997). The socioeconomic status of the respondents in this research is somewhat higher than average in the Netherlands, and women are also over-represented. In addition, as might be expected, participants in the EcoTeam Program score higher on a measure of environmental involvement than the general Dutch population. On a 5-point scale (1 = low environmental involvement; 5 = high environmental involvement), the Dutch population scores 3.4, while participants in the EcoTeam Program score 4.0.

The mailed out questionnaires used in the pre-test and first post-test included questions about all 93 environmentally relevant household behaviors and investments that were included in the EcoTeam Program (e. g., separation of organic household waste, use of the standby function of the television, and installation of energy-saving light bulbs). As will be explained later, the second post-test (two years after participation) included a selection of these behaviors.

In addition, data was collected concerning the amount of solid household waste, and the amounts of natural gas, water, and electricity that were used over a period of two weeks by each participating household. Like the behavioral measures, these natural resource data were collected before participation in the EcoTeam Program as well as on both post-tests.

Modifying Environmentally Relevant Behavior: Short-Term and Long-Term Effectiveness of the EcoTeam Program

In this section the changes in environmentally relevant household behavior are described. We will start with a description of short-term behavioral changes observed among the EcoTeam Program participants. The long-term view will also be described. Finally, comparison of behavioral changes among non-EcoTeam participants (the Dutch population) in the same period will be highlighted.

Short-Term Behavioral Changes

To compare changes in environmentally relevant household behavior, 93 questions on specific behaviors were asked in the pre-EcoTeam Program questionnaire and in the first post-EcoTeam Program questionnaire. Each of the six environmental domains distinguished in the workbook (waste, natural gas, water, electricity, transportation, and consumer goods) was represented by a series of relevant behaviors. The changes in environmentally relevant behaviors are described here for each environmental domain. Changes are defined as statistically significant differences (p at least $<.05$) between pre- and post-measurements, and were tested with t-tests. When a difference appears to be statistically significant, this means that the chance that these differences are caused by normal behavioral fluctuations is acceptably small. In this chapter we only report changes which are statistically significant. In addition, it must be noted that all differences in behavioral performance are improvements (i.e., changes toward more environmental friendliness).

Waste

Three out of four behaviors related to the production of waste changed: an increase in the separation of organic waste, of old textile products, and of chemical waste was ascertained. The number of people who composted organic waste at home remained unchanged.

Natural Gas

Two out of 13 behaviors improved. The position of furniture in relation to the sources of heat was changed to improve heating efficiency and thermostats were checked to insure proper functioning.

Electricity

Eight out of 15 behaviors improved. Turning off lights in unoccupied rooms, switching off the television set instead of having it on standby, using the washing machine only when it is fully loaded, and cooking as efficiently as possible were numerous daily behaviors that changed. Energy-efficient lights were installed by a larger number of people and in greater amounts. The same was ascertained for the installation of time switches on electric devices for more efficient use.

Water

Ten out of 18 possible behaviors improved. The frequency by which individuals left the tap running while washing hands, washing the dishes, and brushing teeth decreased. Shower time and the frequency of washing the car were reduced. Participants also made four small investments in water-saving devices for their taps, toilets, and shower heads. Finally, a significant number of leaking taps were repaired.

Transportation

Five out of 17 behaviors changed. More cars with catalytic converters were in use after the EcoTeam Program, average speed on highways was reduced and for distances up to five kilometers, the use of public transportation and cars decreased while the use of bicycles increased.

Consumption

In a series of 26 items, participants had improved on 18. Of these behavioral changes, eight referred to food consumption, five to the use of cleaning products, two to the kind of paper that is used, one to the refusal of plastic bags in shops and two to product repair instead of buying new products.

Although the precise figures are not presented here (see Staats and Harland 1995), it appears that the size of the behavioral changes is modest. Rather than radical behavioral changes, the effects must be interpreted as improvements of already environmentally friendly trends. Additionally, behavioral changes that would have led to considerable investments of money or effort or that would have led to a considerable decrease of comfort hardly occurred during participation in the EcoTeam Program. This might be the reason why most of the behaviors concerning natural gas use and transportation did not improve. Another reason for the relatively few improvements concerning the use of natural gas might be related to the season in which the current participants followed the program. The larger part of the program was followed during spring and summer. Almost all behavior concerning natural gas use were related to spatial heating, and so were not relevant to focus on at that time of year.

The aforementioned changes are summarized in Table 14-1. The table makes clear that directly after participation in the EcoTeam Program, many frequently performed behaviors improved, while two out of eight behaviors that only involved a one- time effort changed. Also some small investments were made while large investments were not.

TABLE 14-1 *Type and Number of Changed and Unchanged Environmentally Relevant Behavior Directly After Participation in the EcoTeam Program (Absolute Numbers)*

	Changed	Unchanged	Total
Frequently occurring	36	27	63
One time behavior	2	8	10
Small investments	8	6	14
Large investments	0	6	6
TOTAL	46	47	93

Long-Term Behavioral Changes

After critically reviewing the research on the effectiveness of behavioral interventions Dwyer et al. (1993) conclude that many interventions can lead to substantial improvements in pro-environmental behavior, but that the long-term effectiveness of most of the interventions is meager and needs more research attention. The inclusion of the second post-test meets this need. In the following paragraph, we provide this long-term view on the effects attained with the EcoTeam Program. We first focus on behaviors that had improved and then on behaviors that had not improved shortly after participation.

Long-Term View on Behaviors That had Improved Shortly After Participation

What happened to behaviors that were improved directly after participation in the EcoTeam Program? A subset of 23 behaviors which had shown an improvement shortly after participation was compared with performance of the same behaviors two years after participation. As stated previously, the differences between behavioral performance just after participation in the EcoTeam Program and performance two years after participation have been tested statistically. The mean score for the group of respondents as a whole, shortly after participation, on each of the 23 questions concerning behavioral performance, were tested against the mean score of each question two years after participation. The results are described in Table 14-2. An example might be to query whether the improvement in the behavior "separation of organic waste," which was observed after participation, has been maintained. It appears that two years after participation, the group of former EcoTeam participants separated organic waste slightly more frequently (M =

6.74) than shortly after their participation (M = 6.68). This improvement is not large enough to be statistically significant, thus this behavior must be assumed unchanged. This result does, however, mean that the improvement which had taken place during participation in the EcoTeam Program has been maintained up to two years after participation.

TABLE 14-2 *Behavior Change Maintenance of Behaviors That Had Improved Shortly After Participation*

	Mean (st.dev) shortly after participation	Mean (st.dev) 2 years after participation	N	t-test
Separate deposit of organic waste	6.68 (1.16)	6.74 (0.96)	141	n.s.
Separate deposit of textile	6.77 (0.59)	6.68 (0.85)	137	n.s.
Shutting off lights in unused rooms	5.61 (1.27)	5.63 (1.20)	137	n.s.
Showering time (< 3 min.(1) – > 20 min(5))	2.03 (0.74)	2.02 (0.71)	144	n.s.
Speed on highways with a 120 km speed limit (km/h)	111.6 (10.8)	110.7 (12.4)	110	n.s.
Public transportation use for short distance (<5 km)	1.37 (0.75)	1.58 (1.22)	51 [1]	n.s.
Bicycle use for short distance (<5 km)	5.37 (1.56)	5.37 (1.49)	107	n.s.
Eating organically grown products	3.96 (2.25)	4.02 (2.18)	121	n.s.
Eating frozen vegetables	2.41 (1.29)	2.42 (1.29)	146	n.s.
Eating canned vegetables	1.85 (0.93)	1.9 (0.95)	140	n.s.
Repairing instead of buying products	5.67 (1.17)	5.74 (1.13)	147	n.s.
Using unbleached toilet paper	5.22 (2.34)	4.92 (2.36)	145	n.s.
Using unbleached writing paper	4.37 (2.10)	4.04 (2.17)	136	n.s.
Refusing plastic bags from shopkeepers	5.40 (1.39)	5.45 (1.31)	144	n.s.
Frequency of eating meat for dinner (per week)	4.88 (1.45)	4.86 (1.43)	128	n.s.
How much meat per dinner (in grams)	87.4 (31.6)	94.9 (43.8)	121	n.s.
Car use for short distance (<5 km)	2.81 (1.48)	3.06 (1.51)	103	n.s.
Turning off the tap while brushing teeth	5.94 (1.46)	5.91 (1.51)	138	n.s.
Turning off the TV set (instead of stand-by mode)	5.55 (2.07)	5.88 (1.87)	121	*

TABLE 14-2 *Behavior Change Maintenance of Behaviors That Had Improved Shortly After Participation (Continued)*

Saving laundry until the washing machine is full	6.34 (0.84)	6.54 (0.83)	145	**
Using detergents in refill packaging	5.15 (2.22)	6.00 (1.63)	141	***
Turning off the tap while washing hands	4.27 (1.92)	4.76 (1.61)	147	***
Turning off the tap while doing dishes	6.16 (1.22)	6.39 (1.00)	139	**

NOTE n.s. = non significant difference; levels of significance: * = $p<.05$; ** = $p<.01$; *** = $p<.001$, M = mean (i.e. mean score for the group of respondents); st. dev. = standard deviation (dispersion); N= number of valid responses; t-test= statistical test of the difference. Unless indicated otherwise, answers were given on 7-point scales with endpoints 1 (never) to 7 (always).

Focusing on the Table as a whole, it appears that 18 of the 23 behaviors remained improved. Moreover, 5 behaviors have improved even further. None of the improved behaviors have returned to pre-EcoTeam Program level. It thus appears that participation in the EcoTeam Program leads to long-term behavioral improvements.

Long-Term View on Behaviors That Had Not Improved Shortly After Participation

What happened with behaviors that were *not* improved directly after participation in the EcoTeam Program? Two years after participation, a subset of 13 behaviors was compared with behavioral performance shortly after participation. Results are presented in Table 14-3 which shows that, at that time, ten behaviors are still not improved compared to the situation shortly after participation. Three behaviors, however, that had not been improved during participation, do show an improvement two years after participation.

TABLE 14-3 *The Situation Two Years Later of Behaviors That Had Not Improved Shortly After Participation*

	Mean (st.dev) shortly after participation	Mean (st.dev) 2 years after participation	N	t-test
Temperature setting in the house (in Celsius degrees)	18.3 (1.5)	18.2 (1.6)	120	n.s.
Temperature setting of boiler (in Celsius degrees)	69.3 (11.2)	66.1 (12.4)	34	n.s.
Frequency of taking a bath (seldom(1) – daily(6))	1.68 (1.13)	1.64 (0.96)	80	n.s.

TABLE 14-3 *The Situation Two Years Later of Behaviors That Had Not Improved Shortly After Participation (Continued)*

	Mean (st.dev) shortly after participation	Mean (st.dev) 2 years after participation	N	t-test
Diminished water use for toilet flushing (no(1) – yes(2))	1.14 (0.34)	1.11 (0.37)	133	n.s.
Car-pooling	2.14 (1.63)	2.07 (1.67)	105	n.s.
Walking short distances (< 5 km)	3.37 (1.71)	3.65 (1.66)	68	n.s.
Means of transportation to go to work/school [2]		-		-
Most often used means of transportation	-	-	102	n.s.
Amount of fuel that the car uses (km per litre)	14.01 (2.60)	13.66 (2.53)	77	n.s.
Bringing own shopping bag while going shopping	6.73 (0.73)	6.71 (0.74)	146	n.s.
Using white (bleached) coffee filter bags	2.15 (2.16)	1.31 (0.99)	136	
Taking a shower (seldom(1) – daily or more frequently(4))	2.81 (1.13)	2.68 (1.10)	145	
Composting organic waste	4.19 (2.48)	3.91 (2.46)	121	

NOTES n.s. = non significant difference; levels of significance: * = *p*<.05; **=*p*<.01; ***=*p*<.001. Mean (i.e. mean score for the group of respondents); std. dev. = standard deviation (dispersion); N= number of valid responses; t-test= statistical test of the difference. Unless indicated otherwise, answers were given on 7-point scales with endpoints 1 (never) to 7 (always).

Similarly to what was observed in respect to the short-term results, it appears that behaviors that are costly (in terms of money, effort, or in the acceptance of a reduction in comfort) do not seem to be easily improved, even two years after participation in the EcoTeam Program.

In general, the results so far show that no fallback from formerly improved behaviors has occurred. Moreover, we have seen some autonomous improvements after participants had completed the EcoTeam Program. Overall, the conclusion of the results so far is positive: *Participation in the EcoTeam Program leads to an improvement of environmentally relevant behaviors.* Unlike effects attained with other behavioral modification instruments (Dwyer et al. 1993), these improvements seem to endure two years after participation.

Long-Term Effects on Environmental Investments

In order to determine whether environmentally relevant investments were made after participation in the EcoTeam Program had ended, questions about five minor and five major investments were included in this study. Table 14-4 shows that two years after participation, seven investments remain unchanged, and three investments are made more often than shortly after participation. Two of the smaller investments that had already been improved during participation (installation of energy-saving light bulbs and water-saving showering heads) were further increased, two years later. Additionally, contrary to what was observed during participation, former participants eventually decided to augment one of the major household investments: the installation of double-paned windows.

In general, Table 14-4 shows that two years after participation, former participants did invest more in some environmentally relevant appliances and thus increased the sustainability of their households.

TABLE 14-4 *Environmentally Relevant Investments Shortly After Participation and Two Years Later*

	Mean (st.dev) shortly after participation	Mean (st.dev) 2 years after participation	N	t-test
Minor investments:				
Any energy-saving light bulbs (no(1) – yes(2))	1.82 (0.38)		1.88 (0.32)	147
Number of energy-saving light bulbs	4.78 (4.06)	5.36 (3.55)	106	n.s.
Foil behind heating radiators (nowhere(1) – everywhere(4))	1.62 (0.82)	1.73 (0.97)	118	n.s.
Insulated central heating pipes (nowhere(1) – everywhere(4))	2.97 (1.15)	2.92 (1.07)	107	n.s.
Water-saving 'shower head' (no(1) – yes(2))	1.55 (0.50)		1.64 (0.48)	145
Major investments:				
Central heating boiler (regular/ gas-saving /extra gas-saving)[4]	—	—	118	n.s.
Energy source water boiler (electricity/natural gas/solar-energy)			—	51
Insulated walls (no(1)- partly(2)- yes(3))	2.22 (0.92)	2.25 (0.89)	139	n.s.

TABLE 14-4 *Environmentally Relevant Investments Shortly After Participation and Two Years Later (Continued)*

	Mean (st.dev) shortly after participation	Mean (st.dev) 2 years after participation	N	t-test
Toilets with a toilet dam (number of toilets)	0.76 (0.82)	0.86 (0.83)	123	n.s.
Double paned windows (nowhere(1)-everywhere(4))	2.96 (0.95)		3.07 (0.93)	140

Notes   n.s. = non significant difference; levels of significance: * = *p*<.05; **=*p*<.01; ***=*p*<.001; = investment had been improved already shortly after participation. M = mean (i.e. mean score for the group of respondents); std. dev. = standard deviation (dispersion); N= number of valid responses; t-test= statistical test of the difference.

Environmentally Relevant Behavior of the Dutch Population

The changes in the behavior of EcoTeam participants have most likely been caused by the EcoTeam Program. However, the possibility exists that the behavioral improvements that are established during participation in the EcoTeam Program are part of a national trend and caused by other factors (e.g., by nationwide information campaigns or new infrastructure that facilitates pro-environmental behavior). To be able to attribute the behavioral changes to the actual cause, we compared the improvements in a subset of environmentally relevant behaviors as performed by EcoTeam participants with changes in the same behaviors within the Dutch population.

Eleven behaviors and two household investments were measured in national surveys in the same periods as our pre- and post-test (Kruijk and Couvret 1994; 1995) and formed the basis for this comparison. Overall, the Dutch population has become more environmentally friendly on four out of eleven forms of behavior, but has worsened on three. In contrast, EcoTeam participants have improved on ten out of eleven behaviors, while the eleventh remained unchanged. In the same line, EcoTeam participants invested in environmentally relevant appliances at a higher rate than the general Dutch population within the same period of time. The long-term view is comparable with these observations. It appears that the environmentally relevant behavior in the Dutch population has largely remained stable up to the time of our second post-test (Couvret and Van de Valk 1997). This comparison clearly indicates that the EcoTeam Program can be considered the cause of the behavioral improvements among its participants.

Short- and Long-Term Changes in the Household Use of Natural Resources

Improvement of pro-environmental behavior is a very valuable result of the EcoTeam Program. Nevertheless, the results of the program should eventually lead to a decrease in the use of natural resources, and thus to a more sustainable household. To what extent does participation in the EcoTeam Program lead to savings in environmental resources? The parameters used in this study are the consumption of gas, electricity, and water and the production of household waste. Quantitative data on these four categories have been collected on the pre-EcoTeam Program measurement (before participation) and on both post-EcoTeam Program measurement (directly after and two years after participation). The participants were asked to keep record of their use of gas, electricity, and water, and the weight of solid waste they disposed. Each measurement was made during a two-week span. The comparisons between the three measurement periods will be presented in this section. It should be noted that the number of respondents that provided data for these analyses is lower than the number of respondents that provided data for the behavioral analyses.

Table 14-5 shows the short-term results. All data are presented as average consumption or production (for waste) per household member. The data clearly indicate that substantial savings have been achieved. The weight of waste produced and the amount of natural gas consumed show the most striking effects. Savings of electricity and water are more modest but significant.

As noted before, an important remaining question is whether short-term savings persist on the long run. In their overview, Dwyer et al. (1993) have shown that short-term savings very often fall back to earlier levels. For this reason, these quantitative analyses were repeated two years after participation. The 150 respondents were again requested to record their use of natural resources. As was noted above, it appeared that the response to that request dropped notably. As a result of this, the results of the people that *did* respond differ slightly from the short-term picture above. More generally, the lower numbers of respondents require caution in making generalizations of the reported results to other generations of EcoTeam participants.

The data indicate that two years after participation improvements have been achieved or maintained in all four domains (Table 14-6). Household waste and natural gas decreased the most. Savings of electricity and water are more modest, but still impressive. The conclusion of this section is that EcoTeam participants have persisted in reducing their consumption of environmental resources.

An important additional remark should be made concerning the use of natural resources in the Netherlands. It appears that contrary to the reported savings above, average energy use by the Dutch households shows a small increase in the period in which the data about the EcoTeams were collected (Weegink 1997a; 1997b).

TABLE 14-5 *Comparison of Environmental Effects Before and Shortly After Participation in the EcoTeam Program*

	Consumption prior to participation	Consumption directly after participation	Savings (in %)	N
Waste (kilograms per person per day)	.31 (.27)	.22 (.22)***	27.6%	92
Gas (m³ per person per degree day)	.32 (.19)	.24 (.18)***	23.1%	144
Electricity (kwh per person per week)	27.7 (15.7)	25.8 (15.2)*	6.8%	153
Water (m³ per person per week)	.89 (.40)	.84 (.39)*	4.9%	132

NOTES *=p<.05, **=p<.01,***=p<.001) Presented are averages and standard deviations (between brackets), the percentage difference (N= the number of persons available for statistical analysis).

TABLE 14-6 *Comparison of Environmental Resources as Used Before, Shortly After, and Two Years After Participation in the EcoTeam Program*

	Consumption prior to participation	Consumption shortly after participation	Consumption 2 years after participation	N
Waste (kilograms p.p. per day)	.216 (.15)	.153 (.12) = - 29%**	.145 (.12) = - 32%**	37
Natural gas (m³ p.p. per week)	.299 (.21)	.237 (.18) = - 21%***	.248 (.18) = - 17%***	77
Electricity (kwh p.p. per week)	27.2 (15.4)	25.9 (15.6) = - 5% ns	25.1 (14.3) = - 8%*	83
Water (m³ p.p. per week)	.854 (.38).830 (.38) = - 3% ns	.796 (.33) = - 7%*		75

NOTES *=p<.05, **=p<.01,***=p<.001, ns = not significant change. It presents averages and standard deviations (between brackets) and the differences in percentage points between the pre-EcoTeam Program data and both post-EcoTeam Program data (N = the number of persons available for statistical analysis

Conclusion

In this chapter the questioned was posed as to what extent the EcoTeam Program is effective in leading households to a greater extent of sustainability. In answering this question, we presented a short- and long-term view. The following short-term effects of the EcoTeam Program were reported and were additionally tested for their durability: improved environmentally relevant behavior, increased environmental investments in the household, and quantitative savings on environmental resources. Effects that were obtained just after the EcoTeam Program was completed were striking, with almost half of all behaviors showing an improvement. Moreover, it appears that these improvements were maintained up to two years after the intervention ended. Thus, in contrast to effects of most other behavioral change interventions, effects achieved with the EcoTeam Program seem to be lasting, and for some themes these effects even continue to improve after participation has ended. These results provide strong support for the effectiveness of the EcoTeam Program. The program appears to be an instrument that leads its participants to a more sustainable way of organizing their households.

The EcoTeam Program in Other Countries

Can these results be generalized to apply to other countries? It appears that the EcoTeam Program attracts people in such culturally diverse countries as Belgium (Flanders), Denmark, Finland, Iceland, Ireland, Korea, Norway, Poland, Russia, Slovenia, Spain, Sweden, Switzerland, the United Kingdom, and the United States. In all, approximately 3,000 teams have followed or are following the EcoTeam Program. The Global Action Plan International organization has had some positive results about effects of the EcoTeam Program. However, an almost total lack of systematic quantitative research into the effectiveness of the EcoTeam Program in these countries mentioned prevents hard conclusions from being drawn about the success of the EcoTeam Program there. This thwarts international comparison of the results reported in this chapter. Fortunately there are, to the best of our knowledge, two exceptions. Research has been carried out both in the United Kingdom and in Switzerland. The results from these studies seem promising. In the United Kingdom a substantial proportion of participants reported having improved several daily habits as was studied by Global Action Plan in the United Kingdom. However, the program in the United Kingdom has been used differently than in other countries. In the United Kingdom people participate individually instead of in teams. Additionally, participants do not work themselves through an integrated "workbook," but order so-called Action Packs which cover one theme (e.g., energy or water) at a time. This prevents direct comparison with the results reported in this chapter. It is interesting to note that the program in the United Kingdom attracted 20,250 participants until 1999, a number which is much larger than in other countries (e.g., approximately 10,000 in the Netherlands. The U.S. and the Netherlands had the greatest number of participants in the original program).

The Swiss EcoTeam Program is comparable to the Dutch version. There have been several interesting research efforts which are summarized by Bruppacher in the following section of this book. Participants carried out the behavioral suggestions given in the program. However, participants followed suggestions concerning some categories (waste, consumption, and water) in greater numbers than suggestions concerning others (transport and energy). This latter phenomenon was also observed in our research, and might be linked to the costs (i.e., time, money, effort, loss of comfort) that are related to behavioral improvements concerning these categories.

The state of affairs—just presented—illustrates that we still lack the necessary knowledge for judgments regarding the effectiveness of the EcoTeam Program in other countries. It also prevents us from understanding the potential effects of cultural aspects on the effectiveness of the EcoTeam Program in the Netherlands.

Further Research

The preceding explanation regarding the restricted view of the international effects of the EcoTeam Program clearly requires more international research. Such research could reveal the generalization of the results presented in this chapter. However, there is another domain of research which, in our view, is even more strongly needed. Until now only a limited number of people have participated in the EcoTeam Program. In order to make use of the effectiveness of the EcoTeam Program, future development of the EcoTeam Program should be aimed at acquiring larger numbers of participants. From 1991 until 1999, approximately 10,000 households in the Netherlands participated in the EcoTeam Program. Although impressive in itself, this number definitely lags behind initial predictions made by the Global Action Plan in the Netherlands. One way to enhance participation might be to lower the threshold for participation. This threshold seems to be fairly high and appeared to be related to the team aspect of the program as has been indicated in a study among non-participants (Staats and Herenius 1995). Decreasing the emphasis on the team aspect might lower the threshold and enhance participation. Development of a program which is to some extent individualized, might be a way to enlist a substantially larger number of people for the EcoTeam Program. Such a program should make it possible, but not necessary, for participants to meet each other, while the program otherwise remains as it is. It might be expected that such a program would also attract people who are not normally very environmentally involved. Research into the effectiveness of a more easily accessible EcoTeam Program, could make an important contribution to future developments of the EcoTeam Program.

Notes

[1]   The number of valid answers is very small because this question was only answered by people who owned a car on both measurements and who answered this question on both occasions.

[2]   This and the following subject consisted of answering categories that could not be transformed into numerical data ("public transportation"/ "car"/ "bicycle"/ "walking"/ "otherwise"). The differences in performance of these behaviors are statistically tested and appeared to be non-significant as can be seen in the right hand column.

[3]   The number of valid answers is very small because this question was only answered by people who owned a car on both measurements, who in addition answered this question on *both* occasions, and who also went to work or school on a daily basis.

[4]   This and the following subject consisted of answering categories that could not be transformed into numerical data. The differences in performance of these behaviors are statistically tested and appeared to be non-significant as can be seen in the right-hand column.

References

Ajzen, I. and T. J. Madden. 1986. Prediction of goal-directed behavior: Attitudes, intentions and perceived behavioral control. *Journal of Experimental Social Psychology* 22: 453-474.

Bandura, A. 1977. Self-efficacy: Toward a unifying theory of behavioral change. *Psychological Review* 84: 191-215.

Couvret, E and J. Van de Valk. 1997. *Tabellen Milieugedragsmonitor 2-7. (Monitor on environmental behavior Tables 2-7)*. Amsterdam: NIPO.

De Boer, J. and P. Ester. 1985. Gedragsbeïnvloeding door voorlichting en feedback. (Influencing behavior by information and feedback). *Nederlands Tijdschrift voor de Psychologie* 40: 87-95

De Young, R. 1993. Changing behavior and making it stick. The conceptualization and management of conservation behavior. *Environment and Behavior* 25: 185-205.

Dwyer, W. O., F. C. Leeming, M. K. Cobern, B. E. Porter and J. M. Jackson. 1993. Critical review of behavioral interventions to preserve the environment. Research since 1980. *Environment and Behavior* 25: 275-321.

Fishbein, M. and I. Ajzen. 1975. *Belief, Attitude, Intention, and Behavior. An Introduction to Theory and Research*. Reading: Addison-Wesley.

Geller, E. S. 1989. Applied behavioral analysis and social marketing: An integration for environmental preservation. *Journal of Social Issues* 45: 17-36.

Harland, P., S. Langezaal, H. J. Staats and W. H. Weenig. 1993. *The EcoTeam Program in the Netherlands. Pilot Study of the Backgrounds and Experiences of the Global Action Plan*. Leiden: Center for Energy and Environmental Research, Leiden University. EM/R-93-37.

Harland, P. and H. J. Staats. 1997. *Long-Term Effects of the EcoTeam Program in the Nether-lands: The Situation Two Years after Participation.* Leiden: Center for Energy and Environ-mental Research, Leiden University. E&M/R-97-69.

Harland, P., H. Staats and H. A. M. Wilke. 1999. Explaining pro-environmental intention and behavior by personal norms and the theory of planned behavior. *Journal of Applied Social Psychology.* In press.

Kruijk, M. D. de and Couvret. 1994. *Milieugedragsmonitor, Vierde meting (Monitor on envi-ronmental behavior, 4).* Amsterdam: NIPO

Kruijk, M. D. de and Couvret. 1995. *Milieugedragsmonitor, Vijfde meting (Monitor on environ-mental behavior, 5).* Amsterdam: NIPO.

McGuire, W J. 1985. Attitudes and attitude change. In G. Lindzey and E. Aronson, eds., *The Handbook of Social Psychology* Vol. 2., pp. 258-276. New York: Random House.

Staats, H. J. and P. Harland. 1995. *The EcoTeam Program in the Netherlands. Study 4: A Longi-tudinal Study on the Effects of the EcoTeam Program on Environmental Behavior and Its Psy-chological Backgrounds. Summary Report.* Leiden: Center for Energy and Environmental Research, Leiden University. E&M/R-95-57.

Staats, H. J. and S. G. A. Herenius. 1995. *The EcoTeam Program in the Netherlands. Study 3:The Effects of Written Information on the Attitude and Intention Towards Participation.* Summary report. Leiden: Center for Energy and Environmental Research, Leiden Univer-sity. E&M/R-95-48.

Staats, H. J. and W. H. Weenig. 1995. Milieupsychologie in de praktijk: Het EcoTeam pro-gramma in Nederland (Environmental psychology in practice: The EcoTeam program in the Netherlands). *De Psycholoog* 29: 323-26.

Staats, H. J., A. P. Wit and C. W. H. Midden. 1996. Communicating the greenhouse effect to the public; Evaluation of a mass media campaign from a social dilemma perspective. *Jour-nal of Environmental Management* 45: 189-203.

Weegink, R. J. 1997a. *Basisonderzoek Aardgasverbruik Kleinverbruikers BAK '96 (Research on the use of natural gas by households).* Arnhem: EnergieNed. juni 1997.

Weegink, R. J. 1997b. *Basisonderzoek Elektriciteitsverbruik Kleinverbruikers BEK '96. (Research on the use of electricity by households).* Arnhem: EnergieNed. juni 1997.

Weenig, W. H., M. E. Gijsman and S. Langezaal. 1994. *The EcoTeam Program in the Nether-lands. Study 2. A Study into the Effects of Three Different Kinds of Feedback.* Summary report. Leiden: Center for Energy and Environmental Research, Leiden University. E&M/ R-94-58.

Wit, A. P. and H. A. M. Wilke. 1995. "Omgevings-en sociale onzekerheid in sociale dilemma's. (Environmental and social uncertainty in social dilemmas). In N. K. De Vries et al., eds. *Fundamentele Sociale Psychologie,* pp. 137-52. Tilburg: Tilburg University Press,

Global Action Plan Research in Switzerland

Susanne Bruppacher

Development of the Program

The association Global Action Plan (GAP) Switzerland was founded in 1993. By October, 1998, 164 EcoTeams (approximately 1,000 households) had worked through the program. Over the last five years the Swiss GAP manual was revised several times, and the latest version dates from 1997. This version introduced a major change in that participants no longer kept records of their consumption, but estimated it by means of a lifestyle questionnaire before and after the program. In contrast to findings from the Netherlands (Weenig et al. 1994), keeping records of consumption was perceived as rather inconvenient and as a consequence, the GAP office received a considerable amount of incomplete data.

Research

Since 1995, two institutions have carried out research on the effects and characteristics of the program in Switzerland. A first study was undertaken on behalf of the Swiss Federal Office of Energy (Graf 1997), and a second series of pilot studies is currently being conducted in two projects within the scope of the Swiss Priority Program "Environment" by the Swiss National Science Foundation. [1] The main aspects being surveyed are demographics (the kind of people attracted to the program), motivation for participation, success in implementing the suggested actions, and experiences (successes, difficulties) of participants while working through the program.The first study on behalf of the Swiss Federal Office of Energy (Graf 1997) evaluated the following four energy conservation programs to determine their effectiveness in fostering environmentally sound behavior of individuals: a lifestyle test by the World Wildlife Fund (WWF) (N=753), the GAP program for private households (N=310), and two organizations by medical doctors that base their membership fee on the amount of energy consumption (N=74 / N=495).[2] A questionnaire was mailed to individuals to collect information on demographics and motivation for participation. The data shows that in all four samples, participation was mostly attributed to individual world views. Women and young people are more willing to invest time, whereas men prefer to invest money in order to act in an environmentally responsible way. Several differences between the GAP program as an action-oriented program and the other programs were also found: only GAP participants also considered motives associated with economic and social rewards to be relevant, e.g. contact with neighbors. They reported significantly higher energy savings and were more positive about their actions. The GAP program mainly attracts families with school children in suburban areas. Most of the time women initiate participation in the program. Half of the participants were recruited by face-to-face contact, and most teams are formed in their own residential area. The team is considered an important but not dominant factor for success. Control and competition within a team is not positively viewed, which can be taken as an indicator that the program builds on the intrinsic motivation of the participants. Participants also reported difficulties in recruiting new teams after completion of the program. They are not willing to act as missionaries in order to spread the program.

Within the scope of the Swiss Priority Program for the Environment, a series of pilot studies for an interdisciplinary research project have been conducted. These studies examine the preconditions of individual environmentally responsible behavior, especially the restrictions and limitations of such behavior.

Känel et al. (1998) evaluated feedback from eight teams where suggestions for actions were actually undertaken. Afterward they presented the results to team counselors and conducted fifteen semi-structured interviews with them on the factors they perceive which foster or impede the implementation of these concrete suggestions for actions made by GAP. The results of the team feedback show that as far as garbage and consumption were concerned, participants succeeded to a great extent. Measures concerning water conservation were only partly implemented. However, most suggestions concerning transport and energy conservation were not carried out. These two areas seem to be the most difficult for stimulating behavior change. The interviews revealed that the suggestions most participants failed to carry out involve one or more of the following problems: they require rather costly investments and considerable sacrifice of time or comfort, which can affect well-being; they are difficult to carry out because of inadequate or missing infrastructure; they require cooperation with other people who have little or no interest in cooperating. Participants readily implement changes that imply only small additional expenditures of time, money, and personal effort, and also one-time investments or efforts. Social networks seem to have both considerable negative and positive influence on performance in the GAP program. For example, models, support from the social environment, and the group aspect of the program (cooperation, sharing of experiences, sense of community) are perceived to foster the success of the program. Negative attitudes of family members toward action suggestions, fear of social control, and loss of prestige by outing oneself as "Green" represent obstacles. The experiences of the team counselors suggest that it is absolutely necessary for local authorities to take responsibility for providing sufficient infrastructure and small financial incentives in order to achieve a more sustainable lifestyle in the whole population. The program reaches only a certain part of the population, which in addition usually has already adopted a sustainable lifestyle in comparison to the average.

Two further pilot studies have recently been carried out and are currently being evaluated. In June 1998 a "Zukunftswerkstatt" (Jungk and Müllert 1989), a problem-oriented workshop with GAP participants and representatives of local authorities (N=25), was held to reveal the various constraints individuals face when working through the GAP program in order to suggest possible ways of overcoming these constraints and to identify community-based interventions that are experienced as effective or ineffective in fostering ecological behavior. From September 1997 until September 1998 all new GAP households received a questionnaire by mail (N=49) with questions about attitudes toward the environment, environmental politics, environmental behavior, and psychological constructs that were expected to relate to attitudes such as perceived control on environmental problems and the like.

The analysis of all the material has not been completed, but first results confirm the importance of social factors such as cooperation of household members and support by the residential community (e.g. by providing infrastructure) as necessary preconditions for ecological behavior.

Conclusion

Overall, the GAP program is positively evaluated by Swiss participants. Moderate behavior change takes place, especially in the areas of waste and consumption, but there is still room for improvement concerning transport and energy. It can be concluded that the program mostly attracts people who have already adopted quite a sustainable lifestyle in comparison to the average population. For this reason, participation often resulted in less savings than expected. The individuals who would make the greatest behavioral improvements by participating in the program are not reached. The strategy that each EcoTeam recruits new teams by personal contact has also only proved moderately effective, in that each team recruits about 1.1 new teams, which cannot spread the program over the whole country within a reasonable time according to GAP aims. It seems sensible at this time to take advantage of and foster the current trend of local community involvement in Local Agenda 21 processes. GAP activities in communities are so far quite promising. In the community of Ittigen, a suburb of the Swiss capital, Bern, which has 11,000 inhabitants, several environmental measures were able to be implemented. Ittigen has also become a GAP partner community. These processes are currently being evaluated by the research project mentioned within the Swiss Priority Program "Environment." In addition to the in-depth interviews designed to follow up on the findings of the pilot studies and to show how specific changes in social and infrastructural conditions lead to a facilitation of suggested actions by the GAP manual, an expert group has been formed to create a system dynamics model for Ittigen following a method proposed by Vennix (1996).

Notes

[1]   Project No. 5001-48832 "Environmentally Responsible Behavior in Community Settings: Theoretical Analysis and Empirical Investigation of Overcoming Barriers to Change," which is part of the Integrated Project No. 5001-48826 "Strategies and Instruments for Sustainable Development: Foundations and Applications, with Special Consideration of the Community Level."

[2]   Ärztinnen und Ärzte für Umweltschutz, Ärztinnen und Ärtze für soziale Verantwortung (Medical Doctors for Environmental Protection and Social Responsibility).

References

Graf, E.O. 1997. *Energiesparaktionen*. Bern: Bundesamt für Energiewirtschaft.

Jungk, R. and N. R. Müllert. 1989. *Zukunftswerkstätten: Mit Phantasie gegen Routine und Resignation*. München: Heine.

Känel, E., B. Magun, R. Öhri and A. Sanchez. 1998. *Umweltverantwortliches Alltagshandeln beim Global Action Plan: die Bedeutung sozialer Netze*. Bern: IKAÖ.

Vennix, J. A. M. 1996. *Group Model Building: Facilitating Team Learning Using System Dynamics*. Chichester: John Wiley and Sons.

Weenig, W. H., M. E. Gijsman and S. Langezaal. 1994. *The Ecoteam Program in the Netherlands. Study 2: A Study Into the Effects of Three Types of Feedback*. Leiden: Centre for Energy and Environmental Research, Leiden University.

Chapter 15

Sustainable Lifestyle Assessment

Dimitri Devuyst and Sofie Van Volsem

Summary

International institutions, governments, NGOs, and individual citizens currently realize that efforts to make our societies more sustainable should be accompanied by a thorough examination and changing of our lifestyles. Every day individuals and households make important decisions and take part in routine activities which have consequences for the sustainable development of our societies. Therefore, there is a need to develop tools that assess whether our daily life decisions result in more sustainable lifestyles. Sustainable Lifestyle Assessment (SLA) can take a self-assessmentapproach or a "research" approach. The self-assessment approach aims to encourage individuals or households to evaluate the sustainability of the decisions they make in their own lives. The research approach is the scientific examination and comparison of different lifestyles; this information can, in turn, be used by individuals and governments to make decisions that can make life both more rewarding and less impacting on the environment. SLA needs to take into account values, social networks, the hierarchy of decisions, the perception of individuals, and the cultural context in which we live. In this chapter, a seven-step approach to a personal or household SLA is proposed. Life-cycle assessment, household metabolism research, Eco-footprint analysis, and social science methods are discussed as scientific ways to approach SLA.

Lifestyles and Sustainability

In the past, environmental management often focused on the role of industry and their production processes. Today, there is a growing awareness that much can be done by consumers. High-consumption lifestyles—common in many developed countries—result in the use of enormous amounts of energy and material in contrast to lifestyles in developing countries. A vivid example of this intensive use of resources is: the average person in the Northern Hemisphere consumes 80 times as much energy as a person in

Africa south of the Sahara, and the World Trade Center in New York City consumes more energy annually than the African country Niger (Reisch 1998).

With the help of NGOs such as Global Action Plan (see Chapter 14), people are beginning to realize that they can direct their lifestyles in a more sustainable way by making minor changes. Sustainable Seattle (1998) gives many examples of the way in which small changes in an individual's lifestyle can make a difference in the global environmental impact of a community. Table 15-1 provides some examples.

TABLE 15-1 *Indication of the Environmental Consequences of Individual Daily Decisions*

Individual decision	Gain for the environment
Eating 100g less beef per week	Saves about 2,000 l water/week required for cattle raising, feeding, processing, and transportation (equals daily water use of seven people) and makes an additional 3.5 meals worth of grain available for human consumption (saves 500g grain)
Driving the car one day less per week	Reduces distance traveled by 34 km/week (average) or 1,768 km/year
Replacing a vehicle with one that is 4 km/l more efficient	Reduces gasoline consumption on average by 6.5 million l/year for a community of 20,000 people
Reducing water consumption (fix leaks, low-flow shower heads, five minute showers, low-flow toilets)	Saves 228 l/day per household (compared to one person's water use of 304 l/day)
Buying one kg organic foods per week	Reduces pesticides and herbicides use by 440 g/year
Using refillable glass bottles for milk	Reduces plastic in solid waste stream by 1.8 kg/year

SOURCE Sustainable Seattle (1998)

The awareness that the individual and the community (being the sum of many individuals) can contribute to a more sustainable future has led to the development of the first simple instruments for Sustainable Lifestyle Assessment (SLA) which will be further discussed in this chapter.

Defining a Sustainable Lifestyle

Christensen (1997) defines "lifestyles" as a cluster of different habits that to different extents are embedded in societal infrastructures. Habits considered in this context relate to housing, transportation, heating and electricity, and provisions. Each habit can be accomplished in different ways, where some ways have a greater environmental impact than others.

A sustainable lifestyle is defined in this chapter as the sum of all habits that together can be identified as a distinct "way of living" of a human being, which guarantees a basic quality of life that can be maintained indefinitely by a certain population and therefore remains within the carrying capacity of the ecoregion considered. This definition contains two important components which deserve critical thought: a) What is a basic quality of life? and b) Is it possible to reduce the impacts of human life on the surroundings and still maintain this basic quality of life?

Exploring Basic Quality of Life

Maslow (1967) made the following list of the basic needs of human life:

* The first essential need is that the physiological needs are met—that we are able to satisfy hunger and thirst
* The second is that we feel secure and have a place for shelter and are able to keep ourselves warm
* The third is that we feel we belong to a group or society
* The fourth is that we are free to express our individual identity in some way
* The fifth is that we live in an environment that allows us to experience a sense of self-fulfillment

From this it becomes clear that our "basic" well-being is not dependent on the consumption of huge amounts of energy or material goods. Maintaining the roof over our heads, heating in winter, and consuming a nutritious meal when we get hungry requires an energy input. Fulfilling our need to feel secure, to belong to a group, to express our individual identity and self-fulfillment, however, is to a large degree independent of energy and material inputs. In other words, the current energy and material needs of people in developed countries are mostly based on the cultural context in which they live. As Reisch (1998) states, culture provides the fundament for values, attitudes, beliefs, and behavior and therefore directly influences consumption. As a result, changing to a sustainable lifestyle will be a slow and gradual process. The following box shows, for example, that changing lifestyles to be more sustainable—in this case through telecommuting (or *teleworking*) is not easy and requires commitment of all parties concerned. Such changes could only be accomplished when linked to sound institutional and social arrangements and an eco-intelligent infrastructure.

Example of Problems Encountered When Trying to Introduce a More Sustainable Lifestyle: Promoting Teleworking in Flanders, Belgium

Working at home or teleworking (using a computer and modem linked up with the office) is currently not a very popular alternative to the daily commute to the office in Flanders. Both employers and employees look at this possibility with a lot of suspicion. Teleworking is seen as a possible solution for persons who would like to combine work with family commitments. It could also solve problems of congestion on the roads and result in less energy consumption, since there is no or a reduced need for traveling. Teleworking is not taking off in Flanders because:

* ❧ employers fear that they will lose control over the work done by employees

* ❧ employers fear that corporate culture will get lost

* ❧ employees fear the loss of informal social contacts

* ❧ employees do not have the necessary infrastructure and space at home

* ❧ employees fear missing chances for promotion if they are absent from the office

* ❧ employees fear less favorable terms of employment (a self-employed status instead of a real contract, earning less).

Making telecommuting a viable option to office work will require changes in the attitudes of both employers and employees. The Ministry of Employment and the labor unions could also play a critical role through the development of legislation, telecenters (providing local office facilities for people who prefer not to work at home but wish to avoid the cost, time and inconvenience of commuting) can make sure that teleworkers retain the same employee status and rights as the other employees.

Growing evidence for progress in this field is illustrated in the Status Report on European Telework 1998 (European Commission 1998). The Belgian Teleworking Association, founded in 1994 and restructured towards the end of 1997, has professionalized its administrative and management activities and is extending its membership. A new law on home working came into force in March 1997. One telecenter was opened in 1996, with plans for 20 centers across Belgium, and a not-for-profit membership organization for individual teleworkers and home based workers— Home Based Business—was formed in 1997.

Another alternative to telecommuting would be a switch from working eight hours for five days a week to shifts of thirteen hours per day, three days a week. This would reduce commuting by 40 percent—with the nice side-effect that the utilization of premises and machinery would be almost doubled and family life would profit (Reisch 1998).

It should be clear that the concept of "basic quality of life" is closely related to the concept of "well-being." We could even assume that a basic quality of life of a person can only be met when this person can experience an overall feeling of well-being. A feeling of well-being is very personal and can therefore not be regulated or explicitly introduced by an authority. An identical set of living conditions may lead to a feeling of well-being in one person and not in another. The result is that it is impossible for a government to introduce one uniform sustainable lifestyle that would be satisfactory for everyone. Governments can, however, create conditions that would facilitate people to make a number of choices which steer their lifestyles in a more sustainable direction, while at the same time still staying within the boundaries of their ability to satisfy their feelings of well-being.

Population Size and Consumption Levels

The other important question deals with the possibility to reduce the impact of human life so that it can be sustained indefinitely and still result in an acceptable quality of life. This question is directly linked to the size of the human population and on the resource consumption level of each individual. The earth can only absorb an impact up to a certain level—this is known as *carrying capacity*—and Holmberg (1992) indicates that there is a growing consensus that this level is being approached at accelerating rates. Urgent action is required to reduce population growth, mainly in the Southern Hemisphere, and also to reduce or eliminate patterns of overconsumption, mainly in the Northern Hemisphere, all of which contribute to the sustainability crisis (Holmberg 1992). Therefore, powerful measures should be made to stabilize the population in the Southern Hemisphere (including improved maternal and child health care, improved education for women and girls, and focus on family planning) and combined with actions to reduce excessive consumption and waste of resources in the Northern Hemisphere.

At the individual level it is impossible to say which level of consumption is sustainable and which is not, but it is possible to steer the lifestyle of people in a more sustainable direction. An SLA can help individuals, institutions, and authorities to evaluate whether changes in lifestyle also lead to a more sustainable lifestyle.

A Movement Toward More Sustainable Lifestyles

In North America and Europe, local movements have emerged that try to encourage people to take action in relation to sustainable consumption, lifestyle simplification, and the introduction of alternative economic systems which place a higher value on community services or family caring than is the case today. In addition, governments, international institutions and academics are examining measures to reduce unsustainable patterns of production and consumption and measures that can lead our way of life in a sustainable direction. In this section we will examine what is meant by *sustainable consumption, lifestyle simplification,* and *Local Employment Trading Systems.* A link with sustainable lifestyles is also made.

Sustainable Consumption

An increasing number of awareness-raising actions are currently being introduced as a reaction to overconsumption and extreme forms of materialism in Western societies. For example, in 1992, Canadian citizens began organizing a "Buy Nothing Day" to protest against the constant call in the developed world to consume and the contrasting lack of basic necessities for many in the developing world. This idea of a "Buy Nothing Day" has now also spread to Europe. Of similar note is a growing trend that calls for a simplification of our lives or parts of our lives, such as a simplification of the Christmas festivities.

Chapter 4 of Agenda 21 addresses "changing consumption patterns," which encourages the development of national policies and strategies to introduce change in unsustainable consumption patterns (UNCED 1992). As a result, the issue of sustainable consumption and production patterns is now high on the international environment and development agenda. For example, the OECD held an international Experts Seminar on sustainable consumption and production patterns, in collaboration with the Massachusetts Institute of Technology (MIT) in December 1994. What was significant about the approach to sustainable consumption and production that emerged from the meeting was the emphasis placed on highlighting opportunities rather than threats, and on broadening perceptions of individual responsibility and choice rather than narrowing horizons (OECD - MIT 1994). Also, private organizations such as the World Business Council for Sustainable Development (WBCSD) and NGOs such as Consumers International (CI) have taken action. WBCSD established a task force on sustainable consumption and production and the CI presented its "Meeting Needs, Changing Lifestyles" campaign at the 1997 World Consumer Rights Day (CI 1997). The so-called Rio+5 meeting, the nineteenth Special Session of the UN General Assembly which gathered to review progress achieved over the five years that have passed since the Rio Earth Summit, considered changing consumption and production patterns as one of the areas requiring urgent action (UNDPCSD 1997).

Elster (1989) states that we are confronted by two alternative models: *homo economicus* and *homo sociologicus*. The former is supposed to be guided by instrumental rationality, whereas the behavior of the latter is dictated by social norms. What is needed is a *homo socio-economicus*, which points to the fact that—at least with regard to consumption behavior—we are far from autonomous. Our consumption decisions are only to a limited extent determined by innate preferences, and to a very large extent by what may be called "lifestyle effects." Individuals have a complex relationship with their reference group, because consumption norms serve functions of inclusion and exclusion (Elster 1989).

Given the number of products available on the market today, one can wonder to what extent is the consumer well-informed of the existence of substitute products, the characteristics of these products, and the impact they might have. It is clear that if an alternative product cannot be easily identified and recognized by consumers, it is useless to hope that there will be a mass movement to buy such a product. Any consumption choice requires a certain effort. Environmentally and socially friendly products that are distributed only in certain specialized shops are less accessible and are an obstacle to purchasing, which the consumer can overcome only at the cost of an additional effort (Godeau 1998).

Hansen and Schrader (1997) linked sustainable consumption to the notion of an ethically founded consumer responsibility. The consumer is considered a citizen who takes into account his morals and values when deciding on buying something.

To make consumption more sustainable, a number of questions need to be asked before an item is bought, such as: a) Do I need it? b) Do I want it? c) How long will it last? d) Can I get along without buying it? e) Are the resources which went into it renewable or nonrenewable? f) Will I be able to fix it? g) How will I dispose of it when I'm done using it? h) Have I researched to get the best cost and quality? i) Am I willing to maintain it? (New Road Map Foundation 1995).

As Reisch (1998) states, rules to guide sustainable consumption can be formulated. These rules should be seen as acts of self-commitment by the individual consumer, based on free will. They are also heuristic in nature and all the limits of the personal, social, infrastructural, and institutional factors influencing consumption apply. Such rules can include: a) decreasing purchases (buy less, use products longer, buy secondhand; repair broken materials); b) reflecting on product quality (consider ethical and environmental qualities of products); c) reflecting on needs (do not buy because of habit, boredom, or frustration); and d) adapting of the lifestyle (adapt your lifestyle to the degree of your environmental and social conscience).

Lifestyle Simplification

Lifestyle simplification is linked to sustainable consumption but has a wider scope. Here, the idea is not only to change the way we buy and consume goods, but to change the way we live in a more far-reaching manner. In North America, the idea of lifestyle simplification has been promoted by an increasing number of organizations, such as The New Road Map Foundation and the Center of a New American Dream and by Internet sites, such as

"The Simple Living Network," which presents tools and examples for a more conscious, simple, healthy, and restorative lifestyle. Several authors have also been very active in presenting possible ways of simplifying life in books using a practical, non-scientific approach, such as *Your Money or Your Life* by Joe Domiguez and Vicki Robin or *Simplify Your Life* by Elaine St. James. The idea of lifestyle simplification is to get rid of all the clutter in people's lives, based on conscious decisions on what is really important in life. The process can both concentrate on the external and internal areas of life. External areas deal with aspects such as finances, careers, social lives, and the routines of our general lifestyle. Inner simplicity means among other things tuning in to the love of family and friends, the wonders of nature, creating joy in our lives, and the serenity and clarity that come from silence and quiet contemplation (St. James 1995). Lifestyle simplification directs us away from a complicated life focused on gaining and maintaining material wealth and is therefore meaningful from a sustainable lifestyle point of view.

LETS - Local Employment Trading Systems

A new and simplified lifestyle has led to proposals for alternative economic models. Local Employment Trading Systems (LETS) are grassroots initiatives for community economic development, based on the development of a local "money-system." LETS is based on the idea that money is nothing more than a means of measuring the value of goods or services and that we can make our own local currency. The result is that skills, knowledge, and materials that sit idle for lack of money can be put to productive work (Burman 1997). At the same time certain tasks that are currently not highly valued (not well paid), but are important for a healthy community can be revalued through LETS. Time dollars can be considered a form of LETS. The Time Dollar Institute was established in 1995 by Edgar Cahn in Washington, DC, to promote the idea of time as a currency. The concept is to reward the time people spend in services (such as caring for others) and reward these with time dollars, which can then be used to purchase other services within the system. The time dollar can be considered a "social currency," designed to reward altruism and support community-service exchange. It was originally created to give the elderly a means of paying for volunteer services, but has currently expanded in the U.S. into community-wide, computer-based "time banks": integrated into health care systems, colleges, churches, state and federal social agencies, neighborhood security patrols, food banks, and the like. A time dollar is earned for one hour of volunteer work and can be spent on one hour of volunteer service.

Tools for Assessing the Sustainability of Lifestyles

Q.   *How can individuals be introduced to the process of sustainable development?*

A.   By letting individuals themselves make their own lifestyles gradually more sustainable, taking small steps at a time.

Q.   *How can they do this?*

A.   By being able to properly analyze and evaluate the impacts of their own lifestyles.

There is a need to develop tools that can help in assessing the sustainability or unsustainability of our lives. Currently, initiatives in the field of sustainable lifestyles mostly focus on sustainable consumption. The OECD Report on Sustainable Consumption and Production (OECD 1997) calls for a broadly accepted set of indicators and benchmarks to measure and monitor progress. Indicators of progress may be qualitative, but quantitative targets and measures should be developed wherever possible. Governments, private sector associations, NGOs, and international institutions should routinely address consumption and production patterns in their assessments of environmental quality trends; for example, through environmental performance reviews and self-assessment programs.

The following priority areas were identified by the OECD - MIT Expert Seminar (1994):

* Extending information collection and analysis: we should provide a baseline against which to assess "unsustainable" behavior

* Applying tools for modifying behavior: social instruments should be applied, particularly at the community level, to help change behavior

* Measuring, communicating and improving performance: clear indicators should be chosen to measure progress towards more sustainable consumption and production

* Generating momentum through practical demonstration: successful demonstration projects should be supported and publicized

Also, Chapter 4 of Agenda 21 (UNCED 1992) states that nations must develop policies and strategies to encourage changes in unsustainable consumption patterns. Such policies and strategies should help individuals and households to make environmentally sound purchasing decisions. Agenda 21 asks governments and international organizations, together with the private sector, to develop criteria and methodologies for the assessment of environmental impacts and resource requirements throughout the full life cycle of products and processes. Results of those assessments should be transformed into clear indicators in order to inform consumers and decision-makers. Governments should also encourage the emergence of an informed consumer public and help individuals and households to make environmentally informed choices by providing information on the consequences of consumption choices and behavior, as well as by making consumers aware of the health and environmental impact of products.

The Treaty on Consumption and Lifestyle (ProSus 1992), which is meant to promote reflection and debate among social movements and NGOs leading to commitments for action on sustainable consumption and lifestyles, also calls for conducting a self-assessment of our own lifestyle choices in light of the elements of the treaty and making personal commitments toward change. In addition, the treaty calls for the provision of educational methodologies that focus on values clarification and moving beyond blame to constructive action. In this sense methodologies should reawaken us to the reality that quality of life is based on the development of human relationships, creativity, cultural and artistic expression, spirituality, reverence for the natural world and celebration of life, and is not dependent upon increased consumption of material goods (ProSus 1992).

Sustainable Lifestyle Assessment

The introduction of a new concept called "Sustainable Lifestyle Assessment" (SLA) would certainly fulfill part of the previous recommendation. SLA aims to generate information on the consequences of life choices and to encourage individuals, households, and authorities to take more conscientious decisions. These decisions should be based on a train of thought in which different alternative options are weighed against each other, while also taking into account the sustainability consequences of the alternatives.

SLA can be defined as a process in which the way of life of individuals or specific groups of people is subjected to a structured and in-depth examination. The aim is to predict, analyze, and evaluate the impacts of all long-term, mid-term, and short-term—or routine—decisions made over a lifetime, making use of sustainability goals as a point of reference. The sustainability goals should integrate the environmental, social, and economic aspects of life and lead to an overall improvement of the quality of life that can be maintained indefinitely by future generations.

SLA should be applied before a decision is made and should lead to information that can help in making a final decision in which all possible options have been made explicit and have been weighed against each other on the basis of sustainability criteria. It should stimulate individuals and authorities to think seriously and creatively about alternative and more sustainable life paths.

A distinction should be made between SLA that can be performed by the individual and SLA as a scientific research tool. SLA as a self-assessment instrument for the individual or household should be both very simple and clear in directing people and families toward a more sustainable lifestyle. SLA instruments for scientific purposes to be used by researchers should give scientifically correct information on the difference between various life-styles. This information can be very useful for policy-makers and authorities to develop measures to encourage sustainable lifestyles. This second type of SLA can be more complex, although policy-makers and the public in general should be able to understand the reasoning behind the calculations.

The "Self-Assessment" Approach

People are currently not taught to consciously plan their lives and assess what changes can be made to make life more fulfilling or sustainable. They are not encouraged to think creatively about alternatives to the lifestyles which are currently prescribed by society and religious, moral, ethical, philosophical, social, and economic institutions. We would most likely be better equipped to do a self-assessment of our lives and come up with rewarding alternatives if we were trained to use problem-solving techniques and analytical methods such as those described in the following box. Also, getting acquainted with techniques for creative thinking (such as random input, problem reversal, applied imagination, lateral thinking, use of draw-ings, and many others), time management skills, planning skills, and com-munication skills would help in performing a personal or household SLA.

Examples of Problem Solving Techniques and Analytical Methods

Brainstorming

Brainstorming is a method for developing creative solutions to prob-lems. It works by focusing on a problem, and then deliberately coming up with as many deliberately unusual solutions as possible and by push-ing the ideas as far as possible. During the brainstorming session there is no criticism of ideas - the idea is to open up as many possibilities as pos-sible, and to break down preconceptions about the limits of the prob-lem. Once this has been done the results of the brainstorming session can be analyzed and the best solutions can be explored either using fur-ther brainstorming or more conventional methods.

Critical Path Analysis

Critical Path Analysis is an extremely effective method of analyzing a complex project. It helps to calculate the minimum length of time in which the project can be completed, and which activities should be pri-oritized to complete by that date. Where a job has to be completed on time, critical path analysis helps persons to focus on the essential activities to which attention and resources should be devoted. It gives an effective basis for the scheduling and monitoring of progress.

Decision Trees

Decision trees are excellent tools for making financial or number based decisions where a lot of complex information needs to be taken into account. They provide an effective structure in which alternative decisions and the implications of taking those decisions can be laid down and evaluated. They also help to form an accurate, balanced picture of the risks and rewards that can result from a particular choice.

Force Field Analysis

This is a method used to get a whole view of all the forces for or against a plan so that a decision can be made which takes into account all interests. Where a plan has been decided on, force field analysis makes it possible to look at all the forces for or against the plan. It helps to plan or reduce the impact of the opposing forces, and strengthen and reinforce the supporting forces.

PMI - Plus/Minus/Interesting

This is a valuable development of the 'pros and cons' technique which has been used for a long time. PMI is a basic decision-making tool. When faced with a difficult decision, the idea is to simply draw up a table headed 'Plus', 'Minus', and 'Interesting'. Users of this technique should write down all the positive points of taking the action under 'Plus', all the negative effects under 'Minus' and the extended implications of taking the action (positive or negative) under 'Interesting'.

SWOT Analysis

This is an effective method of identifying Strengths and Weaknesses, and to examine the Opportunities and Threats of certain decisions. Often carrying out an analysis using the SWOT framework will be enough to reveal changes which can be usefully made.

SOURCE MindTools (1998)

We are talking here about all kinds of decisions we face in our lives: to have our own children or not and how many; to adopt children or sponsor children in need; to live in a particular location (close to other family members, close to work, accessible or not by public transportation) and in what type of dwelling; to do what kind of paid and/or volunteer work; to do which kind of studies or training; to get involved in the local community in what way; to go on holiday (frequency, location, mode of transportation, purpose); to eat what types of food (organically grown, vegetarian or otherwise); to commute to work by what mode of transportation; to travel how far each year by foot, bike, public transport, car, high speed train, boat, plane; to use pesticides in the house or not, and so on. One of the goals of SLA is to promote the explicit articulation of values during the personal decision-making process.

Increasingly, NGOs are developing SLA systems based on questionnaires linked to a scoring system. These are not real in-depth studies of life and often take the format of a quiz. They are valuable as tools to help people to start thinking about their lifestyle. They do not predict sustainability impacts of lifestyle options, but help people understand where they are at a certain moment. It is important to establish the baseline situation before a person or household begins a process of developing a more sustainable lifestyle. This auditing approach is most interesting if we repeat the assessment periodically and compare results to examine how much progress has been made. In Flanders (Belgium) Global Action Plan uses a Lifestyle Assessment approach as a marketing tool for its EcoTeams program. It widely distributes a two-page test which consists of twenty questions assessing the degree of environmental sensitivity of the lifestyle of individuals.

More detailed is the "How Earth-Friendly Are You—A Lifestyle Self-Assessment Questionnaire," developed by the New Road Map Foundation (1995), the "Living More Lightly Profile" of the Institute for Earth Education (1998), and the "EarthScore—Your Personal Environmental Audit & Guide," by EnviroAccount (1993). The latter is the most elaborate and deals with fourteen different issues: household energy - general; household energy - winter; household energy - summer; water; transportation; consumerism - durable goods; consumerism - food and agricultural products; consumerism - paper and forest products; toxins; waste, packaging, single-use items, and recycling; environmental advocacy; respect for the land; livelihood; and family planning. It is characteristic of all of these tests that they focus on the environmental effects of our lifestyles, shedding some light on the social and economic effects.

An assessment system that is more socially inspired is the "Peaceful Lifestyle Assessment" of World Peace One (1992). World Peace One is an organization whose mission is to help people look at their lifestyles and make changes that improve the quality of life for all on the basis of a peaceful lifestyle. Table 15-2 gives examples of the topics discussed in the Peaceful Lifestyle Assessment.

Examples of fully developed scientific methods for the assessment of the impact of the lifestyle of individuals, making use of a predictive impact assessment approach, could not be found and should therefore be introduced in the future.

The Research Approach

Scientific approaches used today to assess the sustainability of lifestyles includes Life-cycle Assessment (LCA), Household Metabolism, Eco-Footprint Analysis, and social sciences-based analysis of individual behavior and attitudes. These methods are used by scientists, and the results should be useful for decision-makers of the government or individuals. The following are examples of scientific-based SLA.

TABLE 15-2 *Examples of Assessment Issues in the Peaceful Lifestyle Assessment*

Major topics	Examples of Assessment Issues
Health	I get enough sleep at night
	I maintain a healthy weight and avoid excess calories in my diet
	I avoid smoking, excess alcohol and drugs
Environment	I avoid buying disposable items
	I buy organic produce and foods whenever possible
	I avoid unnatural household, lawn and garden chemicals
Psyche	I plan each week to do things I enjoy
	I feel that I have unconditional worth and value
	I do not overburden myself with too many responsibilities or activities
Political	I vote in every election that I can
	I study candidates for their decision-making ability and their past voting records
	I actively support candidates that promote the long-range good of all
Personal Relationships	I spend enough time on the important personal relationships of my life
	I make an effort to really listen to others, being alert to their experiences and feelings
	I am able to resolve conflicts nonviolently
Economy	I live within my means
	My job is necessary to the production of a good or service that enhances the quality of life
	I contribute money to charities that promote self-sufficiency among the disadvantaged
Nonpersonal relationships	I do not talk negatively about others (gossip)
	I confront unethical or dehumanizing behavior in the groups I belong to
Society	I do about five hours a week of volunteer work
	I affirm my children through positive comments several times a day
	I teach my children to use a systems approach; to "leave the trail better than you found it"

SOURCE World Peace One (1992)

Making Use of Life-Cycle Assessment (LCA)

An interesting scientific study in this context was carried out by Christensen (1997). He examined the environmental impacts from cradle to grave of different lifestyles, based on LCA. This is an instrument which is used to identify, quantify, and evaluate the environmental consequences associated with a product, process, or activity from "cradle-to-grave" on a scientific and systematic basis (Nierynck 1998). Christensen (1997) considered four types of families to be examined. All families consisted of two adults and two children. The major differences between the families were the number of cars (none, one, or two), the distance traveled by these cars (between 0 and 40,000km), the type of house (ordinary or low-energy), the way heat and electricity is generated (oil burner; district heating system supplied by a coal-fired power plant that co-generates heat and electricity; or solar heating and windmills), and the food intake (less meat than average, average amount of meat, more than average amount of meat, organically grown or non-organically grown foodstuffs). Preliminary results of the Christensen (1997) study shows that the most "resource consuming" family uses more than eight times the amount of energy in comparison to the most "green and ecologically minded" family. Transportation is the component of lifestyle with the widest differences in environmental impact. Car use is considered by Christensen (1997) as the single most important factor in the creation of non-sustainable lifestyles.

Calculating Material and Energy Flows - the Household Metabolism Approach

A study by Villaluz and Frostell (1998) developed an "Environmental Profile for Individuals" that makes it possible to calculate material and energy flows directly influenced by the personal lifestyle of individuals during one year. This instrument is also called Substance Intensity Analysis. The model calculates the material and energy fluxes of different individuals in order to compare the fluxes between the different lifestyles and thus obtain a better understanding of the interrelationships between material fluxes, lifestyles, and consumer behavior. Each individual is asked to answer sixty questions covering six aspects: food, clothing, shelter, transport, work, and holidays.

Evaluating the "household metabolism" is the goal of the HOMES project. HOMES, the acronym of *Household Metabolism Effectively Sustainable*, is a cooperation between the Dutch universities of Groningen and Twente. Questions asked in this research project are:

➔ What are the opportunities for and constraints to achieving substantial reductions in the future use of resources by households?

- By what technical, economical, spatial, behavioral, and/or administrative policy options, instruments, and methods could such reductions be achieved?
- What reductions in the use of resources by households can be (maximally) realized if the identified technical, economic, spatial, behavioral, and administrative policy options, instruments and methods are (optimally) implemented? (Noorman and Schoot Uiterkamp 1998)

The Ecological Footprint Analysis Approach

The ecological footprint analysis, such as discussed in more detail in Chapter 13, is traditionally applied at a city-wide level. It can also be used to assess impacts of households. Walker and Rees (1997) examined the ecological implications of the household-level choice of dwelling type. The ecological footprint analysis translates the total ecological impact associated with different housing types into the area of productive land required to support associated resource consumption. Ecological footprint calculations were made for four dwelling types: detached single-family house, townhouse, apartment in small building (so called walk-up), and apartment in high-rise building. The authors found out that occupants of detached houses have the largest housing-related ecological footprints. The smallest eco-footprints are for residents of high-rise and walk-up apartments at 60 percent and 64 percent, respectively, of the value obtained for occupants of standard detached houses.

Analyzing Behavior and Attitudes of Individuals

The analysis of the behavior and attitudes of individuals in relation to consumption patterns can give interesting insights into the way people make personal decisions. It is based on social sciences methods and gives a more qualitative result. The following box gives an example of an SLA making use of such an analysis approach. Also, social science-based methods, such as surveys and life histories can be used. Life histories consist of biographical material assembled about particular individuals and gives much information about the development of beliefs and attitudes over time (Giddens 1997).

Households and Electricity Consumption

The Danish Institute of Local Government Studies prepared a report concerning the relationship between way of life and electricity consumption, based on standard interviews with about 800 respondents. The analysis covered three dimensions: conditions of life, orientation of life, and social relations, as three dynamic interacting aspects of a person's way of life. Conditions of life are objective parameters, such as income, connection with the labor market, and housing conditions. Orientation of life includes the values, preferences and ways of thinking that develop in the course of a person's life. Social relations describe the social network and codes of conduct within the network.

The analysis shows that the factors that most affect electricity consumption are associated with the respondent's conditions of life, whereas his or her orientation of life - including values and thinking on environmental issues - has only limited effect. Also, the existence of codes of eco-friendly conduct within a person's social network has no significant effect on electricity consumption.

This contrasts with the relationships found between way of life and ecologically aware consumption in general. The consumption of eco-friendly products is governed primarily by the codes of conduct, and the attention paid to environmental issues, within the social network. In addition, the analysis shows that those people who do not know the size of their electricity bill have a relatively larger electricity consumption, but also a relatively larger consumption of green products. This indicates a discrepancy in pattern of behavior between electricity consumption and consumption of green products. An explanation for this could be that electricity consumption is basically invisible and tends to be forgotten by the consumers. Moreover, serious reductions in consumption of electricity would likely interfere with the routine of daily life.

On the one hand, those with the best financial resources also have the largest electricity consumption. On the other hand, they also have the best educational resources, the highest degree of environmental awareness and the most motivation. This indicates that there could be a potential for changing behavior with respect to use of electricity within this group. A first step must be to make people aware of electricity consumption and its importance.

SOURCE Pedersen and Broegaard (1997)

Case Study: Personal Decision-Making on Travel Behavior

In 1996 the OECD organized two workshops on individual travel behavior: one on "values, welfare and quality of life" (OECD 1996a) and one on "culture, choice and technology" (OECD 1996b). Both workshops raised a number of issues that are very relevant to SLA. Many of the findings, although focused on travel behavior, are applicable in a broader SLA context.

The first workshop dealt mostly with values. It was generally felt that an individual's travel behavior was more likely to result from habit and circumstances than to reflect fundamental values, since values have potential for affecting behavior only when: a) the behavior is one that requires a deliberate decision; b) specific values are explicitly articulated during the decision-making process; and c) the behavior is not constrained by external factors. For example, values seem to have relatively little influence on day-to-day decisions relating to commuting to work, running errands, and in those cases where there is no real choice. Less habitual and less constrained actions such as car selection, housing location decisions, vacation travel, and voting seem more likely to be consciously influenced by values (OECD 1996a).

In the case where values matter in determining behavior, some values may matter more than others. This is highly dependent on the context and can change unpredictablybecause individuals' values are not necessarily coherent and are often in contradiction with each other. Individual choice is often the result of an internal negotiation between conflicting values. The example given in the OECD report (OECD 1996a) makes this very clear: individuals might value environmental protection, on the one hand, and freedom and flexibility, on the other. Depending on the circumstances, either one or the other—or neither—of these values might play a role in their decision. The result is that in those cases where values matter, there seems to be no clear link between certain values and certain behavior.

An important workshop conclusion in the SLA-context is that information can contribute only to a certain extent to changing behavior. Structural constraints, such as existing infrastructure and technology, represent a major obstacle to information-led behavior change (OECD 1996a).

The second workshop focused on the following six topics: decisions and networks, the hierarchy of choice, the perception of individuals, culture, information, and learning (OECD 1996b).

Examining individual travel behavior shows that structural factors play a major role. For example, the use of a transport mode, such as the car, is dependent on a broad web of social interactions including car manufacturers, fuel supply networks, repair and maintenance facilities, driving schools, car washes, roads and traffic control systems. In the same way, the factors giving rise to travel activity can be described by an interconnected network of influences ranging from the spacing, location, and density of the built environment and the cultural significance of face-to-face contact to time management strategies and lifestyles (OECD 1996b).

Because individual travel behavior is embedded in specific technical-social-organizational networks, it may be difficult for individuals to perceive alternative patterns of behavior. People may feel that their behavior is the only one possible given their current context and therefore feel they have "no choice" except to do as they do. "Choice" seems to be an important term here. To choose means to pick one alternative in preference to others. This implies that individuals have information regarding different alternatives before choosing among them. This means that information about alternatives must exist and be available to individuals (OECD 1996b).

The workshop participants stressed the importance of networks in conditioning the travel behavior of individuals. For example, the simple act of walking to a shop is dependent on a number of relationships that give rise to shoe production and distribution, pavement construction, social conventions regarding rights-of-way when meeting other pedestrians, shop location, and other such factors. Linked to this network concept is the fact that choices are not equal in influencing the travel behavior of individuals. There is a hierarchy of choice, meaning that higher-order choices will influence a whole range of secondary decisions. For example, the choice of car ownership leads to new travel behavior and often halts the formation of alternate mobility patterns, or the location of the home may increase or decrease the need to travel (OECD 1996b).

Perception is also considered important in individual travel behavior. Perception is the result of social and cultural influences as well as individual life histories and experiences, and can affect travel behavior on at least three levels: a) the level at which individuals perceive needs that can be satisfied by travel; b) the level at which people perceive the range of alternatives to specific choices; and c) the level at which people perceive the relative qualities and/or disadvantages of different travel modes. In many cases, individuals overestimate the benefits of their current choice and underestimate the capacity of alternate modes to satisfy their needs. For example, adults who have never learned to drive may perceive car use as costly, difficult, and dangerous; people who have rarely taken public transportation or have rarely cycled may not perceive these modes as suited to their lifestyle because of the perceived disadvantages. Both sets of users feel that they have "no choice" than to act as they do, and yet the first-hand experience of alternate modes might reveal the viability and acceptability of alternative travel behavior (OECD 1996b).

Culture also plays an important role in determining the status, image, and acceptability of different types of travel behavior. Furthermore, information gathering, processing, and assimilation is at the heart of non-instinctive individual travel behavior. Individuals are, however, not always receptive to new information nor are they equally receptive to all forms of information. The type of information, how it is made available, and when it is made available all have important implications for the decision made. Information can also be used in a learning process. Over time, the behavior of individuals is shaped by vast amounts of information from different media and through day-to-day experience (OECD 1996b).

Reflections on Personal or Household Sustainable Lifestyle Assessment Methodology

The self-assessment approach to SLA currently lacks an instrument that can be used to predict and evaluate the impact of specific personal or household decisions on the overall sustainability of a lifestyle. Such an instrument could follow a seven-step approach.

Step 1: Description of the Issue

First, a description should be made of the issue on which a decision has to be made.

Step 2: Type of Decision

The hierarchical nature of decisions should be taken into account. Therefore, the following aspects should be examined and described:

> ✦ The sequence of decisions: *Is there a logical order in which a series of decisions should be made?*
> ✦ The degree of reversibility/irreversibility of decisions: *Can decisions be reversed and in which time frame (short-term, medium-term, long-term)?*
> ✦ The extent to which decisions help frame others or exclude others

The more important the decision is, the more effort needs to be put into the analysis of the particular SLA case.

Step 3: Development of a List of All Possible Options

In this step all the possible options that can lead to the desired outcome need to be described on the basis of creative thinking. The attainability, attractiveness, or desirability of the options do not have to be taken into account. The only requirement is to make a list and describe as many alternatives as possible.

Step 4: Examination of the Different Options

For each option the following aspects should be examined:

> ✦ Description of the different options
> ✦ Comparative description of the environmental consequences of the different options
> ✦ Comparative description of the social consequences of the different options (including who it will affect and in what way it will change social relations; possible health effects, etc.)
> ✦ Comparative description of the economic consequences of the different options (including the cost of the different options)

❖ Description of the forces and actors which are in favor and those not in favor for each option

❖ Description of the structural constraints for each option

❖ Description of the degree of uncertainty (which important information is not available at the time of the decision) for each option

❖ Overview of the personal perception of people concerned (are they personally convinced that the options are feasible?)

❖ Description of moral and ethical values of the people concerned, and the acceptability of the options within their own culture and community

❖ Description of the personal feelings of the people concerned—let their "hearts speak" and indicate if they have a strong personal feeling of what should be done

❖ Indicate the source of the information on which this SLA is based. Were books, Internet sites, experts, family members, friends, and colleagues consulted?

Step 5: FEALWELL Analysis

An overview should be made of the findings of the assessment for all the options and the "Friendliest Environmental Alternative Without Excessive Losses in Lifestyle," the FEALWELL, should be examined. The idea here is to give a description of the option that is, on the one hand, most in line with sustainability principles and, on the other hand, still desirable for those involved.

Step 6: Discussion

The different options should be discussed with a friend who is at the same time neutral toward the outcome of the decision. The same should be done with people who will be immediately affected by the decision. They will all be able to give insight into the problem.

Step 7: Decision

A final decision should be taken on the basis of all the information. It should be clear that this decision is not necessarily the most sustainable one because in a certain case, for example, the opinion of a loved one may be considered more important than the amount of energy saved. It is, however, very useful to have gone through this process and make all options and their pros and cons explicit.

For regularly recurring decisions, SLA should be repeated regularly since modern life is dynamic and sudden changes and opportunities present themselves without warning. These changes may create possibilities and choices not available before or they may also limit possible options.

We also need to introduce to idea of life phases—people go through a number of distinct periods throughout their lives. The stages of the human life course are social as well as biological in nature (Giddens 1997). Stages such as childhood, adolescence, young adulthood, mature adulthood, and old age are mostly biologically determined, while the formation of a family, moving, changing jobs, having children, divorce, having children leave the house, retirement, or passing away of a family member have mostly social consequences. Changing from one life phase to another is a time when a lot of decisions have to be made, decisions which may have long-term consequences. At these times SLA becomes very important.

Application of Sustainable Lifestyle Assessment

SLA applied by individuals or households will only work if people can easily receive high quality and understandable information on the environmental, social, and economic impacts of all kinds of decisions. Therefore, the self-assessment approach to SLA will only work when supported by the results of the research approach. Scientists looking into the sustainability of lifestyles should make an effort to present results in a way which is usable by authorities and individuals. Companies should also be required to provide a standard set of data giving an indication to consumers on the sustainable nature of the products or services they sell. Authorities have the responsibility of disseminating this information and should train citizens to perform an SLA. For example, individuals could be taught specific behaviors in school in order to make it a habit.

Moreover, SLA could be integrated in scientific disciplines such as Family Resource Management (FRM). FRM is a science based on the philosophy that the quality of life of individuals and families is dependent on the procurement, use, and conservation of adequate resources. The quantity and quality of resources available and the ability to use them effectively varies. Individuals and families are confronted with daily challenges in producing and managing resources to optimize their well-being. They need education for skill development in: problem solving, decision-making, goal setting, and prioritizing, resource allocation, budgeting, and record keeping, using credit, planning for future financial security, interpreting economic trends, shopping skills, coping with marketplace practices, exercising their consumer rights and responsibilities, using consumer laws, and the coordination of legal, financial, housing, and health care decisions for retirement planning and estate planning (NC State University 1998). FRM could play a very important role in teaching families how to manage resources in a more sustainable way. It appears to be an appropriate way to teach individuals and households how to make more sustainable decisions. Currently Sustainable Family Resource Management seems to have received only limited attention from experts in this scientific discipline.

The implementation of SLA will be most effective when it is included in a broader "sustainable household action plan." As shown in the following box, SLA should be part of a new way of managing households that includes, for instance, the development of a sustainable household vision, specific household actions, and monitoring of progress.

Sustainable Lifestyle Assessment as Part of a "Household Agenda 21"

Agenda 21 has been applied at local level by municipalities, towns and cities, following the "Local Agenda 21" approach. Agenda 21 could be taken to an even more local level by introducing it in household management. A so-called "Household Agenda 21" (HA21) could help households to develop a more sustainable lifestyle and consciously introduce the "sustainability" concept in the "household decision-making process". School children could be taught how to make use of a HA21 throughout their lives and households could be encouraged to develop their own HA21 by local authorities introducing tax relief for those households which can prove to meet certain sustainability standards. Persons would still be free to live the lives they want, but those who aim for a more sustainable lifestyle could be financially or otherwise rewarded. HA21 is not implemented today. The program which currently comes closest to a HA21 is the EcoTeam program of Global Action Plan. A HA21 is, however, more ambitious: while the EcoTeam focuses mostly on the day to day environmental impacts , the HA21-approach also wants participants to look into long-term decisions and wants to include examining the social and economic aspects of lifestyle decisions. One cannot expect each household to develop its own HA21-methodology and process. Such a system should be developed and introduced by a public authority and follow a prescribed approach which is logical, clear and simple. Households should be supported and guided in practice. A HA21 could include following six steps.

Step 1. Development of a vision

A vision for the future should be made in which the major choices in life are delineated. Here we discuss with all members of the household how we vision ourselves in the long-term future. All members of the household will have their own goals, but a common ground should be found on which all can work together. The vision should indicate where we would like to go in relation to the sustainable development of the household. The vision could also lead to a clear statement of commitment of the household and could be revised regularly.

Step 2. Setting the baseline situation

Perform an audit of the activities of the household, examining environmental, social, and economic impacts.

Step 3. Development of an action plan

The members of the household should examine which very specific actions can be taken today and in the near future to steer the household activities in a more sustainable direction.

Step 4. Sustainable Lifestyle Assessment

A household sustainable development decision-making support system should be introduced. Decisions should be made being aware of their consequences on the sustainability of the household and the community.

Step 5. Monitoring

Members of the household should keep track of progress made through the periodical auditing of the household activities. Sustainability indicators and specific targets could help households to evaluate progress during their monitoring efforts.

Step 6. Reporting

All households participating in a HA21 should be asked to report in a standard way to a central public authority. The results can be used to reward households which reach certain goals and can be used by governments to examine where households would need extra support.

While it is very important that individuals and households take responsibility for their actions, it is also of utmost importance that governments deal with existing structural constraints. Public authorities should introduce policies and take actions which encourage people to live a more sustainable life. This could be, among others, policies or actions which encourage people to live close to work, use public transportation more frequently, and make time to invest in the local community.

Conclusion

At this time little research has been carried out on the application of impact assessment principles to the lives of individuals or households. The self-assessment approach to SLA is therefore largely undeveloped and limited to actions by NGOs. The scientific world should not only focus more on its own scientifically based SLA studies, but should also get involved and examine methodologies for future use by individuals and households. SLA will only be efficient when people are trained in using the instrument and sufficient and correct information is made available so that people can make a more sustainable lifestyle decision.

One of the ideas of a self-assessment approach to SLA is to introduce more rational decision-making in the personal and household spheres, encouraging people to become informed about the consequences of decisions, and focusing on a sustainable development of our lifestyles. Introduction of rationality in our personal decisions should not result in the pushing aside of our personal feelings. On the contrary, it should simultaneously encourage us to really get in touch with what we truly feel, and become less influenced by commercial messages and social pressures. SLA should lead us to open up to new ideas that increase the quality of life for ourselves and those around us, while at the same time lead to a reduced environmental impact.

SLA should promote creative thinking in individuals and authorities. SLA may encounter opposition because it may result in recommendations which are not congruous with societal, cultural, or religious traditions. Sustainable development will, however, only be successful if we do not cling with desperate tenacity to our old habits, but rather try to be open-minded about new suggestions for positive lifestyle changes for ourselves and for future generations.

More Information

On the Internet:
> The Simple Living Network: http://www.simpleliving.net
> World Peace One: http://trfn.clpgh.org/orgs/wp1/

References

Burman, D. 1997. Enhancing community health promotion with local currencies. The local employment and trading system (LETS). In M. Roseland, ed., *Eco-City Dimensions*. Gabriola Island: New Society Publishers.

Christensen, P. 1997. Different lifestyles and their impact on the environment. *Sustainable Development* 5: 30-35. New York: John Wiley & Sons.

CI, Consumers International. 1997. *Meeting Needs, Changing Lifestyles*. Consumers International.

EnviroAccount. 1993. *EarthScore—Your Personal Environmental Audit & Guide*. Lafayette, CA: Morning Sun Press.

Elster, J. 1989. *The Cement of Society: A Study of Social Order*. Cambridge: Cambridge University Press.

European Commission. 1998. *Status Report on European Telework—Telework 98*. August 1998, Brussels.

Giddens, A. 1997. *Sociology, third edition*. Cambridge: Polity Press.

Godeau, A. 1998. Responsible consumption as a lever for sustainable development: A critical analysis. *International Conference Beyond Sustainability, Integrating Behavioral, Economic and Environmental Research, 19 - 20 November 1998*. Amsterdam: Netherlands Organization for Scientific Research.

Hansen, U. and U. Schrader. 1997. A modern model of consumption for a sustainable society. *Journal of Consumer Policy* 20, 4: 443-468.

Holmberg, J. 1992. *Making Development Sustainable. Redefining Institutions, Policy, and Economics*. International Institute for Environment and Development. Washington DC: Island Press.

Institute for Earth Education. 1998. *Living More Lightly Profile*. Greenville, WV: The Institute for Earth Education.

Maslow, A. H. 1967. A theory of metamotivation: The biological rooting of the value-life. *Journal of Humanistic Psychology* 7: 93-127.

MindTools. 1998. *Problem-Solving Techniques and Analytical Methods*. Website http://www.psy-www.com/mtsite/index.html (1 Oct. 1998).

NC State University. 1998. Family resource management resource guide. *Department of Family & Consumer Sciences*. Website http://www.ces.ncsu.edu/ depts/fcs/frmguide/intro.html (3 Dec. 1998).

New Road Map Foundation. 1995. *How Earth-Friendly Are You? A Lifestyle Self-Assessment Questionnaire*. Seattle: New Road Map Foundation.

Nierynck, E. 1998. Life-cycle assessment. In B. Nath, L. Hens, P. Compton and D. Devuyst, eds., *Environmental Management in Practice* Vol.1. London: Routledge.

Noorman, K.J. and T. Schoot Uiterkamp, eds. 1998. *Green Households? Domestic Consumers, Environment and Sustainability*. Littlehampton: Earthscan.

OECD - MIT. 1994. *Expert Seminar on Sustainable Consumption and Production Patterns*. 18-20 December 1994. Paris and Boston: Organization for Economic Cooperation and Development and Massachusetts Institute of Technology.

OECD. 1996a. *First OECD Workshop on Individual Travel Behavior: "Values, Welfare and Quality of Life," Final Report., 18-19 March 1996*. Paris: Organization for Economic Cooperation and Development.

OECD. 1996b. *Second OECD Workshop on Individual Travel Behavior: "Culture, Choice and Technology" Final Report., 17-19 July 1996*. Paris and Brighton: Organization for Economic Cooperation and Development and University of Sussex.

OECD. 1997. *Report on Sustainable Consumption and Production*. Paris: Organization for Economic Cooperation and Development.

Pedersen, L. H. and E. Broegaard. 1997. *Households and Electricity Consumption. An Analysis of Behavior and Attitudes Towards Energy and Environment*. Institute of Local Government Studies - Denmark.

ProSus, Program for Research and Documentation for a Sustainable Society. 1992. *Treaty on Consumption and Lifestyle*. Website http://www.prosus.nfr.no/andre-doc/alternativ-agenda/Consumption_Lifestyle.html (27 Nov. 1998).

Reisch, L. 1998. *Sustainable Consumption: Three Questions About a Fuzzy Concept, Working Paper 13*. Copenhagen: Research Group "Consumption, Environment, and Culture," Department of Marketing, Copenhagen Business School.

St. James, E. 1995. *Inner Simplicity. 100 Ways to Regain Peace and Nourish Your Soul*. New York: Hyperion.

Sustainable Seattle. 1998. *Indicators of Sustainable Community 1998*. A Status Report on Long-Term Cultural, Economic, and Environmental Health for Seattle/King County. Seattle: Sustainable Seattle.

UNCED, United Nations Conference on Environment and Development. 1992. *Agenda 21: Program for Action for Sustainable Development*. New York: United Nations Conference on Environment and Development.

UNDPCSD, United Nations Department for Policy Coordination and Sustainable Development. 1997. *Program for the Further Implementation of Agenda 21*. New York: United Nations Department for Policy Coordination and Sustainable Development.

Villaluz, M. and B. Frostell. 1998. "An Environmental Profile for Individuals." *International Conference Beyond Sustainability, Integrating Behavioral, Economic and Environmental Research, 19 - 20 November 1998*. Amsterdam: Netherlands Organization for Scientific Research.

Walker, L.A. and W.E. Rees. 1997. "Urban Density and Ecological Footprints: An Analysis of Canadian Households." In M. Roseland, ed., *Eco-City Dimensions. Healthy Communities, Healthy Planet*. Gabriola Island: New Society Publishers.

World Peace One. 1992. *Peaceful Lifestyle Assessment*. Pittsburgh: World Peace One.

Linking Sustainability Assessment to a Vision for a Sustainable Future

Dimitri Devuyst

Sustainable Urban Development and Sustainability Assessment

The Western lifestyle is characterized by high consumption levels, extravagant use of natural resources, and excessive production of waste, a widening gap between the rich and poor, and rapid growth of the global human population, all of which pose a major problem for the future survival of our species. The earth can only absorb an impact up to its carrying capacity, and scientists indicate that this level is being approached rapidly. Sustainable development means "development that meets the needs of the present generations without compromising the needs of future generations" (WCED 1987). It aims to develop societies in which humans can maintain or improve the health, productivity, and quality of their lives in harmony with nature. Although such a goal can count on broad support from all corners of the world and all parts of human societies, putting it into practice remains a major challenge.

As Lester Brown puts it:

> We do know what is happening. The question for us is whether our global society can cross the social threshold that will enable us to restructure the global economy before environmental deterioration leads to economic decline. (Worldwatch Institute 1999)

In other words, human beings have the knowledge to foresee that environmental, social, and economic trouble is waiting if we maintain our current development path. It is, however, uncertain whether or not humanity will have the necessary dynamics to turn the situation around and develop in a sustainable way.

Sustainability assessment aims to steer societies in a more sustainable direction by providing tools that can be used either to predict impacts of various initiatives on the sustainable development of society or to measure progress toward a more sustainable state.

This book deals with sustainability assessment at the local level. At this level, one can find a wide variety of human settlements ranging widely in size, development level, main functions, and local characteristics. When examining the "local level," reference is often made to urban areas. This can be explained by the growth in urban settlements and the fact that the twenty-first century will be the first time in our history that the majority of humanity will be living in cities. We should not forget that sustainable development of metropolitan regions cannot be separated from the sustainability of rural and natural regions, since the sustainability of our cities is inextricably linked to the integrity and sustainability of the rural hinterland.

Moreover, sustainable development at the local level cannot be detached from international, national, and subnational initiatives. The importance of the national level is shown, for example, when examining the introduction of Local Agenda 21in cities. O'Meara (1999) reports on a 1997 survey which showed that 82 percent of known Local Agenda 21 initiatives were concentrated in 11 countries with national campaigns sponsored by government or country-wide municipal associations. In other words, the local level often depends on the higher authorities to realize sustainable development initiatives.

The successful shift of human settlements to a more sustainable state is dependent on a number of critical factors. For example, the presence of people in positions of local authority who stand behind the principles of sustainable development and who are able to motivate is an important catalytic factor in the introduction of measures for more sustainability. In addition, for a successful introduction of sustainable development, it is very important for local authorities to be open-minded and consider new ways of governing, including new participatory approaches that involve the local population. Citizens and their leaders should also be willing to examine the introduction of innovative policies, plans, and projects in which holistic approaches are developed and traditionally separated sectors are brought together in finding creative solutions for sustainability issues.

One of the most successful initiatives for the introduction of sustainable development at the local level is the establishment of a Local Agenda 21 process, which is a translation of the Agenda 21 UNCED action plan to the local level (UNCED 1992). To be most effective in introducing sustainable development, the Local Agenda 21 process should include, among others, a new dialogue between authorities and citizens, efforts to integrate language and values of sustainable development in key planning documents, and the monitoring of all municipal activities with respect to values, goals, and targets for sustainable development. Also, sectoral integration should be attempted, the precautionary principle should be applied, North-South cooperative arrangements should be introduced, and evaluation of progress made should be reported.

The eco-city approach to sustainable development in urban areas is an example of how new ways of thinking about urban planning and urban design can lead to the innovative integration of apparently disconnected ideas in relation to transportation, health, housing, energy, economic development, natural habitats, public participation, and social justice. By doing this more sustainable communities can be created.

Lessons To Be Learned from Past Impact Assessment Practices

Sustainability assessment aims to check whether new policies, plans, or other initiatives are in line with principles for sustainable development. Sustainability assessment will function most effectively if it can rely on a clear assessment framework. Such a framework can take the form of a local community vision for sustainable development or a local sustainable development policy. While we can attempt to develop "standard visions" for sustainable urban development, in the end each local authority will have to develop its own vision, based on the specific sustainability problems with which that area is faced. These issues may be influenced by, for example, the size, economic status, and the stage of development of the city.

Two types of tools for sustainability assessment can be distinguished: a) tools to be used in the decision-making process, and b) tools for setting a baseline and measuring progress. Sustainability assessment tools to be used in the decision-making process follow an impact assessment approach, whereas tools for setting a baseline and measuring progress use an audit approach.

The impact assessment approach to sustainability assessment aims to predict the sustainability consequences of certain initiatives before a decision is made on their introduction. This means that the impact assessment approach has an advantage over the audit approach because of its preventive nature. The audit approach measures effects when the activity is already in operation, possible damage has been done, and the possibilities for taking corrective measures have become more limited. The major advantage of the audit approach is that impacts can be measured and possible consequences do not need to be predicted. Since techniques for the prediction of sustainability impacts are not yet fully developed, results are easier to obtain and more reliable when following the audit approach.

The impact assessment approach to sustainability assessment is currently not widely practiced or applied on a regular basis. We are still in an experimental phase. However, this is also the time to learn from experiences with related instruments, such as EIA.

On the basis of experience with EIA in Canada, Gibson (1993) discusses eight principles which form the main design requirements that should be met by impact assessment processes. Sustainability assessment systems will also benefit from the introduction of these principles:

✦ The assessments should follow an integrated approach

✦ The assessments should clearly and automatically apply to planning and decision-making on all undertakings that may have significant environmental effects

✦ The assessments should be aimed at identifying best options and should focus on the examination of alternatives

✦ The assessment requirements should be specified in legislation and be mandatory and enforceable

✦ The assessment and decision-making process must be open, interactive, and fair. The system must be designed to ensure evenhanded treatment of all parties

✦ The conditions of approval must be enforceable and must be followed up by the monitoring of effects

✦ The assessment process must be designed to facilitate efficient implementation

✦ The assessment must be accompanied by clear and well-accepted criteria for judgments. To some extent these criteria must depend on local environmental conditions and community values

In the Netherlands there is a sense among high-placed environmental officers that EIA has in the past focused too much on paperwork and that the EIA procedure is not yet linked closely enough to the decision-making mechanisms. There appears to be a need to study how the often irrational decision-making process works, taking into account the practical side of the story and the way that decision-makers function. This should be the starting point for trying to influence and steer decision-making in a more environmentally friendly way. The Dutch are thinking about a paperless EIA system (Berkenbosch and Herlaar 1998). This type of lesson from actual EIA practice needs to be taken into account in developing sustainability assessment systems.

In the application of sustainability assessment we also have to make sure that the following three conditions are met:

1. Sustainability assessments should be performed by people who are trained to do this work. They should be advocates of the sustainable development philosophy, aware of the meaning of the concept "sustainable development" and how it is translated into practice. They should be trained in motivating people to consider introducing initiatives for sustainable development.

2. Sustainability assessments should stimulate planners, policy-makers and decision-makers to consider nontraditional ways of reaching long-term sustainable goals.

3. People preparing sustainability assessment studies should be provided with information on "good practice" in relation to the introduction of sustainable development principles in all kinds of policies, plans, and other activities. On the basis of a database of international "good practice" cases, the sustainability assessment process can lead to the introduction of innovative ideas in the formulation of new initiatives.

Considering sustainability assessment as an impact assessment tool also means looking into the relationship between sustainability assessment and Strategic Environmental Assessment (SEA). SEA examines the environmental impacts of policies, plans, and programs and is considered a most promising technique to tackle limitations of project level EIA systems. Many impact assessment professionals offer SEA as the most effective tool to look into sustainability considerations of higher decision-making levels. In this book a proposal is made to develop sustainability assessment tools which go further than SEA in the holistic examination of environmental, social, and economic aspects, fully integrating sustainability principles in the assessment process.

The goal of the impact assessment approach to sustainability assessment is to influence the decision-making process and to introduce sustainability principles in that process. This goal is very ambitious because of the complexity and high level of confusion which is characteristic of decision-making.

When introducing sustainability assessment, we have to take into account the possibility of a negative reaction of administrative staff and decision-makers, who may not be motivated to use the instrument to its fullest potential. Therefore, it is important to see sustainability assessment not only as a tool to be used by public officials or public authorities. Sustainability assessment systems should be designed in such a way that they can be used by NGOs or political opposition parties. Results of sustainability assessment studies performed by watchdog groups can be released to the media, and in this indirect way sustainability assessment can sensitize the general population and encourage public authorities to take into account sustainable development principles in their decision-making.

Recommendations for the Introduction of Sustainability Assessment Systems at the Local Level

A Need for Help from Higher Authorities

Research shows that it will be difficult for local authorities to manage sustainability assessment on their own because: a) the smaller municipalities especially do not have the necessary staff members (they may not have an officer responsible for sustainable development) and b) many local authorities do not feel confident that they have the necessary knowledge. Many municipalities therefore often seek external advice.

A Need for Sustainability Assessment Training

Many local authorities in many countries around the globe have never been directly involved in impact assessment exercises, and many are unaware of the meaning of concepts such as sustainability assessment or sustainable development. This shows that there is a need for training in sustainability assessment and hence, training manuals should be developed.

A Need for a Simple and Flexible Sustainability Assessment System

A sustainability assessment system for municipalities should be as flexible as possible because:

→ This group of authorities is very variable: cities, small rural municipalities, industrial centers, coastal towns, and provincial authorities all have their own characteristics, problems, and needs
→ The range of policies, plans, and other initiatives developed at the local level is very broad
→ Different municipal authorities have different experiences with impact assessment

A sustainability assessment system should be as simple as possible because:

→ The aim of sustainability assessment is to make decisions which are more in line with principles for sustainable development and which are based on correct information, without excessive costs or loss of time
→ Sustainability assessment should give decision-makers the opportunity to consider the sustainability consequences of proposals in a cost-effective way

❖ Sustainability assessment is not an aim in itself. It should only be done if it can add to the quality of the decisions made and should be organized as efficiently as possible

❖ The staff members of municipal authorities are not often qualified to do extensive scientific studies, but could, for example, learn to apply a simple checklist

❖ The public should be encouraged to participate in sustainability assessment. This can be encouraged by a simple system

A Need for Guidelines on Best Practice

The International Association for Impact Assessment and the Institute of Environmental Assessment, U.K., has developed principles of Environmental Impact Assessment best practice. The same principles, presented in Table 16-1, also apply to sustainability assessment.

A Need for a Hierarchical Sustainability Assessment System

A sustainability assessment system for the municipal level cannot be detached from assessments at the national, federal, and/or regional levels of government. Policy-making and planning should start at the highest levels of government and be filled in by policies, plans, programs, and projects at lower levels. It is therefore very important to introduce a tiering approach. Once in place, tiering ensures that environmental implications, issues, and impacts of development decision-making can be addressed at the appropriate level(s) and with the degree of effort necessary for informed choice (Sadler and Verheem, 1996).

A Need for Further Research

The introduction of sustainability assessment at the local level needs to be guided by additional scientific research. Examples of possible research projects are:

❖ Examination of policy-making and planning processes at the municipal level and a study of how sustainability assessment can be integrated in these processes

❖ Study of sustainability assessment methodology for the local level, such as community-based approaches to sustainability assessment

❖ Development of sustainability assessment training packages for municipal staff

TABLE 16-1 *Basic Principles of Environmental Impact Assessment That Also Apply to Sustainability Assessment*

Purposive:
The process should inform decision-making and result in appropriate levels of environmental protection and well-being.

Rigorous:
The process should apply "best practicable" science, employing methodologies and techniques appropriate to address the problems being investigated.

Practical:
The process should result in information and outputs which assist with problem solving and are acceptable to and able to be implemented by proponents.

Relevant:
The process should provide sufficient, reliable and usable information for development planning and decision-making.

Cost-effective:
The process should achieve the objectives of EIA within the limits of available information, time, resources, and methodology.

Efficient:
The process should impose the minimum cost burdens in terms of time and finance on proponents and participants consistent with meeting accepted requirements and objectives of EIA.

Focused:
The process should concentrate on significant environmental effects and key issues; i.e., the matters that need to be taken into account in making decisions.

Adaptive:
the process should be adjusted to the realities, issues and circumstances of the proposals under review without compromising the integrity of the process, and be iterative, incorporating lessons learned throughout the proposal's life cycle.

Participative:
The process should provide appropriate opportunities to inform and involve the interested and affected publics, and their inputs and concerns should be addressed explicitly in the documentation and decision-making.

Interdisciplinary:
The process should ensure that the appropriate techniques and experts in the relevant bio-physical and socio-economic disciplines are employed, including use of traditional knowledge as relevant.

Credible:
The process should be carried out with professionalism, rigor, fairness, objectivity, impartiality and balance, and be subject to independent checks and verification.

Integrated:
The process should address the interrelationships of social, economic, and bio-physical aspects.

Transparent:
The process should have clear, easily understood requirements for EIA content; ensure public access to information; identify the factors that are to be taken into account in decision-making; and acknowledge limitations and difficulties.

Systematic:
The process should result in full consideration of all relevant information on the affected environment, of proposed alternatives and their impacts, and of the measures necessary to monitor and investigate residual effects.

SOURCE IAIA and IEA (no date)

❖ Follow-up of specific case studies, carrying out of pilot studies, and sustainability assessment experimental studies

❖ Development of monitoring systems and indicators for sustainable
development to be used in sustainability assessment at the local level

Integrated Assessment and Sustainability Management

The impact assessment approach is certainly not the only way to lead communities towards more sustainable development. More results will be obtained
when impact assessment is combined with other management instruments
and decision-making aids such as the auditing approach. In the end this
might lead to an integrated type of assessment. It is not a coincidence that
integrated environmental assessment is an active and rapidly developing field.
In Europe, a European Forum on Integrated Environmental Assessment
(EFIEA) was recently created. The two main objectives are: a) To improve the
scientific quality of integrated environmental assessment and b) To
strengthen the interaction between environmental science and policy. By fostering a collaborative network of scientists, decision-makers and stakeholders
concerned with complex environmental issues, the EFIEA hopes to improve
the necessary information flows both quantitatively and qualitatively.

Sustainability Assessment as part of the impact assessment family should
be seen as only one element in an extended sustainability management tool-
kit. Other examples of tools which can be adapted for use in sustainability
management are:

❖ Environmental Care Systems, such as the Eco-Management and Audit
Scheme (EMAS) and ISO14,001

❖ Indicators Systems

❖ Norms, standards, and targets

❖ State of the Environment reporting

This approach leads to the integration of tools such as Sustainable Development Indicators, Sustainability Targets, Sustainability Reporting, and Sustainable Lifestyle Assessment in a Sustainability Management System.

Developing standard sustainability indicators and targets which are applicable by any local authority is difficult. The major benefit of nationwide
standardized sustainability indicators would be the possibility of comparative performance measurement, which would allow local authorities to compare results. We should, however, be very cautious about the introduction of
such standardized performance indicators. One of the problems encountered in any assessment is the need to allow for various community objec-

tives. It is clear that no two local authorities are identical in terms of their priority issues.

The basic test for the appropriateness of specific indicators is that they make sense to those affected by them. Because circumstances vary from person to person and place to place, the development and identification of indicators need to be tailored carefully to resonate with people in language that they understand, in a time frame consistent with their horizons, focused on issues that have significant meaning and which can be applied in ways that help produce tangible improvements in their lives.

Sustainable development indicators should target "core" concerns rather than symptoms and should look beyond urban boundaries to the eco-regions to which the survival of the urban areas are linked. Indicators should focus on changes in the well-being of people and the health of nature at both neighborhood and citywide levels, and be tailored to correspond with the mood of the community as a whole. Moreover, they should focus on the priorities set in the local community, measure positive change rather than deficits, address equity concerns, and should include both qualitative and quantitative measures.

Sustainability Reporting has its origins in State of the Environment Reporting. It aims to obtain and interpret information, which can be used to guide actions and decisions. Periodically repeating the Sustainable Reporting process makes it possible to get information on progress made over time. Examples of Sustainability Reporting show that each one uses its own terms and methodology, such as goals, objectives, benchmarks, milestones, and signposts. These are used interchangeably and should therefore be clearly defined.

Sustainability targets are not only important for measuring progress, but are also necessary in an impact assessment process in which clear statements have to be made on the acceptability and significance of predicted impacts. Therefore, sustainability assessment certainly benefits from clear sustainability targets. To date, very little information is available on the effectiveness of sustainability targets. Literature on evaluation studies relating to setting targets is rare. Clearly there is a need for scientific research in relation to setting and evaluating sustainability targets.

This book also introduces a new concept called Sustainable Lifestyle Assessment (SLA), which aims to generate information on the consequences of life choices and to encourage individuals, households, and authorities to take more environmentally conscientious decisions, taking into account the sustainability consequences of the alternatives.

The implementation of SLA will be most effective when it is included in a broader "sustainable household action plan." A proposal is made to develop a "Household Agenda 21" process following the successful Local Agenda 21 approach. A Household Agenda 21 (HA21) could help households develop a more sustainable lifestyle and consciously introduce the concept of sustainability in the household decision-making process. School children could be taught how to make use of a HA21 throughout their lives. Local authorities can encourage households to develop their own HA21 by introducing tax relief for households that prove that they can meet certain sustainability targets. People would still be free to maintain their lifestyles, but those who aim for a more sustainable lifestyle could be financially or otherwise rewarded. Steps in the HA21 process are: development of a vision, setting the baseline situation, development of an action plan, continuous sustainable lifestyle assessment, monitoring, and reporting. The introduction of an HA21 will only be successful if individuals learn to use the instrument itself—hence the important role of educational systems.

While it is very important that individuals and households are trained to take responsibility for their actions, it is also of utmost importance that governments deal with existing structural constraints which reduce possibilities for people to live a more sustainable lifestyle.

Practical examples of sustainability assessment presented in this book, such as Direction Analysis, Ecological Footprint Analysis, and evaluating the long-term effectiveness of the EcoTeam Program show the diversity of sustainability assessment approaches and its usefulness in producing sustainability-related information.

Future Directions for Sustainability Assessment

Sustainability assessment should lead to corrections of initiatives in a more sustainable direction. We should not expect this approach to result in a revolutionary change of society in the short-term. However, sustainability assessment should invite initiators to think about the consequences of their actions for the sustainability of society. Even small steps in the direction of a more sustainable decision-making process should be considered encouraging signs.

Chances are that the introduction of sustainability assessment will be a slow process, comparable to the slow establishment of EIA. EIA has been in practice for nearly 30 years and is still not accepted by many governments. There are, however, encouraging signs that the basic principles of EIA are slowly permeating into governmental decision-making processes, not only in relation to environmental aspects, but also in respect to health and social issues, for example. There is still much resistance to EIA in many countries, but slow progress is being made. If the same slow approach is followed in sustainability assessment, we risk that it will result in sustainability assessment being "too little, too late." Therefore, special attention should be given to developing effective and powerful sustainability assessment systems which can count on the support of our governments, and can also be used by NGOs, academics, corporations, and interested individuals.

It is also important that we start making sustainability assessments now, even if we are not sure if the right methodology or assessment frameworks have been developed. Pilot projects in which policies, plans, programs, and projects are subjected to sustainability assessment (using, for example, the ASSIPAC method presented in this book) will help us to improve our knowledge base and teach us practical lessons as to what works and what does not.

It is also important that sustainability assessment not be labeled as just another bureaucratic government tool. It should be seen as a flexible instrument with a wide range of application in government, corporations, and civil society. We should not force companies or citizens to introduce sustainability assessment, but we should encourage and motivate them to do so, for example, by providing financial or other incentives. We could consider a system whereby corporations and households receive benefits from the government if they reach certain sustainability targets. In such a system sustainability assessment can help in reaching these sustainability targets.

It should be clear that a sustainability assessment system cannot function on its own, but exists to monitor the introduction of sustainability principles in decision-making processes. In other words, sustainability assessment is an instrument that is an aid in implementing a sustainable development policy or vision on sustainable development. Sustainability assessment will only work if the people involved understand its usefulness. A sustainability assessment system can be no better or worse than the political and economic conditions that are set for it.

As long as there are no scientifically established and generally accepted methods to predict sustainability impacts, we will need to revert to more qualitative approaches to sustainability assessment. The lack of quantitative methods for use in sustainability assessment should not stop us from using the tool. A more qualitative approach may already be sufficient in encouraging people to think about and consider more sustainable scenarios to reach their goals. Qualitative approaches to sustainability assessment may also increase the quality of debates on the introduction of sustainable development principles in the functioning of our societies.

Sustainability assessment is most of all a communicative process, improving communication in relation to sustainability issues. Sustainability assessment should be designed to initiate creative and innovative thought processes, which lead to solving current problems of sustainable development.

A Humanist Vision for a Sustainable Future

Today, humanity is faced with important problems, such as: rapid population growth; poverty, hunger and disease (AIDS, malaria, tuberculosis); global warming, destruction of forests, and declining biodiversity. There are still totalitarian regimes which try to stop democratic tendencies in their societies; fundamentalist voices of religious, political, or tribal origin which oppose efforts to solve social problems and create bitter conflicts and wars between communities. Moreover, we should not forget problems such as the concentration of power and wealth in the hands of global corporations which have no democratic representation; high unemployment; and runaway consumption which puts unprecedented pressure on the environment.

We should also be very concerned about the fact that much of what passes today for "sustainable," "sustainability," or "sustainable development" is barely an improvement over the status quo. Many people think we are well on our way to sustainability when we have introduced a municipal recycling program and an improved bus service. Such limited measures fundamentally do not change the problems mentioned in the previous paragraph.

Moreover, tools to perform sustainability assessment alone will not have the strength to change the world and make societies truly sustainable. Their role is supportive, encouraging people to take into account sustainable development principles in their policy-making and decision-making processes. Sustainability assessment tools will consequently only be effective when linked to a strong vision for a sustainable future, not only at the local level, but especially at the global level. Therefore, at the end of this book a vision for a sustainable future of humanity on this planet is presented. This vision is offered to the reader in modesty as an alternative to the present management practices in the world, as a proposal which is open to discussion and constructive dialogue. It is based to a large degree on the *Humanist Manifesto 2000: A Call for a New Planetary Humanism* (Kurtz 1999), which presents a humanist outlook on the future of the world.

The humanist view is basically an optimistic view which states that global problems can be overcome by making use of critical intelligence and cooperative efforts. It rejects nihilistic philosophies of doom and despair and apocalyptic scenarios, believing in the ability of humankind to solve its problems on the basis of continuous scientific and technological progress, the realization of the highest ethical values, and the use of reason and cognition. In other words, humanist ideas and values express a renewed confidence in the power of human beings to solve their own problems and conquer uncharted frontiers (Kurtz 1999).

This optimism stands in contrast to, for instance, the observation that widespread concern about the future of humankind has not yet been translated into effective political action (Wackernagel 1997). World leaders have known about the aforementioned problems for decades, but no fundamental initiatives to solve them have been taken. Therefore, we need to focus in the future on approaches that encourage action and go beyond just documenting warning signs. Consequently, we have to tackle the following political, epistemological, and psychological problems (Wackernagel 1997):

✢ The erosion of the power of political institutions and community networks as a result of the globalization of industrial production

✢ The incapability of traditional science and economic analysis to comprehend the sustainability crisis

✢ The psychologically rooted social behavior that makes people and societies continue their daily routine without making fundamental changes

The following recommendations can be part of the solution and are a vision which can lead to a more sustainable future.

A New Global Ethical Framework

A new ethical framework should take into account the new reality of a globalization of human interactions. *Planetary humanism* seeks to preserve human rights and enhance human freedom and dignity with an emphasis on a commitment to humanity as a whole. The following seven points are the basis of this *planetary humanism:*

1. Respect of dignity and worth of everyone in the world community: we not only have a responsibility to people in our immediate surroundings, but also to people beyond our national boundaries. We are linked morally and physically to each person on the globe.

2. Mitigation of human suffering and increase of the sum of human happiness all over the world.

3. Avoidance of an overemphasis on multicultural parochialism: we should be tolerant of cultural diversity except of cultures which are intolerant or repressive. Moral caring and loyalty should not end at ethnic enclaves or national frontiers. We should support institutions of cooperation among individuals of different ethnic backgrounds.

4. Respect and concern for people should apply equally to all human beings: we should defend human rights everywhere. Each of us has a duty to help mitigate the suffering of people anywhere in the world and to contribute to the common good.

5. These principles should apply not only to the world community of the present time, but also to the future.

6. Each generation has an obligation, as far as possible, to leave the planetary environment a better place.

7. We should do nothing that would endanger the very survival of future generations.

Moreover, a *Planetary Bill of Rights and Responsibilities* is proposed which includes the following major points (Kurtz 1999):

✦ We should strive to end poverty and malnutrition, and to provide adequate health care and shelter for people everywhere on the planet

✦ We should strive to provide economic security and adequate income for everyone

✦ Every person should be protected from unwarranted and unnecessary injury, danger, and death

✦ Individuals should have the right to live in a family unit or household of their choice, consonant with their income, and should have the right to bear or not to bear children

✦ The opportunity for education and cultural enrichment should be universal

✦ Individuals should not be discriminated against because of race, ethnic origin, nationality, culture, caste, class, creed, gender, or sexual orientation

✦ The principles of equality should be respected by civilized communities, including equality before the law, equality of consideration, satisfaction of basic needs, and equality of opportunity

✦ Everyone has the right to live a good life, pursue happiness, achieve creative satisfaction and leisure in his or her own terms, as long as he or she does not harm others

✦ Individuals should have the opportunity to appreciate and participate in the arts

✦ Individuals should not be unduly restrained, restricted, or prohibited from exercising a wide range of personal choices. This includes the right to privacy

A Focus on Empowerment of World Citizens

Wackernagel (1997) identifies a number of psychological phenomena which result in political leaders and citizens continuing to follow unsustainable development paths. First there is "active promotion" and "passive tolerance" of the current condition. Active promotion includes the positive portrayal of unsustainable lifestyles through advertising and other forms of media, for example. Passive tolerance is evidenced by the social denial of the current crisis by industrialized countries in their perseverance in planning for more rather than planning for sustainability. There is also the "boiled frog syndrome," meaning that human beings cannot easily perceive slow changes and long-term implications, comparable to the frog who cannot detect when water in which it is placed is slowly heated to a boiling point. Another problem is "mental apartheid" in which it is easier to see ecological and social problems that are far away, to point fingers or recommend to others what to do. Moreover, individuals often feel insignificant, overwhelmed, and powerless when confronted with the global dimensions of the sustainability crisis.

People should be encouraged to try new, less materialistic lifestyles in which having fewer possessions is seen as a liberation instead of a deprivation and in which fulfillment comes from giving and being with other people rather than taking and withdrawing. It is liberating for people to know that our most precious resource is not our disposable income but rather our life energy, and that building independence does not come from having more, but from needing less (Wackernagel 1997).

Parents, friends, schools, and media can all help in giving children the skills, power, and knowledge to work toward personal growth and growth of the communities in which they live. To be able to tackle global and local sustainability issues, people need to learn how to deal with deep-rooted fears, taboos, challenges, and changes in their lives. Building self-esteem is also very important: only by valuing ourselves rather than imitating consumer-oriented role models can people make sustainable choices (Wackernagel 1997). It is also important to give people a positive mental attitude, in which they are optimistic about the future. All individuals should know that they can make—in their own way—a positive contribution to a changed, more humane and sustainable society.

New Global Institutions

There is a need to create new global institutions that deal with the global problems directly and focus on the needs of humanity as a whole. There is a necessity for the introduction of a World Parliament that represents the people of the world rather than nation-states. This World Parliament should only deal with questions that can only be solved on the global level. A World Security System to resolve military conflicts and a World Court and an International Judiciary with sufficient power to enforce its rulings should also be introduced.

There is a need for an international system of taxation. This could include: a) a tax on the Gross National Product of all nations; b) a tax which ensures that multinational corporations pay their fair share of the global tax burden; c) a tax on international fund transfers. These taxes should be used for economic and social assistance and development, including selective cancellations of burdensome debts of poor countries.

The *Global Sustainable Development Resolution* that was placed before the Congress of the United States in 1999 shows that it is possible for nations to develop comprehensive proposals of legislation which tackle sustainable development issues. The aims of the resolution are for the government to reconstruct the global economy to realize the following goals: a) democracy at every level of government from local to global; b) human rights for all people; c) environmental sustainability throughout the world; and d) economic advancement for the most oppressed and exploited, including women, children, minorities, and indigenous peoples. While the chances of this resolution being passed in the United States are negligible, it does force leaders to speak out on issues such as worker rights, environmental protection, democratic process and accountability (Clifford 1999). Moreover, it gives an example of the type of legislative action that can be taken at the national level.

At the level of global institutions there is also a need for encouraging worldwide distribution of high quality information, free expression of ideas, respect for diversity of opinion, and the right to dissent. Global media should not be dominated by powerful private corporations which are only interested in maximizing ratings. The media should be a forum which reflects the cultural diversity of the world, provides factual or educational information, and encourages a free exchange of ideas.

New Research Approaches

Scientists should give more attention to building capacity for conducting interdisciplinary, collaborative, action-oriented research on sustainability issues (Wackernagel 1997).

New technologies should be developed, critically examined and evaluated in terms of potential risks and benefits to society and the environment. Research efforts should strive for technological innovations that reduce overall human impact on the environment and are affordable to the poor.

New Planning Approaches

Sandercock's (1998) "cosmopolis" approach to planning is a celebration of human diversity and the multicultural community. Following the work of King (1981), a people-oriented definition of development is used. A community becomes more developed as it becomes more diverse, incorporating more cultural and ethnic traditions, and developing the skills and confidence to solve its own problems.

Sandercock (1998) proposes a new framework for urban planning, with an acknowledgement of historical injustice, and an understanding and positive valuation of difference, a respect for and honoring of a different way of life and a desire to preserve that difference. The stories of immigrants, indigenous peoples, people of all colors and cultures, gays and lesbians and others need to be included. It presupposes an ability to communicate with people whose lifestyles and daily rhythms are very different, a genuine openness in negotiation and mediation processes, and requires face-to-face mutual learning, compassion, empathy, and patience.

There is also a need for planning that focuses on the formulation of goals and is less oriented to the preparation of documents and more concerned with the interaction of people, a type of planning that is never apolitical or value-neutral (Sandercock 1998).

New Educational Approaches

One of the fundamental challenges in building a sustainable society is the reform of educational systems. What people think and feel about the world affects what they do as voters, consumers, and resource owners, as government officials, international diplomats, and employees (Roodman 1999).

First of all, education and training should be a fundamental and basic right, not only for all children, but also for adults.

Schools should focus on giving students experiences that they can use later for the benefit of a more sustainable world. The exchange of experiences of people from different backgrounds is very important. For example, programs could be introduced for the exchange of pupils from different countries, communities, cultures, and social backgrounds in order to learn from each other.

One of the main tasks of schools should be to give children and adolescents the capacity to participate in community development. In school, pupils should discover their own unique talents and learn how they can use these talents for the benefit of the local and world community.

Schools should aim to educate responsible, free spirited, inquiring, creative world citizens who care about the long-term health of the world. Pupils should learn how to build trust and security between the members of a society, and they should learn to value cooperation instead of competition.

Roodman (1999) draws our attention to the encouraging fact that mindsets can change quickly in response to education. For example, in developing nations, education campaigns, along with increased availability of family planning and contraception, are one major reason that fertility rates fell remarkably quickly between the early 1960s and the first half of the 1990s.

Basically a humanist vision of a more sustainable future means that (Kurtz 1999):

> In the midst of our diversity and the plurality of our traditions, we need to recognize that we are all part of an extended human family, sharing a common planetary habitat. The very success of our species now threatens the future of human existence. We alone are responsible for our collective destiny. Solving our problems will require the cooperation and wisdom of all members of the world community. It is within the power of each human being to make a difference. The planetary community is our own, and each of us can help make it flourish. The future is open. The choices are for us to make. Together we can realize the noblest ends and ideals of humankind.

References

Berkenbosch, R. and C. Herlaar. 1998. Milieu-effectrapportage zonder papier? *Kenmerken* 5,5: 10-13.

Clifford, D. 1999. The global sustainable development resolution. Regaining democracy. The campaign of the 21st century. Website http://www.users.on.net/rmc/global.htm#INTRODUCTION (3 Dec 1999).

Gibson, R. B. 1993. Environmental assessment design: Lessons from the Canadian experience. *The Environmental Professional* 15: 12-24.

IAIA and IEA, International Association for Impact Assessment and Institute of Environmental Assessment. No date. *Principles of Environmental Impact Assessment Best Practice.* Leaflet of the International Association for Impact Assessment.

King, M. 1981. *Chain of Change.* Boston: South End Press.

Kurtz, P. 1999. Humanist manifesto 2000: A call for a new planetary humanism. *Free Inquiry* 19, 4: 4-19.

O'Meara, M. 1999. Exploring a new vision for cities. In L. Brown, C. Flavin and H. F. French, eds., *State of the World 1999. A Worldwatch Institute Report on Progress Toward a Sustainable Society.* New York: W.W. Norton & Company.

Roodman, D. M. 1999. Building a sustainable society. In L. Brown, C. Flavin and H. F. French, eds., *State of the World 1999. A Worldwatch Institute Report on Progress Toward a Sustainable Society.* New York: W.W. Norton & Company.

Sadler, B. and R. Verheem. 1996. *Strategic Environmental Assessment. Status, Challenges and Future Directions.* Den Haag: Ministry of Housing, Spatial Planning and the Environment of the Netherlands; International Study of the Effectiveness of Environmental Assessment and The EIA-Commission of the Netherlands.

Sandercock, L. 1998. *Towards Cosmopolis. Planning for Multicultural Cities.* Chichester: John Wiley & Sons.

UNCED, United Nations Conference on Environment and Development. 1992. *Agenda 21: Program for Action for Sustainable Development.* New York: United Nations Conference on Environment and Development.

Wackernagel, M. 1997. Framing the sustainability crisis: Getting from concern to action. *Sustainable Development Research Institute, University of British Columbia* Website http://www.sdri.ubc.ca/publications/Wackerna.html (1 Dec 1999).

WCED, World Commission on Environment and Development. 1987. *Our Common Future.* Oxford: Oxford University Press.

Worldwatch Institute. 1999. World may be on edge of environmental revolution. *Worldwatch Institute* Website http://www.worldwatch.org/alerts/990225.html (4 Mar 1999).

Index

CPSIA information can be obtained
at www.ICGtesting.com
Printed in the USA
JSHW050741140622
27051JS00003B/37